EU ENLARGEMENT AND THE ENVIRONMENT

This volume focuses attention on key environmental and institutional changes associated with eastern expansion of the European Union, assessing and challenging prevailing views about the outcomes and processes of this historic development.

Looking at four central themes:

- capacity changes and limitations
- the EU's mixed messages and conflicting priorities
- non-state actor roles and developments
- and the exchange of ideas and information

the volume shows that enlargement will change the EU, not just make it bigger, and that EU officials and programs are improving aspects of environmental policy in CEE countries even as they are making others less sustainable.

This is a special issue of the journal *Environmental Politics.*

JoAnn Carmin is an Assistant Professor in the Department of Urban Studies and Planning at Massachusetts Institute of Technology. Her research examines on the role of civil society actors in environmental policy, planning, and sustainable development initiatives both in the United States and in Central and Eastern Europe. **Stacy D. VanDeveer** is an Assistant Professor in the University of New Hampshire's Department of Political Science. His research interests include international environmental policymaking and its domestic impacts, the connections between environmental and security issues, and the role expertise in policy making.

Statement of Aims

Environmental Politics is concerned with four aspects of the study of environmental politics, with a primary, though not exclusive, focus on the industrialised countries. First, it examines the evolution of environmental movements and parties. Second, it provides analysis of the making and implementation of public policy in the area of the environment at international, national and local levels. Third, it carries comment on ideas generated by the various environmental movements and organisations, and by individual theorists. Fourth, it aims to cover the international environmental issues which are of increasing salience. Its coverage of the developing world does not reach beyond this to the affairs of individual countries, partly because of the journal's chosen focus and partly because of the number of existing journals dealing with development. *Environmental Politics* is sensitive to the distinction between the goals of conservation and of a radical reordering of political and social preferences, and aims to explore the interface between these goals, rather than to favour any one position in contemporary debates.

EU ENLARGEMENT AND THE ENVIRONMENT

Institutional Change and Environmental Policy in Central and Eastern Europe

Edited by JoAnn Carmin and Stacy D. VanDeveer

LONDON AND NEW YORK

First published 2005
by Routledge
2 Park Square, Milton Park, Abingdon, Oxon OX14 4RN

Simultaneously published in the USA and Canada
by Routledge
270 Madison Ave, New York, NY 10016

Routledge is an imprint of the Taylor & Francis Group

Typeset by Elite Typesetting Techniques Ltd,
Eastleigh, Hampshire, UK
Printed and bound in Great Britain by Antony Rowe Ltd,
Chippenham, Wiltshire

British Library Cataloguing in Publication Data
A catalogue record for this book is available
from the British Library

Library of Congress Cataloguing in Publication Data

ISBN 0-415-34788-2 (hbk)
ISBN 0-415-35186-3 (pbk)

ENVIRONMENTAL POLITICS/ROUTLEDGE RESEARCH IN ENVIRONMENTAL POLITICS

Edited by Matthew Paterson

Keele University and Graham Smith, University of Southampton

Over recent years environmental politics has moved from a peripheral interest to a central concern within the discipline of politics. This series aims to reinforce this trend through the publication of books that investigate the nature of contemporary environmental politics and show the centrality of environmental politics to the study of politics per se. The series understands politics in a broad sense and books will focus on mainstream issues such as the policy process and new social movements as well as emerging areas such as cultural politics and political economy. Books in the series will analyse contemporary political practices with regards to the environment and/or explore possible future directions for the 'greening' of contemporary politics. The series will be of interest not only to academics and students working in the environmental field, but will also demand to be read within the broader discipline.

The series consists of two strands:

Environmental Politics addresses the needs of students and teachers, and the titles will be published in paperback and hardback. Titles include:

GLOBAL WARMING AND GLOBAL POLITICS
Matthew Paterson

POLITICS AND THE ENVIRONMENT
James Connelly & Graham Smith

INTERNATIONAL RELATIONS THEORY AND ECOLOGICAL THOUGHT TOWARDS SYNTHESIS
Edited by Eric Lafferière & Peter Stoett

PLANNING SUSTAINABILITY
Edited by Michael Kenny & James Meadowcroft

DELIBERATIVE DEMOCRACY AND THE ENVIRONMENT
Graham Smith

EU ENLARGEMENT AND THE ENVIRONMENT
Institutional change and environmental policy in Central and Eastern Europe
Edited by JoAnn Carmin and Stacy D. VanDeveer

Routledge Research in Environmental Politics presents innovative new research intended for high-level specialist readership. These titles are published in hardback only and include:

1 THE EMERGENCE OF ECOLOGICAL MODERNISATION
Integrating the Environment and the Economy?
Stephen C Young

2 IDEAS AND ACTIONS IN THE GREEN MOVEMENT
Brian Doherty

3 RUSSIA AND THE WEST
Environmental Cooperation and Conflict
Geir Hønneland

4 GLOBAL WARMING AND EAST ASIA
The Domestic and International Politics of Climate Change
Edited by Paul G. Harris

5 EUROPE, GLOBALIZATION AND SUSTAINABLE DEVELOPMENT
Edited by John Barry, Brian Baxter and Richard Dunphy

6 THE POLITICS OF GM FOOD
A Comparative Study of the UK, USA and EU
Dave Toke

7 ENVIRONMENTAL POLICY IN EUROPE
The Europeanization of National Environmental Policy
Edited by Andrew Jordan and Duncan Liefferink

8 A THEORY OF ECOLOGICAL JUSTICE
Brian Baxter

CONTENTS

Acknowledgements

In the spring of 2002, a group of scholars met at the Woodrow Wilson International Center for Scholars in Washington, DC to discuss the environmental implications of EU enlargement for Central and Eastern Europe. Participants presented views of how EU enlargement would alter environmental institutions, policies and practices in the CEE region and across Europe. At the conclusion of the conference, it was evident that a variety of views and perspectives about the relationships of EU bodies to candidate countries had emerged. At the same time, the discussions identified numerous questions about the future of international relations, policy diffusion and environmental leadership in the region. Since the conference, the questions, concerns and insights endured, serving as the basis for this volume.

We are indebted to numerous individuals and organisations for supporting this project. At Woodrow Wilson International Center for Scholars, Martin Sletzinger, Director of East European Studies, and Geoffrey D. Dabelko, Director of The Environmental Change and Security Project, helped secure funds and ensured the availability of administrative support for the conference held in Washington, DC. James R. Bohland and C. Theodore Koebel, both from Virginia Tech, provided conference funds through the School of Public and International Affairs and the Metropolitan Institute. Robert Jenkins and Gary Marks, from the University of North Carolina at Chapel Hill, also made conference funds available through the Center for Slavic Eurasian, and East European Studies and the Center for European Studies. Earlier versions of some of this volume's contributions were included in the conference proceedings, entitled *EU Enlargement and Environmental Quality: Central and Eastern Europe and Beyond*, produced by the Woodrow Wilson International Center for Scholars.

We are grateful to the Center for Environmental Solutions at Duke University, the Department of Political Science at the University of New Hampshire, and the Department of Urban Studies and Planning at Massachusetts Institute of Technology for providing support that made the completion of this volume possible. We also thank Marcie Anderson and Jodi Vandervort for providing administrative assistance, Betsy Albright for compiling the index that accompanies the book version of this volume, and Barbara Hicks and Miranda Schreurs for reading and commenting on drafts of the introductory and concluding contributions. In addition, Stacy extends his thanks to Michael Harding for providing constant support and encouragement.

We both appreciate the detailed and constructive feedback we received on each of the contributions from a complement of anonymous reviewers. We also are grateful to Andrew Dobson, Slav Todorov and other members of editorial staff at *Environmental Politics* and Frank Cass Publishers for their assistance throughout the review and publication process. Finally, we thank the contributing authors for their cooperation, hard work, commitment to the project, and their abiding interest in the ongoing, historic transformations across Europe.

JOANN CARMIN and STACY D. VANDEVEER
September 2003

Foreword

The stage is set for eight Central and Eastern European states to join the European Union in 2004. Although the final stages of the EU entry negotiations appeared to many to be unduly taken up by haggling over minutiae, the amalgamation of the majority of the European people into a European Union based on the principles of democracy and free enterprise is a cause for rejoicing. The years ahead, like the two decades past, will involve considerable expenditures by the existing member states as they help create the conditions and infrastructure necessary to promote rapid growth and to close the gap in living standards between the old and new parts of the Union.

The next phase of EU enlargement includes the three Baltic States (Estonia, Latvia and Lithuania), the four Visagrad states (the Czech Republic, Hungary, Poland and Slovakia), one former member of Yugoslavia (Slovenia) and two Mediterranean countries (Cyprus and Malta). Accession negotiations continue with Bulgaria, Romania and Turkey, while future phases of enlargement might also include other states in the Balkan Peninsula. When complete, enlargement will bring the population of the EU to around 500 million and constitute the largest economic market in the world. There is little doubt that this powerful economic bloc will become a larger player on the global political stage. The nature and pace of the EU's progress towards greater political integration will be on the table for discussion as the new members take their place at the table. While political integration may be variable, in keeping with the EU's core policies of promoting social and economic cohesion, it is reasonable to expect that the rate of economic growth in new member states will be comparatively rapid.

The presence of accelerated economic growth raises important questions and poses important challenges for the EU overall. First, the new member states come, by and large, from a background of environmental neglect that was part and parcel of the Soviet system. In the process of catching up with the other member states, issues arise as to whether this growth can be secured without pro rata increases in environmental pollution and ecological devastation. Second, at the moment of its enlargement, the EU is the recognised leader of international concern on global environmental issues. Therefore, it is also essential to consider the effects of enlargement on the evolution of environmental policy within the EU and beyond the Union.

EU Enlargement and Environmental Quality: Four Arguments

As the idea of EU membership for the newly independent democracies in Central and Eastern Europe gained ground in the early 1990s, the groundswell of political opinion did not give rise to detailed assessment or debate about the economic, social and environmental implications of enlargement. For many, the 're-unification' of Europe was seen as necessary and desirable, and it was generally taken for granted that it would serve the best interests of all parties. These assumptions were based on a rather superficial analysis driven by an overriding emotional commitment to the newly liberated countries and to bridging the Cold War rifts across the continent. Today however, with the first wave of new members poised to join the Union and after a decade of concern about sustainability, the question whether enlargement is sustainable must be addressed.

The assumption that EU membership of CEE countries would have beneficial impacts on the quality of the environment in Europe was based on four arguments. First, admission into the European Union obliges candidate countries to adopt and implement the entire body of EU environmental legislation that has been built up over 30 years. It was generally assumed that replacing the previously ineffective and unimplemented laws of the Soviet era with a comprehensive body of regulation would produce significant improvements in environmental quality.

Examination of this argument in the light of subsequent experience suggests that it will take more than implementing the EU body of environmental regulations to ensure environmental protection. In 1998, after conducting extensive research, the European Environment Agency (EEA) determined that environmental quality in the EU was declining despite years of the implementation of EU regulations in all member states. Environment quality was unchanged in only four of the twelve sectors examined; in all the others (except the impact of natural and technological disasters) environmental quality was in decline. This study demonstrated that regulation may be a necessary condition for coping with the environmental consequences of increasing prosperity, but it is by no means sufficient. Thus the implementation of EU environmental regulations in CEE countries was not, by itself, going to produce the desired results. This view has been borne out in practice. In the first six or so years after the fall of communism, industrial production in the region declined and environmental quality improved. Now that production is increasing, and a 'consumer society' with its incentives to produce ever-increasing amounts of waste has taken hold, improvements in environment quality have stalled. A number of contributions to this volume demonstrate both the accomplishments and the limitations of extending EU environmental policy to the candidate countries.

The new member states will participate along side the existing 15 countries in implementing the EU strategy for achieving sustainability, launched at the Gothenburg Summit in June 2001 as a response to the *Environment Quality Assessment* published by the EEA. This is an interesting and unique exercise, the effectiveness of which will ultimately depend on the capacity of European politicians to introduce implementation measures, many of which will not be popular. The jury is still out.

A second argument suggesting that EU enlargement would improve environmental quality was made on the basis that replacing the central planning system of economic organisation with a market economy would result in the substitution of old gas-guzzling technology for clean and green technology. Yet, the positive effects of technology and markets are not entirely clear-cut. On the one hand, there is hope that with many investment decisions considering the environment and with a high rate of technology substitution, the growth in CEE may be less environmentally damaging than much of the previous investments that have taken place in the EU. On the other hand, increased growth in the region is already giving rise to increasing pollution in some sectors, with transport being the most obvious. At the moment, the prospect of accession to the EU has given rise to major shifts in trade patterns and transport needs. Additionally, with growing prosperity, the number of automobiles in use has more than quadrupled and Prague, Budapest and Warsaw seem to be in a state of permanent gridlock. WIFO, the Austrian economic research institute, predicts that the tonnage of transport passing through Austria will increase sevenfold by the date of accession. To this must be added the 'modal shift' from rail to road, which the EU has done little to reverse in its own territory or as a by-product of the investment supports it has supplied to the accession countries.

In some countries, improvements in air quality as a consequence of reductions in emissions from the energy sector have been significant. However, these gains have been offset by the slow adoption of new approaches towards the prevention and minimisation of waste. On balance, the mix of tight regulation, new technology, and the policy directions coming as a result of the EU's sustainable strategy may have overall beneficial effects on environmental quality both in the region and in the wider EU. This remains to be seen.

A third argument maintained that environmental protection would remain a political priority in CEE countries, since concern for the environment was high on their political agendas at the time of liberation. However, the significance of the environment in the eyes of the citizens in the region is no longer a priority. During the 1990s, efforts were made by numerous governments to promote increased public and political awareness of environmental issues. For example, the Regional Environment Center, initiated and financed by the US and the European Commission, had a significant impact on public awareness of the environment and on the development of environmental NGOs. The Environment for Europe process, launched in Dobris in 1991, also contributed to the growing capabilities of officials at all levels to come to grips with environmental policy.

The final argument is derived from the fact that environmental quality in Europe is not nationally defined, but instead, is determined by conditions in many European countries. Therefore, the argument was advanced that improving environmental quality in CEE states would, other things being equal, raise environmental quality throughout the whole of continental Europe.

Future Evolution of EU Environmental Policy

The obligation to implement the EU regulatory regime as a condition of membership has undoubtedly been an important incentive to CEE officials. Indeed, it is interesting to compare the approach to environmental issues in CEE with those of other ex-Soviet countries (Russia, Belarus, Ukraine) where there is no such incentive and far less environmental policy development. Here then is the big imponderable – will citizens of these new democracies support improved environmental quality in the face of other more immediate and material priorities – including aspects of the lifestyle being 'imported' from the rest of the Union? This question brings us to a vitally important aspect of the overall impact of enlargement – the extent to which the new member states will be either a progressive or a restraining element in the development of future environmental policy. This question is taken up in this volume.

Environmental concern and its integration into the European Community was driven by a relatively small number of member states. Through successive amendments to the Treaty of Union, Germany, Denmark and The Netherlands were joined in the most recent (1995) enlargement of the EU by Sweden, Austria and Finland in promoting 'sustainable growth' as a goal of the Union. As other member states developed positive approaches to the environmental issues, the present Treaty of Union enjoins the observance of a series of environmental principles and, perhaps most important, the integration of environmental concerns and requirements into *all* Community policies.

Policy decisions are taken in the EU by Ministerial Councils where each minister from each member state casts one vote, the outcome being determined by the achievement of a 'qualified majority'. In recent years, voting in the Council of Environment Ministers has shown itself to be quite 'green'. What will be the influence of this perspective on the votes of the new member states? Quite clearly, some are environmentally proactive and can be expected to have a 'green' influence in an enlarged Union. For others, cost implications may figure substantially in their voting. Countries with considerably lower national incomes per capita than those of current member states will likely concentrate scarce budgetary resources on industrial/economic investment, rather than on environmental infrastructure. Thus, national voting is likely to be heavily influenced by the cost implications of legislative proposals.

In the final analysis, three factors are likely to influence legislative decisions: (1) the nature and cost implications of proposals; (2) whether this cost burden falls on the public purse or on private enterprise; and (3) the way in which Poland and eventually Romania decide to vote, since in combination they will hold 40 per cent of the votes of the new member states. In any case, the EU provides funds to assist new member states in meeting the costs of building their environmental infrastructures; this is already happening during the accession process and I see it being strengthened in the future.

Increasingly, one must also take into account the influence of the European Parliament. Most of the proposals in the environmental sector require a 'co-decision' that gives Parliament a key role in determining the outcome. The European Parliament has traditionally been sympathetic to environment issues. This seems unlikely to change with the arrival of parliamentarians from the new member states. On balance, this suggests that the EU will continue to adopt strict environmental regulations, although the process will become increasingly complex. It follows that the EU can be expected to continue playing a leading role in global environmental politics. The commitment made in at the 2002 Johannesburg World Summit on Sustainable Development by Russian President Putin to support the EU position in favour of Kyoto Protocol ratification – and European officials' subsequent efforts to promote Russian ratification – give further indication of the increasingly influential role played by the EU in global environmental problem solving.

Contribution of this Volume

Enlargement to the east may have some deleterious impacts on the quality of the European environment in the short term. I expect to see such impacts increasingly offset by the cumulative positive effects of stricter regulation and the rapid spread of newer and greener technology in the new member states. Further, the addition of the CEE states will only serve to strengthen the influence of the EU in international environmental negotiation, an influence that is likely to remain environmentally concerned and supportive of international cooperation to solve common problems.

As Deputy Director General of the European Commission's Environment Directorate General, I was closely involved in the process of preparation for enlargement. On the heels of this involvement I was invited to address a group of distinguished participants at a conference held in Washington, DC at the Woodrow Wilson International Center for Scholars in March 2002. This volume and its constituent contributions follow from the conference presentations and discussions.

The timing of this volume could hardly be better. As a whole, the contributions direct attention to the environmental benefits ensuing from the extension to the CEE region of EU environmental regulations, while examining important questions and issues regarding the impacts that of accession on environmental policymaking and institutions in the EU and in Central and Eastern Europe. This volume examines the roles played in accession processes by international institutions and actors as they relate to domestic CEE agents. It also calls attention to the influence of relationships between government officials and a wide variety of civil society actors. The contributors have a wealth of expertise and knowledge about the EU and the CEE region and contemporary challenges facing both. This volume provides balanced and valuable perspectives to those seeking greater understanding of the institutional and environmental implications of Europe's historic 'enlargement to the east' – the penultimate step in the reunification of our continent.

TOM GARVEY

Former Deputy Director General for Environment, European Commission
Member of the boards of the Regional Environmental Centres

INTRODUCTION

Enlarging EU Environments: Central and Eastern Europe from Transition to Accession

JOANN CARMIN AND STACY D. VANDEVEER

The 2004 enlargement of the European Union (EU) from 15 to 25 member states is an enormous step in the historic process of European economic, political and cultural integration. It was, quite literally, unimaginable 15 years ago. The notion of expanding the EU to include Central and Eastern European (CEE) countries followed the collapse of the communist systems in the region in 1989. Many justifications for EU enlargement have been offered, but it is clear that advocates of the idea viewed CEE membership in the EU as an important way to help stabilise the region's new political and economic systems, assist Europeans to compete in a globalising economy, and improve CEE and continental European environmental protection and quality. Although proponents maintain that EU enlargement has numerous benefits, critics assert that it will hinder environmental quality and lower environmental standards throughout all of Europe.

Since the collapse of communist rule, many aspects of the environmental policy agendas in CEE states have been influenced by the desire to join the EU. Between 1994 and 1996, a number of countries applied for EU membership, and in 1998 accession negotiations were started with the Czech Republic, Estonia, Hungary, Poland and Slovenia. Later, negotiations were opened with Bulgaria, Latvia, Lithuania, Romania and Slovakia, resulting in ten CEE countries formally engaged in negotiations for EU membership. As laid out in the so-called Copenhagen criteria, membership in the EU requires the adoption, implementation and enforcement of the *acquis communautaire* – the body of EU law and regulations. The *acquis* consists of 31 thematic chapters, each detailing laws, regulations, norms and standards. Environmental law and regulations constitute one such chapter.

Transposing the environmental chapter of the *acquis* requires that candidate countries adopt framework legislation, measures on international conventions, biodiversity protection, product standards, and provisions to ensure reductions in national, transboundary and global pollution [*Europa, 2002a*]. Accession negotiations were provisionally closed in December

2002 for eight CEE countries (the Czech Republic, Estonia, Hungary, Latvia, Lithuania, Poland, Slovakia and Slovenia) and two Mediterranean countries (Malta and Cyprus). The accession treaty was signed in April 2003. By September, all ten of these countries had approved entrance into the EU (most by referendum) with the expectation that they would join the EU in May 2004, in time to participate in the European Parliament elections being held that year.

As the subtitle notes, this volume focuses on changes in environmental policy and related institutions as driven by eastern enlargement of the EU. Five general questions, all responses to aspects of contemporary scholarly and popular debates about EU enlargement, motivate this work: (1) what are the likely impacts of eastern enlargement on EU institutions?; (2) how has the EU accession process shaped environmental policies and practices in CEE countries?; (3) do CEE states have the capacities required to implement the environmental *acquis* and, if not, what must they do to build them?; (4) how does accession shape opportunities for domestic agents and actors in CEE countries?; and (5) how does harmonisation and implementation of the *acquis* affect environmental quality and sustainability in CEE countries?

To contextualise these questions and the subsequent contributions, this introduction discusses some key concepts and issues associated with the past 15 years of 'transitions' away from the communist systems and the process of EU expansion. We briefly review a number of political and environmental changes that have taken place in CEE states and societies since 1989 and discuss the roles and impacts of foreign assistance in these transitions. Next we summarise key aspects of the 'Europeanisation' turn in contemporary social science research. Because of its relevance for this volume, the discussion of Europeanisation pays particular attention to prevailing ideas about how the EU influences state-level actors and decisions. We conclude with an overview of the volume's organisation and a brief discussion of how the contributions collectively augment conventional views of eastern enlargement of the EU.

Fifteen Years of Changing Environments

As the communist regimes gave way in the late 1980s and early 1990s, the challenges faced by CEE states and societies in their 'transitions' away from state socialism and towards new democratic political institutions and capitalist economic systems began to come into focus. Decades of state socialism left a legacy of entrenched and inefficient bureaucratic institutions. Just as political institutions were deeply ingrained, so too were values and beliefs about how government should function and its role in society. State socialism also contributed to the presence of serious environmental problems and challenges, many of which required immediate attention and remediation. Of the many interconnected 'transitions' away from the communist systems experienced by CEE states and societies, the political and environmental transitions are the focus here. These transitions are connected with each other, just as they overlap with many aspects of the post-communist economic transitions.[1]

Political Transitions

Transitions away from authoritarianism predate the 1989 collapse of Soviet-style state socialism in the CEE region. Examinations of these changes, such as the decolonisation and colonial independence movements and movements away from authoritarian rule in Latin America, suggest that transitions towards more democratic forms of government entail dramatic, indeed revolutionary, change for government officials, citizens, the private sector and civil society institutions alike. To be successful, governmental and societal actors must develop and adapt to new laws, organisational forms and social institutions [e.g. *O'Donnell, Schmitter and Whitehead, 1986; Huntington, 1991*]. Studies of earlier 'waves' of democratisation generally focused on domestic level politics and institutions. In contrast, CEE transitions demonstrate the impact that international actors, including international organisations, non-governmental organisations, and other states can have on state-level politics and institutional development [*Linden, 2002; Pridham, Herring and Sanford, 1994; Whitehead, 1996*]. The accession of CEE countries to the EU has provided an avenue for particularly strong external influence on the path of democratisation.

Democratisation across the 'post-communist' CEE region is quite varied, with some new democracies judged as well consolidated and others experiencing either stagnation or reconsolidation of authoritarian rule [*McFaul, 2002*]. By the mid-1990s, EU (and NATO) officials were repeatedly asserting that countries without well-consolidated democratic rule would not be admitted as new members [*Wood and Yesilada, 2004*]. Consequently, invitations to join the EU have been extended to those states regarded as well on their way towards realising consolidation, but not to those in which democratisation efforts have been stalled or reversed.

The transitions of state and societal institutions away from authoritarian rule and the consolidation of democratic forms of government have important implications for environmental politics and policy. Critical state factors affecting environmental governance in the transition away from state socialism include legal and constitutional provisions regarding rights of assembly and expression, state authority and ability to regulate private enterprises and enforce contracts, and the organisational structure and competencies of local, regional and national government bodies. Important societal institutions include legislation governing the creation and operation of non-governmental organisations, citizens' access to environmental information, and public involvement in decision making.

One aspect of CEE democratisation bearing directly on environmental governance is the decentralisation of political authority. Decentralisation can encompass a wide array of 'ways to divest the central government of responsibility to outside organisation' [*Yoder, 2003: 263*] including shifting authority towards sub-national public sector actors such as local and regional governments and delegating powers to public authorities at any level of government [*Hicks and Kaminski, 1995*]. These bodies can be administrative bureaucracies or elected bodies.

EU financial assistance and investment programmes encourage and reinforce many types of decentralisation and regionalisation through their funding criteria [*Yoder, 2003*]. To date, however, decentralisation has had mixed results. Several CEE states have reorganised sub-national governance numerous times and at multiple levels. Consequently, the names, competencies, and jurisdictional authorities and boundaries have often undergone wholesale change. Such changes have left uncertainty and inexperience in their wake at local and regional levels. Because key environmental policy functions – monitoring, inspections, enforcement, permitting and licensure – are assigned to these new or reformed sub-national bodies by national environmental law and regulations, policy implementation often suffers when agencies lack the capacity to perform their assigned tasks [*Ecotech, 2001a*].

Political changes in the wake of the demise of state socialism also had a significant impact on the rights and responsibilities of citizens. Throughout the region, new political parties formed and democratic elections took place. People were guaranteed the freedoms of speech and association. Non-governmental organisations began to proliferate. In the communist era, most social and cultural organisations were controlled or closely monitored by the government [*Wolchik, 1991*]. However, these groups could now form and act independently, a change that plays an important role in fostering democratic practice and stability.

In the years leading up to the fall of the communist regimes, many civil society initiatives had an environmental theme. While important in their own right, discontent with the state of the environment and environmental protection offered citizens opportunities to criticise government institutions and ultimately helped destabilise the CEE communist states [*Singleton, 1987; Jancar-Webster, 1993; Vari and Tamas, 1993; Tickle and Welsh, 1998*]. The intensity of environmental concern led to a surge in the formation and activity of citizens' groups in the years prior to and immediately following the fall of the regimes. For a time, it seemed that civil society actors and politicians would work collaboratively with government officials and that an emphasis would be placed on environmental remediation and institutional development. However, as other priorities began to dominate the political agenda in the early 1990s, especially those related to the hardships brought on by the economic, political and cultural transitions, environmental issues moved to the margins [*Fagin, 1994; Slocock, 1996*].

For environmental organisations and movements, as well as for the promotion of civil society more broadly, greater political openness was accompanied by greater access to transnational and international sources of aid and influence. Environmental organisations had new opportunities to participate in international conferences and to build far-reaching networks. While aid from EU and other international sources was available during this period, it placed pressure on these groups to adopt patterns of action and organisation that closely mimicked those found in the West [*Carmin and Hicks, 2002*]. As these countries moved into the consolidation period, activism became further professionalised [*Jancar-Webster, 1998*] and sources of aid more limited. The EU was one of the few remaining donors in the region [*Carmin and Hicks, 2002*], but many of these funds were dedicated to specific types of environmental projects rather than to civil society initiatives.

Environmental Transitions

In the wake of the collapse of the communist systems, Western media and scholars chronicled the tremendous environmental damage and the environmental challenges facing transition states and societies [*DeBardeleben, 1993; DeBardeleben and Hannigan, 1995; Pryde, 1995; Vari and Tamas, 1993; McCuen and Swanson, 1993; Simons, 1990*]. While severe degradation was present, these reports and studies often focused on environmental 'hot spots' or relied on data collected at points of heightened pollution [*Pavlínek and Pickles, 2000*]. They also tended to focus on environmental damages, while ignoring or downplaying more positive environmental practices and conditions in the region. For example, many countries had relatively high levels of recycling and low levels of automobile use [*Gille, this volume; Pavlínek and Pickles, this volume*]. Furthermore, these reports ignored or bracketed the presence of numerous protected natural areas and landscapes.

With the pivotal role played by ecological issues in the overthrow of the communist regime, many anticipated that environmental concern would remain prominent and that CEE countries would become environmental leaders [*Pavlínek and Pickles, 2000; Beckmann, Carmin and Hicks, 2002*]. Although this potential was not fully realised, significant resources and energy have been dedicated to the development of new environmental policies and laws, the formation of political institutions, and the remediation of past environmental degradation. In terms of environmental policy development, CEE states such as Poland and Czechoslovakia were early movers. By 1992, these countries had begun to strengthen their environmental ministries and passed laws that expanded government authority to regulate environmental quality.[2] By 2000, such developments had moved well beyond the environmental lead states, with changes taking place in counties such as Bulgaria, Croatia, Lithuania, Slovakia and Romania [*Kruger and Carius, 2001; UNECE, 1998, 1999, 2000a, 2000b, 2001*].

Though environmental policy institutions have been strengthened substantially across the CEE region, a large portion of reductions in pollution emissions in the 1990s were a result of economic restructuring [*Archibald et al., this volume; HELCOM, 1998; Selin and VanDeveer, forthcoming*]. Broadly speaking, the economic transitions away from state socialism can be characterised by three overlapping processes: privatisation, liberalisation and institution building,[3] all of which are intended to produce more efficient, more effective and more innovative markets and market actors. Like many aspects of the political transition, much of the economic transition has important impacts, both positive and negative, on environmental politics, policy and quality. Privatisation and liberalisation, for example, resulted in the closure of many inefficient and polluting industrial facilities, helping to reduce some types of waste and inefficiencies engendered by the subsidies and corruption that characterised state socialist economic management [*Archibald et al., this volume*]. On the other hand, because these same economic processes are promoting the adoption of Western-style practices, such as material consumption, they are contributing to unsustainable environmental outcomes commonly found in Western Europe and North America [*Legro and Auer, 2004; Gille, this volume; Gille, 2000; Pavlínek and Pickles, this volume; Pavlínek and Pickles, 2000*].

Significant investments have been made in building environmental policy institutions. However, bringing about broad environmental policy reforms of the type required in the CEE region to harmonise with EU policies is costly. Recent reports suggest that the total cost of adopting the environmental *acquis* will range from 80 to 100 billion Euro, requiring that candidate countries spend an average of two to three per cent of their gross domestic product (GDP) to implement the environmental *acquis*. There are trade-offs between the high costs of implementing the *acquis* and reductions in expenditures and costs that will be achieved within candidate countries as a result of improved environmental quality and human health [*Ecotech,*

2001b]. For example, it has been estimated that by 2010 there will be a reduction in particulates by over 1.8 million tons. This translates into lower medical costs since improvements in air quality are expected to cut significantly the number of premature deaths and contribute to an overall decrease in cases of chronic bronchitis. Similarly, anticipated reductions in landfill methane emissions and improvements in urban wastewater treatment, surface water, and groundwater aquifers should produce health benefits that will result in savings in medical costs. Additionally, the expense of implementing the *acquis* will be offset by lower costs in areas such as waste collection, treatment and disposal [*Ecotech, 2001b*].

In the late 1990s, assessments by the EU suggested that it was unlikely that the applicant countries could comply with the *acquis* in the short term and that staggered accession would lead all of the countries to reduce their efforts at environmental protection, leading to a deterioration of environmental quality in all of Europe [*European Parliament, 1998*]. More recent reports, however, suggest that the presence of environmental rules and standards in the *acquis* will foster improvements in air and water quality, enhance the efficiency of waste management, and reinforce the protection of natural areas [*Ecotech, 2001b*]. Further, while EU policies will promote improvements in the environmental quality of candidate states, they also will reduce transboundary pollution affecting present member states [*Europa, 2002b*].

International Assistance, Capacity Development, and the Environmental *Acquis*

Environmental transition, like the more general political and economic transitions, engenders needs for different skills, information and knowledge as well as different organisational structures and social institutions than were common in the communist era. In an effort to promote the requisite changes, international assistance from various national, intergovernmental and non-governmental sources flowed into the CEE region following the collapse of the socialist systems. Many international assistance programmes with environmental components explicitly sought to develop environmental capacities of state actors and institutions, NGOs, private sector actors and domestic publics.

Assistance away from Communism

The regime changes in 1989 were accompanied by a rush of international actors seeking to assist CEE states and societies in their transitions away from state socialism [*VanDeveer and Carmin, forthcoming*]. Bilateral assistance was offered by most West European states, the United States,

Canada and Japan and by intergovernmental organisations such as the World Bank and IMF as well as a myriad of non-governmental organisations. With stories of severe environmental degradation receiving significant play, many donors focused on providing financial assistance for pollution remediation and the development of environmental institutions [*Baker and Jehlièka, 1998*].

An estimated 3.5 billion ECU was invested in environmental remediation and protection by international governments between 1990 and 1995 [*Kolk and van der Weij, 1998*]. While a significant portion of these investments came from loans through multilateral development banks, bilateral assistance efforts also were common. Germany and Denmark led the list of European donors, dedicating approximately 392 and 118 million ECU respectively to the environment. The United States committed approximately 231 million ECU to environmental issues during this same period [*Kolk and van der Weij, 1998*]. Intergovernmental donors also provided support for environmental issues. For instance, the World Bank provided loans totalling US$788 million for environmental projects between 1990 and 1994 [*Connolly, Gutner, and Bedarff, 1996*]. In addition to providing financial support, states, national and multilateral governmental agencies, private foundations, NGOs and private firms from around the world provided scientific, technical and policy guidance during this period [*Baker and Jehlièka, 1998; Gutner, 2002*].

International actors did not limit their attention to the development of governmental institutions and organisations. They also provided support for NGOs, democracy promotion and civil society development programmes [see *Carmin and Hicks, 2002; Quigley, 2000; Kolk and van der Weij, 1998*]. For example, EU and US funds were channelled towards environmental education and used to establish NGO funding programmes. They also led to the creation and support of the Regional Environmental Center (REC). Founded in 1990, REC was designed to build the capacities of environmental NGOs through training, education and direct support of environmental initiatives and to encourage CEE states to recognise and work with NGOs [*Jancar-Webster, 1998*]. In the early transition years, assistance was also offered by international environmental NGOs such as the World Wide Fund for Nature (WWF), Friends of the Earth, and the World Conservation Union (IUCN), as well as by international foundations such as the Rockefeller Brothers Foundation and the German Marshall Fund.

Assistance towards EU Membership

The years immediately following the 1989 collapse of communist regimes are distinguished by efforts made to relegate authoritarianism and the environmental damage attributed to state socialism to the annals of history. Since the mid-1990s, CEE officials and civil society actors – as well as international assistance programmes – increasingly have directed their efforts and support towards satisfying the requirements of EU accession. With many donors leaving the region, European intergovernmental organisations and the EU have become the dominant sources of financial and technical support for environmental policy change and remediation.

By the late 1990s, large percentages of EU assistance to CEE countries prioritised 'harmonisation' of CEE policy and practices with EU directives and regulations [see *Carius, Homeyer and Bär, 2000*]. Environment-related examples include the PHARE Twinning programme and LIFE (Financial Instrument for the Environment). PHARE (Poland/Hungary Aid for the Reconstruction of the Economy) was initially designed to assist in the development of democratic institutions and aid the economic transition. PHARE focuses primarily on increasing capacities of public organisations and, over time, it dramatically increased its environmental assistance. LIFE assists EU member states in financing nature conservation and the implementation of Community environmental policies. As aspiring EU members, Estonia, Hungary, Latvia, Romania, the Slovak Republic and Slovenia took advantage of this programme. The consolidation of EU influence in the 1990s extends to civil society assistance as well. PHARE funded CEE environmental NGOs to carry out environmental projects and the EU's 6th Environmental Action Programme, approved in 2002, provides funds to NGOs from the EU15 and candidate countries.

Capacity Development and Accession

As the enormity of the tasks associated with harmonising CEE domestic law, regulation and practices with the *acquis* came into focus, it became clear that CEE state and civil society structures and actors did not have sufficient resources to realise the necessary changes. These lagging capacities were abundantly clear around environmental issues. The term 'capacity building' refers to efforts and strategies intended to increase the 'efficiency, effectiveness, and responsiveness of government performance' [*Grindle, 1997: 5*]. Capacity-building activities have historically focused on the enhancement of regulatory mechanisms, technical capabilities and resource availability. As problems with many technical assistance programmes illustrate, capacity-building initiatives often fail to assess the actual roots of constraints on the performance of individuals and organisations. Instead, they focus on concrete and obvious (to donors) expressions of incapacity such as the absence of certain technologies or procedures or the failure to perform specific functions [*Grindle, 1997*].

Proponents of 'capacity development' suggest the frequent focus of international capacity-building programmes on such factors as the provision of training programmes, information and technologies, is not adequate to ensure public sector capacity [*VanDeveer and Dabelko, 2001; Berg, 1993*]. Instead, to promote good governance, it is essential to take a more integrated approach, considering human resources as well as organisational and institutional capabilities [*Grindle, 1997*]. While capacity development requires well-trained and well-equipped personnel, it is also essential to have effective and efficient governmental and non-governmental organisations and to establish appropriate institutional environments in which these organisations can operate [*Grindle, 1997*].

'Capacity development for the environment' (CDE) applies this more integrated perspective directly to the environmental arena [*Sagar, 2000*]. From this perspective, the capacity to implement the environmental *acquis* in CEE countries not only relies on government capabilities, but also on the combined capacities of civil society and public sector organisations and institutions [*Garvey, 2002; Ecotech, 2001a*]. For example, the effective development and implementation of pollution control and prevention programmes necessitates the clear delineation of legal and regulatory authorities at national and local levels, adequate monitoring and enforcement capabilities, and integrated processes to link scientific and technical information with ongoing legal and regulatory development [*VanDeveer and Sagar, forthcoming; Miller, 1998; VanDeveer, 1998*].

During the decade of EU–CEE accession negotiations and preparations for EU membership, the focus has remained primarily on harmonisation of CEE law and regulations with the *acquis*. Much of the EU's environmentally focused capacity-building assistance was designed to support this legal and regulatory development. Funding was often provided to support the translation of EU directives and regulations into CEE languages, educate and train CEE policymakers as to the requirements of EU policymakers, draft CEE legislation and regulations, and assess the distance or contradictions between existing CEE policy and that required by the EU. As demonstrated by periodic reports by EU and other international bodies on CEE progress towards harmonisation, these efforts have borne fruit in the form of dramatically rewritten environmental law and policy across the CEE region.[4] However, with the focus of these efforts on capacity building, rather than capacity development, the abilities of CEE states, NGOs and firms to actually implement this vast new body of law and regulation at the time of accession remains in question.

'Europeanisation' and CEE Accession

Foreign assistance often directs resources to remediating environmental problems and establishing new environmental programmes, policies and organisations. In CEE countries, many international assistance programmes with environmental components explicitly sought to develop environmental capacities of state actors and institutions, NGOs, private sector actors or domestic publics. While building capabilities, they also serve to diffuse norms and expertise and can thereby shape institutional development [*Linden, 2002*]. Improvements achieved in environmental quality and the formation of new environmental institutions in Central and Eastern Europe have been heavily influenced by foreign aid through traditional capacity-building programmes as well as those that adopt a capacity development perspective. At the same time that these programmes have reshaped CEE environmental policies, they also diffuse Western environmental norms and values [*VanDeveer, 1997*].

A rapidly expanding social science research agenda seeking to define, assess and measure the 'Europeanisation' of domestic policies testifies to both the dramatic expansion of EU competencies in many policy areas and the widespread acceptance of the fact that EU member states and societies are significantly influenced by decisions taken in Brussels [*Börzel, 2002; Jordan and Liefferink, 2004; Knill, 2001; Knill and Lenschow, 2000*]. While this 'Europeanising' turn in social science research has generally been used to understand the relationships between the EU and its member states, a number of findings and lessons from this research can be applied to EU enlargement.

One school of Europeanisation research has focused on the 'top-down' dynamics of EU influence on such factors as member state policy content, policy styles, state structures and processes. Another has examined the movement of various policy competencies 'up' to the EU level, and a third assesses the dynamic interaction of EU and member state bodies and debates [*Jordan and Liefferink, forthcoming*]. The Europeanisation scholarship focusing on the domestic adjustments needed for implementation of EU policies – the more top-down approach – is likely the most relevant to eastern expansion because of the EU requirement that CEE states enact the *acquis*. This perspective of EU–state relations begins with the assumption that member states must change to accommodate EU policy decisions. This is certainly the case for CEE states required to adopt the entire *acquis* – a body of law that CEE officials did not participate in making.

Europeanisation scholarship analyses EU-induced domestic adjustment through one of three general mechanisms or causal pathways [*Knill and Lehmkuhl, 1999*]. The first, a hierarchical institutional model, suggests that subordinate units, such as EU member states, adjust domestic institutions when required to do so by EU policymaking processes. From this point of view, EU decisions prescribe changes that formally subordinate organisations such as member states are supposed to accommodate. Altering domestic opportunity structures is a second pathway through which change can take place. This approach examines how incentives for various actors in domestic politics are altered by international processes such as EU political debate and policymaking. Changing incentives for actors may, for example, result from changes in market incentives within the EU's single market [*Andonova, 2003*]. They might also stem from the ability of domestic environmental officials to leverage EU environmental policy debates or requirements into greater domestic political influence. The third mechanism of Europeanisation sees actors changing more than their strategies and interests; they change their preferences, beliefs and expectations. For example, environmental concern and awareness among citizens or policymakers may be increased by EU policy debates, procedures and dictates. If this happens, actors' preferences regarding policy and environmental quality may be changed [*Knill, 2001; Knill and Lehmkuhl, 1999*].

EU Enlargement and the Environment

In the years following the transition and leading towards accession, CEE countries have focused on supporting negotiations and harmonising law and regulatory policy with EU directives. While these processes have their costs and benefits, as the three pathways suggest, they also serve as conduits for Europeanisation of CEE states and societies. Accordingly, many accounts of CEE transition focus on the socialisation processes stemming from EU influence, giving limited attention to the role of domestic agency and action. Further, most accounts of Europeanisation and capacity development address a wide range of substantive areas [*Cowles, Caporaso and Risse, 2001; Linden, 2002*]. This volume uses environmental governance and action as a means for understanding the profound role that the EU plays in shaping norms and practices and, at the same time, the interactions between external forces and state-level agency and history.

This volume brings together scholars and policy practitioners with a broad range of intellectual training and backgrounds. Because the authors are not bound by a single conceptual framework or set of questions, their diverse training and perspectives provide a basis for broad understanding of the interactions between international influences and domestic agency. The authors were selected for their substantive areas of expertise, their knowledge and experience in CEE countries, and the unique vantage point that they could provide on EU enlargement and the environment. Each was charged with addressing the anticipated impact of EU enlargement on environmental policy and politics, paying particular attention to socialisation and the diffusion of ideas, norms and practices, capacity development for the environment, the role of non-state actors and civil society, and changes in environmental quality. Authors were encouraged to address accomplishments to date and remaining challenges associated with environmental institutions. They were also asked, where appropriate, to assess environmental outcomes.

Many of the volume's contributors draw examples and case material from Hungary, Poland and the Czech Republic. This focus mimics a commonly held belief that this group of countries would be among the first to join the EU. Although these countries appear throughout the volume, a number of contributions also examine them relative to other CEE countries, developing regional insights and trends. As a result, this volume provides depth of analysis of a small group of nations while offering sufficient breadth to afford a comparative perspective.

Organisation of the Volume

Each of the volume's authors address state-level opportunities and tensions arising from the harmonisation and transposition processes. At the same time, each also focuses on a particular aspect of EU enlargement, using a unique lens to understand its implications for the environment. Part I, 'EU Enlargement, Institutions, and Environmental Politics', considers broad institutional changes associated with EU enlargement from practical and theoretical perspectives, drawing attention to changes in EU institutions as well as to those in CEE states. Schreurs examines the effects of the CEE accession on environmental policymaking within the EU. Drawing on evidence from the historical development of EU environmental policy authority and from previous expansions in EU membership, she argues that harmonisation offers opportunities for innovation and collaborative approaches to problem solving. Most importantly, Schreurs asserts that by making investments to remediate pollution and enhance environmental quality in CEE and by promoting aggressive environmental policies, the EU is assuming a regional and global leadership role in environmental protection.

Homeyer brings a more theoretical perspective to this section, using a historical–institutionalist lens to examine the potential impact of enlargement on areas of EU environmental governance. He anticipates that enlargement will affect three distinct EU governance regimes – the internal market, environmental management and sustainability regimes – that comprise and help to reproduce EU environmental policy. He maintains that the interrelated institutional mechanisms associated with each of the regimes will contribute to overall resilience and stability of environmental governance.

While Homeyer examines the impact of change, the final contribution to this section by Jehlička and Tickle examines the presence of institutional capacity to implement the *acquis*. The authors use empirical evidence to support their claim that CEE states have taken on passive and reactive roles regarding EU environmental policy, with national perspectives and priorities now dominated by EU goals and interests. While EU policies have been transposed in earnest and good intentions abound in CEE states, implementation of the body of newly transposed laws and policies will be difficult to realise in the short term owing to capacity limitations. At the same time, Jehlička and Tickle remain sceptical that CEE states' entrance into the EU will greatly slow or reverse the Union's environmental policy expansion, as many Western analysts have claimed. Together, the three chapters in Part I suggest that fears of negative CEE impacts on EU environmental governance expressed by many in Western Europe are overdrawn.

The contributions in Part II, 'Environmental Policy Challenges', examine pressing issues faced by accession countries as they gain EU membership. Using their particular issue areas and country cases, the contributors draw general lessons and conclusions about EU environmental policy and the challenges faced by various actors in an enlarged EU. Kružíková initiates the section with an investigation of legal institutions and the legislative challenges that countries face as they near accession. Using the case of the Czech Republic, she briefly describes how much environmental law has changed since 1990 to become increasingly harmonised with EU law. While the adoption and transposition of EU laws is a necessary step towards CEE accession, this set of challenging tasks has been accomplished within the context of many remaining aspects of the socialist legal order that developed over 40 years. Kružíková maintains that adoption of the laws alone and the accompanying mandate that national law be superseded by Community law pose numerous domestic and international institutional challenges. She suggests that the implementation and enforcement of Community laws creates tensions in domestic legal culture and practices, particularly in the face of numerous capacity limitations within the legal system. Unlike many other analysts, Kružíková argues that some implementation difficulties for CEE states stem from the content and nature of EU law, and not only exclusively from deficiencies in CEE countries.

Gille's discussion of waste and waste minimisation in Hungary argues that the EU sends mixed messages to candidate countries. On the one hand, the EU requires the adoption of waste management and recycling policies.

On the other hand, it is promoting consumption and the use of non-recyclable materials. Understanding issues pertaining to waste is critical since this has been one of the most difficult policy issues for the accession countries to tackle and is one area where extensions for compliance have consistently been granted. Moreover, this case provides an example of how market forces play into environmental decisions and how some environmental conditions may have been more sustainable under the former regime.

Gille's articulation of the presence of mixed messages within a single policy arena is similar to the types of tensions that Beckmann and Dissing demonstrate are present across policy domains. They suggest that the EU has developed rural policies and agricultural programmes that have the potential to build local capacity and enhance environmental quality. At the same time, overall environmental quality and sustainable development are undermined by the goals, practices and incentives supported by many EU funding programmes, including those designed to prepare countries for EU accession. In conclusion, they argue that, although EU rural and agricultural policies are unable to realise the promise of sustainable agriculture, NGOs and foundations are assuming leadership roles in fostering rural sustainability. Nevertheless, such actors face the daunting challenges presented by EU policies and EU-driven incentives that contradict movement towards sustainable agriculture.

Axelrod chronicles political processes and decisions associated with the controversial Temelín nuclear power plant in the Czech Republic. This case illustrates strengths and weaknesses of environmental policy and politics within the EU and between EU members and CEE candidate states. The Temelín case also highlights the EU's role as potential intermediary in bilateral disputes between EU members and non-members. In the process of chronicling the developments related to Temelín, Axelrod argues that the EU plays a powerful role in shaping the norms and behaviour of both member and applicant states. Her piece demonstrates that bilateral disputes, and the EU role in them, can be especially difficult to sort out when they concern issues such as nuclear power plant safety standards, around which the EU has very little regulatory competency and about which its member states differ greatly.

Environmental issues and organisations played critical roles in the overthrow of the communist regime. While all of the volume's authors acknowledge the significance of non-state actors in accession and environmental governance, Part III, 'Civil Society in an Enlarged EU', takes an in-depth look at CEE civil society development and the relationships between these organisations and EU bodies. Hallstrom initiates this section with an investigation of the informal dynamics of environmental politics and governance. Drawing on interviews with EU officials, he maintains that the participation of non-governmental organisations in Brussels, particularly those that do not have specialised scientific or technical expertise, is limited to symbolic gestures and token forms of involvement. Rather than candidate states pushing for greater inclusion of these groups, they tend to reinforce the technocratic, top-down approach to governance common in Brussels. Even though NGOs presently may have limited involvement and influence, Bell maintains that accession will create important opportunities and critical obligations for these groups to contribute to environmental policy and management in the CEE region and across Europe. Her contribution furthers the theme of EU domination of political agendas introduced in Part I. Bell suggests that in the short term, involvement will be oriented to agenda items and issues that are established by the EU. At the same time, she maintains that if NGOs can utilise the opportunities available to them, they will serve important roles in building stronger and perhaps more enduring ties between government and civil society actors in the CEE region.

In the final contribution to Part III, Hicks takes these arguments a step further by suggesting that the EU is a source of diffusion for environmental norms and practices. She presents a framework used to gain greater understanding of the different ways that EU bodies shape the agendas and actions of environmental movements and movement organisations. She maintains that centralised decisions and decision-making processes influence environmental activism. However, this is just part of the story since laws, policies, funding and organisational requirements that place constraints on power and resources also channel the priorities and activities of environmental activists. Hicks limits her examples and analysis to environmental movements and movement organisations. However, taken as a whole, Part III suggests that the entire transition, harmonisation and accession processes are shaping – even defining – the actions and agendas of civil society actors more broadly. This raises questions about whether such groups actually function as an independent ‘third sector’, or whether they have aligned their goals and priorities too closely to those of the EU.

The contributions in Part IV of the volume, ‘Environmental Outcomes: From State Socialism to EU Membership,’ reflect on environmental outcomes achieved to date and highlight numerous challenges faced by the accession states and, by implication, the enlarged EU as a whole. Pavlínek and Pickles demonstrate that implementing the environmental *acquis*, like broad based economic and political reform, is a complex process in countries with a legacy of socialism and environmental struggle. They agree with other authors in the volume who suggest that significant environmental improvements have been achieved, though they maintain that these results stem from both economic reforms and environmental policy action shaped by the *acquis*. They also assert that the break with socialist practices has not been fully realised and that limitations of EU policies and the promotion of unsustainable consumption may have unintended environmental consequences over the long term. The contribution by Archibald, Banu, and Bochniarz focuses on environmental changes driven by market liberalisation and privatisation. They argue that significant pollution reductions have been achieved and that associated health and social gains also have been realised in those states that enacted the quickest and most dramatic economic reforms. Kramer concludes Part IV by highlighting a set of key challenges faced by accession states and, by implication, EU institutions and the other countries remaining in line for EU membership. In so doing, he outlines numerous fiscal, administrative, environmental, democratic, nuclear, and political tasks that lie ahead. The volume's conclusion summarises some of the main points raised by the contributors and draws out lessons and possible inplications of eastern accession to EU membership.

Key Themes and Arguments

Collectively, the contributions in this volume examine environmental initiatives driven by EU policies and programmes and the desire of CEE officials and publics to gain EU membership. They also explore the impacts of the EU on environmental policy and protection, as well as the relationship between government and civil society actors in the policy process. When reviewed as a whole, the authors suggest that CEE states have significant capacity limitations, but are making concerted efforts to address them even in the face of the mixed messages they are receiving as a result of the EU's conflicting priorities. The authors further note the importance of non-state actors, both with respect to their present accomplishments, and, more importantly, as an untapped resource that can benefit CEE states and the EU alike. Finally, the contributions suggest that individual CEE states and NGOs could bring knowledge to the EU, in contrast to the unidirectional dynamics of the accession process that have assumed that CEE states and societies were only recipients of expertise. A more concerted effort to promote a multi-directional exchange of ideas and information between the EU15, accession states, and NGOs and officials in Brussels is likely necessary to realise this joint learning potential.

Some of the points raised by the authors reinforce prevailing arguments in the literature. In particular, they maintain that EU pressures are not only altering environmental policies and incentives, but also are changing values and behavioural norms in individual countries. However, while the

Europeanisation debates centre on EU–member state relations, the authors suggest that external pressures in the race to accession are promoting Europeanisation in applicant states and that all three of the pathways associated with Europeanisation are contributing to the changes in environmental governance and behaviour that have taken place across the CEE region. They observe that the transition and accession processes of the last several years have changed both the strategic environment in which CEE domestic actors operate and the values, beliefs and norms held by some CEE individuals, groups and organisations.

Together, the contributions indicate that accession does not preclude opportunities for independent forms of national and sub-national action in the new member states. CEE states potentially offer perspectives, resources and innovations that could enhance EU policy along important dimensions. In other words, not only are opportunities present for independent state action and the influence of domestic actors, but the potential exists for CEE countries to strengthen EU governance. These views represent different framings of Europeanisation and EU enlargement than have been articulated to date. In effect, the contributions collectively suggest that, although various environmental policy and civil society capacities are limited in CEE states, these countries have the potential to make genuine contributions to EU environmental policy and quality. Further, despite the many challenges associated with eastern accession documented in the contributions that follow, this volume suggests that enlargement presents the EU with numerous opportunities to enhance its leadership role in regional and global environmental politics.

NOTES

1. For general discussions of economic transitions from communist to capitalist economic systems see Aslund [*1999*], Gerber [*2002*] and World Bank [*1996*].
2. See Table 1 in Kružíková (this volume) for an illustrative summary of changes in the Czech Republic.
3. Across the transition period, there have remained tensions between institution building and public and private sector capacity-building programmes and the logic of privatisation, liberalisation and decentralisation. Privatisation essentially involves the transfer of state-owned property to private hands. Liberalisation is a general term often used to refer to the 'freeing' of various economic sectors or transactions from state or monopolistic control.
4. See, for example, REC [*1996, 2000*], the European Commission's series of 'Regular Reports' on enlargement and on progress toward accession [e.g. *European Commission, 2001*], and the series of 'Environmental Performance Reviews' organised under the auspices of the Organisation for Economic Cooperation and Development and the United Nations Economic Commission for Europe [e.g. *OECD, 2000; UNECE, 1996*].

REFERENCES

Andonova, Liliana (2003), *Transnational Politics and the Environment: EU Integration and Environmental Policy in Central and Eastern Europe*, Cambridge: MIT Press.

Aslund, Anders (1999), 'Post-Communist Economic Transformation', in A. Brau and Z. Barany (eds.), *Dilemmas of Transition: The Hungarian Experience*, Lanham, MD: Rowman & Littlefield, pp.69–90.

Baker, Susan and Petr Jehlièka (1998), 'Dilemmas of Transition: The Environment, Democracy and Economic Reform in East Central Europe – An Introduction', in S. Baker and P. Jehlièka (eds.), *Dilemmas of Transition: The Environment, Democracy and Economic Reform in East Central Europe*, London: Frank Cass Publishers, pp.1–28.

Berg, Elliot J. (1993), *Rethinking Technical Cooperation: Reforms for Capacity Building in Africa*, New York: United Nations Development Program.

Beckmann, Andreas, JoAnn Carmin and Barbara Hicks (2002), 'Catalysts for Sustainability: NGOs and Regional Development Initiatives in the Czech Republic', in Walter Leal Filho (ed.), *International Experiences on Sustainability*, Bern: Peter Lang Scientific Publishing, pp.159–77.

Börzel, Tanja (2002), *States and Regions in the European Union; Institutional Adaptation in Germany and Spain*, Cambridge: Cambridge University Press.

Carmin, JoAnn and Barbara Hicks (2002), 'International Triggering Events, Transnational Networks, and the Development of the Czech and Polish Environmental Movements', *Mobilization: An International Journal*, Vol.7, No.3, pp.305–24.

Carius, Alexander, Ingmar von Homeyer and Stefani Bär (2000), 'Eastern Enlargement of the European Union and Environmental Policy: Challenges, Expectations, Multiple Speeds and Flexibility', in K. Holzinger and P. Knoepfel (eds.), *Environmental Policy in a European Union of Variable Geometry?: The Challenge of the Next Enlargement*, Basel: Helbing & Lichtenhahn, pp.141–80.

Connolly, Barbara, Tamar Gutner and Hildegard Berdarff (1996), 'Organizational Inertia and Environmental Assistance in Eastern Europe', in Robert O. Keohane and Marc Levy (eds.), *Institutions for Environmental Aid*, Cambridge: MIT Press, pp.281–323.

Cowles, Maria Green, James Caporaso and Thomas Risse (2001), *Transforming Europe*, Ithaca: Cornell University Press.

DeBardeleben, Joan (1993), *To Breathe Free: Eastern Europe's Environmental Crisis*, Washington, DC: Woodrow Wilson Center Press/Johns Hopkins University Press.

DeBardeleben, Joan and John Hannigan (1995), *Environmental Security and Quality after Communism*, Boulder, CO: Westview Press.

Ecotech (2001a), *Administrative Capacity for Implementation and Enforcement of EU Environmental Policy in the 13 Candidate Countries*, DGENV Contract: Environmental Policy in the Candidate Countries and their Preparations for Accession, Birmingham, UK.

Ecotech (2001b), *The Benefits of Compliance with the Environmental Acquis for Candidate Countries*, DGENV Contract: Environmental Policy in the Candidate Countries and their Preparations for Accession, Birmingham, UK: July.

Europa (2002a), 'Enlargement: Negotiations of the Chapter 22: Environment'. Available at http://europa.eu.int/comm/enlargement/negotiations/chapters/chap22/index.htm.

Europa (2002b), 'Accession Negotiations: State of Play'. Available at http://europa.eu.int/comm/enlargement/negotiations/pdf/stateofplay_july2002.pdf.

European Commission (2001), 'Regular Reports on Progress Towards Accession', Brussels. Available at http://europa.eu.int/comm/enlargement/report2001/.

European Parliament (1998), 'Environmental Policy and Enlargement', Briefing No.17, Luxembourg, 23 Mar., PE 167.402. Or. DE. Available at http://www.europarl.eu.int/enlargement/briefings/17a3_en.htm#6.

Fagin, Adam (1994), 'Environment and Transition in the Czech Republic', *Environmental Politics*, Vol.3, No.3, pp.479–94.

Garvey, Tom (2002), 'EU Enlargement: Is it Sustainable?' in Sabina Crisen and JoAnn Carmin (eds.), *EU Enlargement and Environmental Quality: Central and Eastern Europe and Beyond*, Washington, DC: Woodrow Wilson International Center for Scholars, pp. 53–62.

Gerber, James (2002), *International Economics, 2nd Edition*, Boston: Addison Wesley.

Gille, Zsuzsa (2000), 'Legacy of Waste or Wasted Legacy? The End of Industrial Ecology in Post-Socialist Hungary', *Environmental Politics*, Vol.9, No.1, pp.203–30.

Grindle, Merilee S. (1997), *Getting Good Government: Capacity Building in the Public Sector of Developing Countries*, Cambridge, MA: Harvard University Press.

Gutner, Tamar L. (2002), *Banking on the Environment: Multilateral Development Banks and Their Environmental Performance in Central and Eastern Europe*, Cambridge, MA: The MIT Press.

HELCOM (Helsinki Commission) (1998), *Final Report on the Implementation of the 1988 Ministerial Declaration*, Baltic Sea Environment Proceedings No.71, Helsinki: HELCOM.

Hicks, James K. and Bartomiej Kaminski (1995), 'Local Government Reform and the Transition from Communism: The Case of Poland', *Journal of Developing Societies*, Vol.9, No.1, pp.1–20.

Huntington, Samuel P. (1991), *The Third Wave: Democratization in the Late Twentieth Century*, Norman: University of Oklahoma Press.

Jancar-Webster, Barbara (1993), *Environmental Action in Eastern Europe: Responses to Crisis*, Armonk, NY: M.E. Sharpe.

Jancar-Webster, Barbara (1998), 'Environmental Movement and Social Change in the Transition Countries', in Susan Baker and Petr Jehlièka (eds.), *Dilemmas of Transition: The Environment, Democracy and Economic Reform in East Central Europe*, London: Frank Cass, pp.69–92.

Jordan, Andrew and Duncan Liefferink (2004), *Environmental Policy in Europe: The Europeanization of Environmental Policy in Europe*, London: Routledge.

Knill, Christoph (2001), *The Europeanisation of National Administrations: Pattern of Institutional Change and Persistence*, Cambridge: Cambridge University Press.

Knill, Christoph and Dirk Lehmkuhl (1999), 'How Europe Matters: Different Mechanisms of Europeanization', *European Integration on-line Papers (EIoP)*, Vol.3, No.7. Available at http://eiop.or.at/eiop/texte/1999-007a.htm.

Knill, Christopher and Andrea Lenschow (2000), *Implementing EU Environmental Policy: New Directions and Old Problems*, Manchester: Manchester University Press.

Kolk, Ans and Ewout van der Weij (1998), 'Financing Environmental Policy in East Central Europe', in Susan Baker and Petr Jehlièka (eds.), *Dilemmas of Transition: The Environment, Democracy and Economic Reform in East Central Europe*, London: Frank Cass, pp.53–68.

Kruger, Christine and Alexander Carius (2001), *Environmental Policy and Law in Romania: Towards EU Accession*, Berlin: Ecologic.

Legro, Susan and Matthew R. Auer (2004), 'Environmental Reform in the Czech Republic: Uneven Progress after 1989,' in Mathew Auer (ed.), *Restoring Cursed Earth: Appraising Environmental Policy Reforms in Central and Eastern Europe and Russia*, Boulder, CO: Rowman & Littlefield Press.

Linden, Ronald H. (2002), *Norms and Nannies: The Impact of International Organizations on the Central and Eastern European States*, Lanham, MD: Rowman & Littlefield Publishers.

McCuen, Gary E. and Ronald P. Swanson (1993), *Toxic Nightmare in the USSR and Eastern Europe*, Hudson, WI: Gary E. McCuen Publications.

McFaul, Michael (2002), 'The Fourth Wave of Democracy *and* Dictatorship: Noncooperative Transition in the Postcommunist World', *World Politics*, Vol.54, No.2, pp.212–44.

Miller, Clark (1998), 'Extending Assessment Communities to Developing Countries', ENRP Discussion Paper E-98-15, Kennedy School of Government, Harvard University.

O'Donnell, Guillermo, Phillipe C. Schmitter and Laurence Whitehead (1986), *Transitions from Authoritarian Rule: Prospects for Democracy*, Baltimore: Johns Hopkins University Press.

OECD (Organisation for Economic Cooperation and Development) (2000), 'Environmental Performance Reviews (1st Cycle): Conclusions and Recommendations: 32 Countries (1993–2000)', OECD Working Paper on Environmental Performance, Paris: OECD.

Pavlínek, Petr and John Pickles (2000), *Environmental Transitions: Transformation and Ecological Defence in Central and Eastern Europe*, Routledge: London.

Pridham, Geoffrey, Eric Herring and George Sanford (1994), *Building Democracy? The International Dimension of Democratization in Eastern Europe*, New York: St Martins Press.

Pryde, Phillip R. (1995), *Environmental Resources and Constraints in the Former Soviet Republics*, Boulder, CO: Westview Press.

Quigley, Kevin F.F. (2000), 'Lofty Goals, Modest Results: Assisting Civil Society in Eastern Europe', in T. Carothers and M. Ottaway (eds.), *Funding Virtue: Civil Society Aid and Democracy Promotion*, New York: The Carnegie Endowment, pp.191–216.

REC (Regional Environment Center for Central and Eastern Europe) (1996), *Approximation of European Union Environmental Legislation*, Budapest: REC.

REC (Regional Environment Center for Central and Eastern Europe) (2000), *Europe 'Agreening': 2000 Report on the Status and Implementation of Multilateral Environmental Agreements in the European Region*, Szentendre: REC.

Sagar, Ambuj (2000), 'Capacity Development for the Environment: A View from the South, A View from the North', *Annual Review of Energy and the Environment*, Vol.25, pp.377–439.

Selin, Henrik and Stacy D. VanDeveer (forthcoming), 'Baltic Hazardous Substances Management: Results and Challenges', *AMBIO: Journal of the Human Environment.*

Simons, Marlise (1990), 'Eastern Europe: The Polluted Lands', *The New York Times Magazine*, 29 Apr., pp.30–35.

Singleton, Fred (1987), *Environmental Problems in the Soviet Union & Eastern Europe*, Boulder, CO: Lynne Rienner Publishers.

Slocock, Brian (1996), 'The Paradoxes of Environmental Policy in Eastern Europe: The Dynamics of Policy-Making in the Czech Republic', *Environmental Politics*, Vol.5, No.3, pp.501–21.

Tickle, Andrew and Ian Welsh (1998), *Environment and Society in Eastern Europe*, Essex: Longman.

UNECE (United Nations Economic Commission for Europe) (1996), *Environmental Performance Review of Estonia*, Geneva: UNECE.

UNECE (United Nations Economic Commission for Europe) (1998), *Environmental Performance Review of Lithuania*, Geneva: UNECE.

UNECE (United Nations Economic Commission for Europe) (1999), *Environmental Performance Review of Croatia*, Geneva: UNECE.

UNECE (United Nations Economic Commission for Europe) (2000a), *Environmental Performance Review of Bulgaria*, Geneva: UNECE.

UNECE (United Nations Economic Commission for Europe] (2000b), *Environmental Performance Review of Lithuania: Report on Follow-up, Estonia*, Geneva: UNECE.

UNECE (United Nations Economic Commission for Europe) (2001), *Environmental Performance Review of Romania*, Geneva: UNECE.

VanDeveer, Stacy D. (1997), 'Normative Force: The State, Transnational Norms and International Environmental Regimes', PhD dissertation, University of Maryland, College Park, MD.

VanDeveer, Stacy D. (1998), 'European Politics with a Scientific Face: Transition Countries, International Environmental Assessment and Long Range Transboundary Air Pollution', ENRP Discussion Paper E-98-9, Kennedy School of Government, Harvard University.

VanDeveer, Stacy D. and Ambuj Sagar (forthcoming), 'Capacity Building for the Environment: North and South', in E. Corell, A. Churie Kallhauge and G. Sjöstedt (eds.), *Furthering Consensus: Meeting the Challenges of Sustainable Development Beyond 2000*, London: Greenleaf.

VanDeveer, Stacy D. and Geoffrey D. Dabelko (2001), 'It's Capacity Stupid: National Implementation and International Assistance,' *Global Environmental Politics*, Vol.1, No.2, pp. 18–29.

VanDeveer, Stacy D. and JoAnn Carmin (forthcoming), 'Sustainability and EU Accession: Capacity Development and Environmental Reform in Central and Eastern Europe', in Gary B. Cohen and Zbigniew Bochniarz (eds.), *Sustainability in the New Central Europe*.

Vari, Anna and Pal Tamas (1993), *Environment and Democratic Transition: Policy and Politics in Central and Eastern and Europe*, Dordrecht: Kluwer Academic Publishers.
Whitehead, Laurence (1996), *The International Dimensions of Democratization: Europe and the Americas*, Oxford: Oxford University Press.
Wolchik, Sharon L. (1991), *Czechoslovakia in Transition: Politics, Economics, & Society*, London: Pinter Publishers.
Wood, David M. and Birol A. Yeslada (2004), *The Emerging European Union*, 3rd edition, New York: Pearson Longman.
World Bank (1996), *World Development Report: From Plan to Market*, Washington, DC: World Bank Group.
Yoder, Jennifer (2003), 'Decentralisation and Regionalisation after Communism: Administrative and Territorial Reform in Poland and the Czech Republic', *Europe-Asia Studies*, Vol.55, No.2, pp.263–86.

Part I

EU ENLARGEMENT, INSTITUTIONS AND ENVIRONMENTAL POLITICS

Environmental Protection in an Expanding European Community: Lessons from Past Accessions

MIRANDA SCHREURS

The EU has transformed environmental policymaking for the states of Europe. Over time, but especially since the Single European Act came into force in 1987, the competencies of the European Community (EC) in environmental policymaking have expanded tremendously. There are now close to 300 environmental regulations and directives governing environmental policy throughout the member states. In many environmental policy areas the EU now plays an international leadership role. For this reason it is important to consider what the planned accession of ten Central and Eastern European (CEE) countries may mean for the future of EU environmental leadership and policy performance. How will the accession of the CEE10, where environmental capacity is still weak and environmental problems are serious, affect environmental policymaking within the EU?

This contribution seeks to answer this question from a comparative perspective by looking at past cases of accession to the EU, with particular attention to the accession of the southern European states that are economically less well off than their northern neighbours and where environmental protection has a weaker tradition in national policy [*Pridham, 2002*]. It argues that to date (2003) in the big picture expansion has not proved to be a substantial drag on EU environmental programmes. Instead, on the whole EU environmental regulations have become more stringent and expansive over time. There also appears to have been considerable environmental capacity building across member states, including in the poorer accession states. A variety of changes to EU institutions and norms have helped to strengthen the EU's capacity to protect the environment, mitigate against moves towards least common denominator outcomes, and aid weaker states in coming into compliance with EU regulations [*Walti, 2003*]. Particularly important in this regard are the influence of the relatively strong and increasingly widespread voices of environmental advocates in Europe, the greater variety of environmental policy instruments available to member states that should make implementation more flexible and cost-effective, institutional reform that has addressed the EU's 'democratic deficit', and the importance that environment has taken on as an arena for foreign policy leadership by the EU.

This is not to argue that there are not important differences that remain in environmental policy performance and awareness across the EU15 [*Weale et al., 2000*]. Rather, it is to suggest that integration and harmonisation appear to be contributing to positive normative, programmatic and institutional environmental change and these trends are being reinforced by developments in the EU that are providing greater voice to environmental interests in society and introducing greater flexibility into environmental protection efforts. There are, however, still many problems with the implementation of the EU's extensive body of environmental laws, and not only in the poorer member states. Thus, with the next wave of accession, the EU will be faced by the double challenge of improving the EU15's environmental implementation performance while helping to bring the CEE10 into compliance as well.

This contribution makes four main points related to the question of why EU environmental standards appear to have strengthened over time despite the accession of states with substantially lower economic and environmental standards than the EU average. First, it suggests that on the whole the European Union has helped to strengthen environmental regulations in states where such regulations were weak and policy enforcement limited because of the relatively strong influence of environmental pressures emanating largely from northern Europe. Pushed largely by the environmental concerns of the northern states of Europe – and especially Denmark, Germany, The Netherlands and, since their accession in 1995 Sweden, Austria and Finland – the EU has adopted increasingly stringent environmental policies and programmes covering an ever wider range of environmental problems [*Liefferink and Andersen, 2002*]. This has forced countries such as Greece, Ireland, Italy, Portugal and Spain where environmental movements are less developed – the so-called 'policy takers' – to introduce many stringent environmental regulations that they might otherwise not have introduced for considerably longer periods of time, if at all [*Jordan, Liefferink and Fairbrass, 2004*].

Second, in response to strong criticisms of the EU's 'democratic deficit', there have been important institutional changes that have provided greater voice to societal opinions in decision making. In the past, decision-making powers were heavily centred in the European Council and European Commission, bodies that are not elected. To address this problem, some steps have been taken to enhance the powers of the elected European Parliament where centre-left parties and green parties have been well represented. This has been very important in the environmental policy field and has helped to continue the drive towards a strengthening overall of EU environmental policies. In preparation for the expansion of the EU from 15 to possibly as many as 30 members in the next years, the Treaty of Nice has created a new set of rules that will govern decision making. While it is still too early to predict what the implications of the new rules will be for environmental policymaking in an expanded EU, the addition of a large group of states where the history of environmental protection is relatively weak is likely to alter in important ways the direction of environmental policy in the next decade. The combined percentage of seats held by the relatively environmentally progressive states of Germany, The Netherlands, Sweden, Austria, Denmark and Finland in the EU Parliament will drop from approximately 33 per cent to 25 per cent and their share of EU Council votes will drop from roughly 33 per cent to 18 per cent. While this could weaken the voice of more environmentally oriented states in EU decision-making bodies, as is discussed more fully below, other institutional changes that have occurred in the EU to address the EU's 'democratic deficit' should counter-balance this trend to some extent [see also *Homeyer, this volume; Jehlička and Tickle, this volume*].

Moreover, expansion could result in a more 'pluralistic' approach to environmental decision making. One likely outcome of expansion, for example, is that the efforts that have begun to emerge in recent years among southern European states to get more of their concerns heard within the EU and to address what they see as the 'northern bias' of EU environmental programmes is likely to gain a boost with the accession of the Central and Eastern European states. Already, with accession looming, the CEE states were able to get some of their concerns discussed when the EU invited them to contribute to the development of the 6h Environmental Action Programme, a five-year action plan focused on moving the EU towards sustainable development and more effective and participatory policy implementation.

Third, there are still substantial problems with implementation of the body of environmental policies within the EU15 [*Glachant, 2001*]. The problem is most apparent in the peripheral states but even the most progressive environmental states in the EU15 have been brought before the European Court of Justice for failure to comply with EU directives. Implementation problems stem from a combination of factors: lack of sufficient institutional capacity, lack of adequate political will, high costs and, at times, simply bad policy. One reaction to this problem has been a slow shift in the EU towards the use of new environmental policy instruments. This could have far-reaching implications for environmental policy implementation in the CEE10 and aid them in coming into compliance with directives more smoothly and quickly than has been the case with past accessions. As has been the case throughout the industrialised world, command and control measures have dominated environmental protection efforts within the EU. There is now a slow transition occurring within the Community towards the greater use of market-based and voluntary mechanisms (taxation, voluntary agreements, emissions trading and joint implementation) that are expected to introduce greater flexibility into the system, with the aim of reducing the costs of implementation and increasing rates of compliance. The growing use and availability of a broad mix of policy instruments may aid the CEE states in meeting the stringent EU environmental standards.

Along with this shift towards the use of a larger mix of policy instruments, the European institutions are moving away from treating environmental protection as a separate regulatory sphere [*Lenschow, 2002*]. The EU's five-year Environmental Action Programmes – of which there have now been six – have shifted from a focus on developing new regulations that deal with pollution problems on a sectoral basis to a broader concern with sustainable development, the integration of environmental concerns into other policy areas (such as transportation, building and agriculture), and enhancing compliance with the existing body of environmental law. The most recent Environmental Action Programme, *Environment 2010: Our Future, Our Choice*, has four priority areas: climate change, nature and biodiversity protection, environment and health, and natural resources and waste. The Action Programme calls for greater attention to sustainable development, cooperation among actors in policy implementation, and the use of a wider range of policy tools. This trend also could have positive implications for the CEE10 to the extent that this kind of policy integration is in fact achieved.

Fourth and finally, as the EU has expanded its environmental expertise, it has become an increasingly powerful presence in international and global environmental policymaking. The EU bodies work cooperatively with the nations in the Mediterranean to improve the quality of the environment in the region. It has spent enormous sums aiding the clean-up of toxic sites in Central and Eastern Europe. It is increasingly active in enhancing environmental awareness and promoting conservation in India, China, Southeast Asia, Latin America and Africa. The EU also has taken a relatively active role in addressing various global environmental issues, such as climate change, the promotion of renewable energies and sustainable development.

Some have voiced their fears that expansion to the east will divert the attention of the EU from global environmental problems as member states struggle to pay for environmental clean-up and develop environmental capacity in the new member states. While this may well be a natural and necessary reaction to the costs and challenges of expansion, there will also be occasions where integration of new member states will provide the EU with new possibilities for cooperative environmental problem solving and global environmental leadership. Also, environmental policy is among the most successful of EU foreign policy areas. At a time when the EU has exhibited deep rifts in its foreign policy in relation to the war on Iraq, presenting a progressive and relatively united environmental foreign policy will take on added significance. This may mean that EU officials will work hard to maintain a strong environmental foreign policy.

The Historical Context: European Integration and Enlargement

The unification of Europe has occurred in a series of waves (Table 1). The European Coal and Steel Community (ECSC) was born in 1951. The six founding member states were Belgium, Germany, France, Italy, Luxembourg and The Netherlands. These six countries deepened their economic integration under the Treaty of Rome in 1957, which established the European Economic Community (EEC) and the European Atomic Energy Authority (Euratom).

In 1973, Denmark, Ireland and the United Kingdom joined the European Economic Community. Greece joined in 1981; Spain and Portugal in 1986; and Austria, Finland and Sweden in 1995 (Norway was also offered the right to accede, but the Norwegian population voted in a referendum not to join). It should also be remembered that with the unification of Germany in 1990, the territory of the German Democratic Republic was added to the European Community. With each new accession the dynamics of policymaking have changed considerably.

Accession to the European Community has proved difficult for many states because of both political factors and concerns about the costs of expansion. It took Greece several years before it was finally allowed to join

TABLE 1

EXPANSION OF THE EUROPEAN COMMUNITY/EUROPEAN UNION

1951	Belgium, France, Germany, Italy, Luxembourg, The Netherlands
1973	Denmark, Ireland, the United Kingdom
1981	Greece
1986	Portugal, Spain
1990	The German Democratic Republic (through unification with the German Federal Republic)
1995	Austria, Finland, Sweden
2004	(planned) Cyprus, the Czech Republic, Estonia, Hungary, Latvia, Lithuania, Malta, Poland, the Slovak Republic, Slovenia

the EC in 1981. Spain too eagerly eyed membership in the EC, but was not a serious contender until after the death of Franco and the establishment of democratic political institutions in 1977. Even once a democratically elected government – the first in 40 years – came to power, EC members were worried about the economic implications of Spain's accession. Spain's economy was far less developed than that of the original EC member countries. Thus, although Spain first applied for membership to the EC in 1977, it was not admitted until 1986. A similar story can be told for Portugal, which also applied for membership close to a decade before finally gaining admittance. With its accession in 1986 Portugal became the poorest member of the Community. Since their accession, Greece, Spain and Portugal have often been viewed as the environmental laggards of Europe. In comparison, the accession of Austria, Sweden and Finland as environmentally progressive states is widely regarded to have helped bolster the coalition of the 'green' northern states – Denmark, Germany and The Netherlands.

The addition of Cyprus, the Czech Republic, Estonia, Hungary, Latvia, Lithuania, Malta, Poland, the Slovak Republic and Slovenia in 2004 and the planned later accession of Bulgaria, Romania and, eventually, possibly Turkey will essentially double the size of the EU. This will have enormous implications for the institutional rules of the EU, which will, in turn, affect environmental decision-making procedures.

Economic Disparity across Europe and the Accession of the CEE10

Clearly, one of the biggest challenges in the development of environmental policies within the EU15 has been the differences in the environmental outlooks and capabilities of member states and regions. Despite the EU's stated goal of promoting 'economic and social cohesion', there are still major differences in per capita levels of wealth among EU regions [*Hix, 1999*]. These differences will intensify with the next wave of accession.

In total, there are 211 regions (for example the 11 provinces of Belgium, the nine *Bundesländer* in Austria, and the 20 *regioni* in Italy) among the 15 EU member states. A study issued in 2001 by Eurostat, the Statistical Office of the European Communities, found that differences in regional gross domestic product (GDP) per capita, expressed in purchasing power standards ranged from a low of 42 per cent of the EU15 average for Ipeiros in Greece to a high of 243 per cent for Inner London and 186 per cent for Hamburg, Germany. In 1998, 46 regions were below 75 per cent of the EU average, including 11 of the 13 Greek regions, 5 of the 7 Portuguese regions, 8 of the 18 regions in Spain, 7 of the new *Bundesländer* in Germany, and other regions scattered across the EU. These 46 regions accounted for approximately one-fifth of the EU population [*Theis, Bautier and Simes, 2001a*]. These regional disparities within the EU15 have complicated the process of environmental harmonisation and at times have hindered environmental clean-up efforts. Nevertheless, it is noteworthy that despite these rather large differences in economic wealth, concerns with environmental protection appear to be spreading throughout the EU15. As will be discussed more below, some fiscal and technological support measures have also been introduced to help narrow the environmental gap that exists across regions.

By way of comparison, and in lieu of the planned expansion of the EU, it is important to note that a parallel study of the Central European candidate countries found that per capita GDP in 41 of their 53 regions was below 50 per cent of the EU average in 1998. The range was from 22 per cent in Yuzhen Tsentralen in Bulgaria to 115 per cent in Praha (Prague) in the Czech Republic [*Theis, Bautier and Simes, 2001b*]. This suggests that the challenges facing the EU with this wave of expansion are larger than any experienced to date. There may, therefore, be some limitations to the lessons that can be learnt from past accessions as the next wave of enlargement is much larger than any single previous accession and the economic disparity between the CEE10 and the EU15 much greater. Still, if past experiences are any indication, then it is possible that despite major differences in wealth, integration of poorer member states need not result in a weakening of EU environmental standards or of the EU's international environmental leadership role. As will be discussed further below, there are also interesting examples of poorer southern European member states serving as policy innovators.

Environmental Protection in an Expanding European Community

The development of framework environmental legislation in several key member states in the early 1970s, environmental policy developments in the United States, Japan and at the international level, and the growth of social movements calling for greater protection of the environment, influenced policy development at the Community level. The European Community issued its first Environmental Action Programme in 1973. Initially, environmental regulations were enacted largely out of concerns about trade and competition. They were introduced in large part to deal with differences among member states' environmental policies that could pose barriers to trade and competition. They were also a reaction to concerns about pollution's impacts on human health.

Early efforts at pollution control were sectorally based; they focused on controlling specific types of pollution with little attention to broader ecological system impacts. Environmental regulations and directives were adopted addressing air and water pollution, noise, and nature conservation. Still, at this early date, environment was not yet a high priority issue for the Community [*Liberatore, 1997*]. Enthusiasm for environmental regulatory action, moreover, was dampened by the two oil shocks of 1973 and 1979, which had much of Europe focused on economic growth and recovery.

The real push to prioritise environmental protection began in the mid-1980s as environmental protection became an increasingly important policy concern in several member states – most notably, Germany, The Netherlands and Denmark, and, somewhat belatedly, the United Kingdom. It also coincided with growing awareness of the international and global dimensions of environmental protection as a result of growing European concerns about acid rain in the early 1980s, radioactive fallout from the Chernobyl nuclear accident of 1986, the discovery of the 'hole' in the stratospheric ozone layer in the same year, a toxic waste spill that polluted the Rhine, and growing public awareness of global warming theories by the end of the decade.

New approaches to environmental protection began to be adopted within several member countries. To give some examples, The Netherlands began experimenting with environmental covenants – voluntary agreements among industry, government and citizens' groups in an effort to enhance understanding among actors and improve policy performance [*Liefferink, 1998*]. They also introduced a Green Plan, an effort to imbed environmental concerns into all government planning. Other countries have followed suit introducing Green Plans and making greater use of voluntary agreements [*De Clercq, 2002*]. Numerous states have made use of environmental taxes. The Flemish Region of Belgium introduced levies on the removal of waste (first, in 1986), the pollution of surface water (1990), and the import or export of waste (1994), among other areas. The Walloon Region has followed suit introducing its own system of levies on polluting activities. These financing levies were supplemented in 1993 at the federal level in Belgium with eco-taxes on polluting products [*Deketelaere, 1998*]. Denmark and later Germany introduced legislation requiring utility companies to buy a certain percentage of their energy from renewable sources.

Developments at the national level have pushed the EU in new policy directions as well. Important driving factors were concerns about competitiveness as well as about improving effectiveness and reducing costs of implementation. While resistance by some member states has slowed the introduction of some new instruments at the EU level, substantial shifts are to be seen in the kinds of measures being introduced by Brussels [*Jordan and Wurzel, 2003*]. One example is the introduction of an Environmental Management and Auditing System (EMAS) in 1993. EMAS requires that companies wanting certification produce and publish an environmental plan and subject themselves to review by independent environmental auditors [*Taschner, 1998; Wätzold and Bültmann, 2001*]. Another is the use of eco-labels to identify products based on their environmental performance [*Eiderström, 1998*].

The expanded use of new instruments at the national level has also pushed the EU to develop guidelines for the use of instruments in an effort to prevent distortions among member states that will result in frictions. Thus, for example, the EU has issued guidelines on the use of environmental subsidies to firms and the use of environmental agreements between public authorities and companies. It also has issued a Communication on the use of environmental taxes [*Delbeke and Bergman, 1998*].

Importantly, the heightened priority given to environmental protection both at the national and at the EU levels paralleled the sudden acceleration of Community institution building. European integration was moving on a slow train throughout the 1970s and the first half of the 1980s. There was considerable inertia because of ambivalence about European integration on the part of the three largest member states: France, Germany and the United Kingdom. This changed beginning in the mid-1980s as France and Germany both determined to push forward with the idea of a stronger union in order to enhance their competitiveness *vis-à-vis* the United States and Japan. The unification of Germany in 1990, which made Germany by far the strongest economy in Europe, altered the relative power balance of the big three and helped to mitigate some of the Euro-scepticism that had prevailed in Great Britain under Margaret Thatcher.

Jacques Delors, European Commission President from 1985 through 1992, oversaw the deepening of European monetary, economic, social and environmental union [*Moravcsik, 1991*]. The Single European Act of 1986 was a turning point in the development of a much stronger supra-national Europe. It called for the development of a single internal European market by 1992. The Act also proved a turning point in EC environmental policy. It amended the Treaty establishing the European Economic Community adding a specific Title on the Environment. For the first time environmental protection was added explicitly to the list of EC responsibilities. The Community was to take action 'to preserve, protect and improve the quality of the environment; to contribute towards protecting human health; (and) to ensure a prudent and rational utilization of natural resources'. Moreover, Community action was to 'be based on the principles that preventive action should be taken, that environmental damage should as a priority be rectified at source, and that the polluter should pay' [*European Commission, 1986*]. This established a legal basis that explicitly defined the expectations for action by the European Community related to the environment. Environmental protection was also to be made a component of other Community policies.

The 1992 Maastricht Treaty further upgraded the environment to an explicit 'policy' responsibility of the Community. The very first objective of the EU as stated in the treaty is 'to promote economic and social progress which is balanced and sustainable'. The Treaty later specifically refers to 'sustainable growth respecting the environment'. The Article dealing with the environment adds to the list of Community objectives spelled out in the Single European Act, the promotion of measures to deal with regional or worldwide environmental problems. The Maastricht Treaty also explicitly calls upon the Community to base environmental decisions on the precautionary principle [*Hildebrand, 2002; Wilkinson 2002*].

Responding to criticisms of the weak status of 'sustainable development' in the Maastricht Treaty, the 1997 Amsterdam Treaty amended the Treaty on European Union to specifically call on the EU to 'promote economic and social progress for their peoples, taking into account the principle of sustainable development'. The Treaty of Nice (2001) dealt primarily with issues pertaining to new institutional rules to be adopted with the accession of new member states, but also reaffirmed the Union's commitment to the environment, both in the new member states and globally. Article 6 of the Consolidated Version of the Treaty Establishing the European Community states that environmental protection requirements must be incorporated into Community policies and implementation efforts [*Nugent 2003; Weale et al., 2000; Axelrod, Vig and Schreurs, 2004*].

The range and speed of the greening of EU institutions is in many ways quite astounding. The European Commission's Directorate General Environment, the executive body of the European Community responsible for defining, drafting and implementing regulations, now has close to 550 personnel. As of 2003, 45 out of 626 members of Parliament (7.1 per cent) were part of the Group of the Greens/European Free Alliance. This figure, moreover, does not include the many parliamentary members in other parties sympathetic to environmental causes. The EU has gained the competency to negotiate on behalf of its member states in international environmental negotiations. In several areas – such as climate change, renewable energy development, packaging waste and recycling, the EU has introduced pioneering measures that are attracting attention internationally. Indeed, environmental policy may be one of the EU's most visible areas of strength in an otherwise underdeveloped foreign policy capacity.

The cumulative effect of these developments has been to expand EU institutional capacity related to the environment and to make the EU a strong player in the development of environmental policies both in relation to national governments in Europe and internationally.

While it is difficult to make broad generalisations, on the whole the trend across the EU15 appears to be towards an enhanced priority for environmental protection *vis-à-vis* the traditionally dominant concern with economic growth. Important differences, however, remain in national environmental performance among member states of the EU15, and this is of relevance to the planned expansion in 2004.

The Impacts of EU Accession on National Environmental Performance

Given the disparities across Europe in terms of wealth, it is not so surprising that environmental movements and green parties tend to be strongest among the wealthiest EU member states. Germany has a Green Party that went into a coalition government with the Social Democratic Party beginning in 1998. Since entering Parliament in 1983, its presence in politics has provided a motivating factor for the larger Social Democratic Party and Christian Democratic Party to green their own policy platforms substantially. In The Netherlands, where membership in environmental groups is high, a system of voluntary agreements or 'covenants' among industries, local administrations and environmental groups have formed to facilitate environmental protection efforts and strengthen rates of compliance. In Denmark, where environmental awareness is among the highest in Europe, green consumerism has become a norm. Denmark can be credited with helping to make wind energy be perceived as a viable energy source; 17% of Danish electricity comes from wind power. This would not have been possible without a strong environmental movement. In 1995, Finland became the first country to have a Green Party member join a national government: Minister for Environment and Planning, Pekka Haavisto. These countries have championed many of the environmental policy ideas – including in relation to packaging waste reduction, recycling, renewable energy production, acid rain and climate change mitigation – that eventually have been incorporated into EU environmental law.

At the end of 1998, the Commission adopted a communication to the Council and the Parliament on the accession of Finland, Austria and Sweden. The report found that the three states were able to maintain environmental rules more stringent than the Community rules. Moreover, the report found that their accession had led to a strengthening of environmental norms in the EU [*European Commission, 2000*].

Environmental movements and awareness tend to be less strong in southern Europe. The European Commission prepares an annual report on compliance with Community environmental law. In most cases, action is brought against a member state because of its failure to transpose EU directives into national law. Over time the number of new complaints and infringements related to the failure to implement environmental law has been rising. There were a total of 612 cases in 1999 and 755 in 2000. The largest numbers of complaints in 2000 were registered against Spain, France, Italy and Germany. The Commission brought 39 cases against member states before the Court of Justice and also delivered 122 'reasoned opinions' in 2000 [*European Commission, 2002*].

The challenge of compliance has certainly been hardest for the southern states [*Börzel, 2003*]. Upon its accession to the EU in 1986, for example, Spain had to transpose a large number of environmental directives without any concessions in terms of timing or content with the one exception of the directive on leaded gasoline [*Fernandez, 1998*]. Prior to accession, Spanish authorities had shown little interest in the environment and Spain had but a weak environmental movement and a poor environmental institutional infrastructure. A strong interest in construction and development has caused considerable environmental destruction, but there are signs of what has even been called an 'environmental revolution' [*Weale et al., 2000: 166*]. A decade after accession, environmental awareness appears to be growing. The 1990s have witnessed a growth in local protest events against industrial waste problems [*Jiménez, 2001*]. Spain created an Environment Ministry in 1996 and, during its presidency of the EU, Spain proposed that the EU move towards integrated protection of soil as it does with air and water and also for the ratification of the Kyoto Protocol (Presidency of the European Union, Press Release, 22 January 2002). Spain has shown considerable interest in renewable energy and is now the world's second largest producer of wind energy after Germany (*Reuters*, 26 May 2003). Thus, although there is still a strong prioritisation of development over environment and access to information regarding environmental decision making is not always forthcoming for NGOs, there are signs of growing environmental concern in Spain.

Greece too has been influenced and pressured significantly by the European Union. Greece passed its first broad legislation covering the environment in 1986 several years after accession. Also at this time, the Ministry for Planning, Housing and the Environment, which had been created in 1980, was given real powers. The real motivation behind the strengthening of Greek environmental law was the coming into force of the Single European Act. In 1991 the Greek Supreme Administrative Court created a special section for the enforcement of Community environmental law [*Getimis and Giannakourou, 2001*]. Some progress in environmental improvements have been made. Close to 90 per cent of the population now has wastewater treatment and air pollution has improved somewhat, although the rise in the number of cars on the road is problematic [*National Center for the Environment and Sustainable Development, 2001*]. C. Koutalakis [*2003*] found that Spain and Greece actually have a long history of citizen environmental protest and that when viewed on a per capita basis, have brought some of the largest number of complaints against their governments of any EU member states (except Ireland and Luxembourg) for failure to comply with EU regulations. Greece, however, has the unfortunate distinction of being the first country in the EU to be fined by the European Court of Justice for failure to comply with waste disposal regulations in a case involving Crete [*Dubal, Lah, Monroe and Roberts (2002)*].

It should also be noted that despite the fact that the peripheral states have tended to be policy takers, there have been some instances where the southern European states and the new *Länder* of East Germany have been the source of environmental policy innovation [*Eder and Kousis, 2001*]. In the discussions over the formation of a constitution for a reunified Germany, for example, East German citizens' movements argued for the inclusion of constitutional guarantees for certain social rights, such as housing, employment and social welfare, in the new constitution. While most of their demands went unanswered, the East German citizens' movements in cooperation with their West German counterparts succeeded in having the protection of the environment included as an objective of the German state [*Baukloh and Roose, 2002*].

In a similar vein, in relation to the accession of the ten Central and Eastern European states in 2004, Environment Commissioner Margot Wallström [*2000*] stated:

> The accession countries are facing a considerable challenge to catch up with EU environmental policy, but they have also an important contribution to EU environmental policies … The two environmental contributions that we most often mention when talking about the accession countries are their rich natural heritage – high biodiversity and vast areas of comparative wilderness – and their innovative use of economic instruments. For several accession countries, eco-taxes and environmental fees and charges play an important role in the financing of environmental policy. This is something we have a lot to learn from. A third contribution that is worth reflecting on is the critical perspective of the accession countries in relation to some EU policies (and, in particular) the environmental impact of the Common Agricultural Policy (CAP) and or our transport policies.

Yet it is not always the economically less well off states that are the environmental laggards in the EU. There are times when it is the richer states, those that are often considered to be the vanguards of EU environmental policies that fail to implement EU directives. While Greece, Spain, Portugal and Italy have had substantial difficulties with implementing air, water, waste and nature conservation measures, northern Europe has substantial problems of its own. Northern Europe has lost more biological diversity than have the peripheral nations of the EU and per capita municipal waste generation and energy consumption tend to be considerably higher as well. This suggests that difficulties with environmental policy implementation are not simply a matter of economic wealth or even environmental awareness, but can be the result of a complex array of factors, including protection of domestic economic interests, power politics, cost considerations, the quality of the legislation itself and public apathy. It also suggests that environmental learning can move in multiple directions [*Weale et al., 2000*].

Expansion of the EU to include CEE states will most certainly alter environmental policy priorities and shift attention from the still serious compliance problems that exist within the EU15 to the great challenges of environmental protection in the next group of accession states. Yet it is paternalistic to think that there are not still serious problems among the current member states or that the EU15 have nothing to learn from the policies, practices or ideas of CEE states.

Environmental Decision Making within an Expanding European Union

Along with the expansion of EU environmental authority, there has been a push to democratise decision-making structures, which has helped to provide new avenues for environmental protection advocates to express their concerns and strengthened the system of checks and balances among different levels of the EU as well as among its institutions. The European Union is a unique federal body. Power relations among federal structures, the states, prefectural and local governments have been a constant matter for negotiation. There are regular struggles among these levels of government regarding the degree of centralisation versus decentralisation that is desirable in relation to environmental policymaking and implementation. Subsidiarity, the idea of keeping decision-making authority at the lowest level applicable for addressing problems, is a guiding principle of the EU [*Fruchart, 2002*]. In other words, the Community is to take action only when it can better meet Community environmental objectives than can be done at the level of individual member states. Yet determining what the lowest applicable level is is often a matter of perspective. Thus, for example, there has been much controversy regarding the idea of an EU-wide tax on carbon emissions as many states are unwilling to cede taxation authority to Brussels even though others argue that an EU-wide tax on greenhouse gas pollutants would be the most effective means of reducing emissions. Expansion of the EU by another ten states will most certainly affect debates on subsidiarity.

As the Community has expanded, there have also been changes to EU environmental institutions and policymaking procedures [*Jordan, 2002*]. Beginning in 1974, regular meetings of the heads of government of member states were institutionalised in the European Council. Broad policy directions are established within the European Council while more specific policy decisions are left to gatherings of relevant ministers – the Council of Ministers of the Environment. The environmental ministers of member states meet regularly to address EU-wide environmental concerns and to negotiate common positions on important foreign environmental policy matters. In the past the Council of Ministers of the Environment operated on a unanimity principle, but now it can pass decisions on environmental matters, except those regarding taxation, on a qualified majority voting principle. Qualified majority voting works on the principle that member states' voting weight should be determined by their population size rather than on a one state, one vote principle. These changes have worked to benefit environmental voices.

The European Commission, with its headquarters in Brussels, is responsible for the development of environmental action plans, regulations and directives, as well as their implementation. Much of the real environmental work of the Community occurs in the Commission's Directorates General, which are similar in function to ministries. There has been much criticism of the EU's democratic deficit, partly a problem of the relatively weak authority of the European Parliament (EP). In reaction to these concerns, the Single European Act introduced a 'cooperation principle' into EU decision making. Prior to this time, the EP had limited ability to object to environmental policy proposals coming from the Council of Ministers of the Environment. With the introduction of the cooperation principle the EP could amend Council drafts, but the Council had the final say over these amendments. It could pass them with a qualified majority vote or reject them with a unanimous vote. The Maastricht Treaty further strengthened the powers of the EU with the introduction of a 'co-decision procedure'. Under this procedure if the Council and the EP disagree on legislation, a Conciliation Committee is convened with equal representation from the Council and the EP. The EP has the final say after the Conciliation Committee convenes.

The near doubling of the number of EU member states in the coming years will mean that there will be further major alterations to decision-making rules and coalitional politics. Many of the changes have been mapped out by the Treaty of Nice. The size of the EP will expand from its current level of 626 but will be capped at 732 to prevent paralysis. The EU15 will hold 73.12 per cent of the seats in the expanded Parliament. What share of these votes will be 'green votes' remains to be seen. Voting weights to be used in the Council's qualified majority voting system will also be changed to account for new membership. Table 2 outlines the changes that will be made to member state representation in the EP and the Council as a result of expansion.

TABLE 2

CHANGES IN INSTITUTIONAL REPRESENTATION AS A RESULT OF EU EXPANSION (AN EU OF 27)

Country	Population (millions)	Percentage EU	Weighted Votes in Council				EP Seats		
			Pre-Nice	%	Post Nice	%	Pre-Nice	Post-Nice	% EU 27
Germany	82.03	17.05	10	11.5	29	8.41	99	99	13.52
UK	59.75	12.31	10	11.5	29	8.41	87	72	9.84
France	58.97	12.25	10	11.5	29	8.41	87	72	9.84
Italy	57.61	11.97	10	11.5	29	8.41	87	72	9.84
Spain	39.40	8.19	8	9.2	27	7.83	64	50	6.83
Netherlands	15.76	3.28	5	5.7	13	3.77	31	25	3.42
Greece	10.53	2.19	5	5.7	12	3.48	25	22	3.01
Belgium	10.71	2.12	5	5.7	12	3.48	25	22	3.01
Portugal	9.98	2.07	5	5.7	12	3.48	25	22	3.01
Sweden	8.85	1.84	4	4.6	10	2.90	22	18	2.46
Austria	8.08	1.68	4	4.6	10	2.90	21	17	2.32
Denmark	5.31	1.10	3	3.4	7	2.03	16	13	1.78
Finland	5.16	1.07	3	3.4	7	2.03	16	13	1.78
Ireland	3.74	0.78	3	3.4	7	2.03	15	12	1.64
Luxembourg	0.43	0.09	2	2.3	4	1.16	6	6	0.82
Total EU 15	*375.31*	*77.99*	*87*	*100*	*237*		*626*	*535*	*73.12*
Poland	38.67	8.04				7.83		50	6.83
Romania	22.49	4.67				4.06		33	4.51
Czech Republic	10.29	2.14				3.48		20	2.73
Hungary	10.09	2.10				3.48		20	2.73
Bulgaria	8.23	1.71				2.90		17	2.32
Slovakia	5.39	1.12				2.03		13	1.78
Lithuania	3.70	0.77				2.03		12	1.64
Latvia	2.44	0.51				1.16		8	1.09
Slovenia	1.98	0.41				1.16		7	0.96
Estonia	1.45	0.30				1.16		6	0.82
Cyprus	0.75	0.16				1.16		6	0.82
Malta	0.38	0.08				0.87		5	0.68
Total EU 27	*481.17*	*100*				*100*		*732*	*100*

Source: Developed from Nugent [*2003*].

These changes and any potential further institutional changes that may occur if a European Constitution is eventually adopted will influence environmental policy choices in the years to come [see also *Homeyer, this volume; Jehlička and Tickle, this volume*]. The EU is not likely to adopt many major new environmental policies in the next years. While this may be seen as a negative consequence of expansion to some observers, the EU is likely to focus on improving its performance with policy implementation and environmental policy integration, a problem with which the EU15 have been struggling in the last decade and that is highlighted in the EU's 5th and 6th Environmental Action Programmes. There will also be a continued expansion in the range of policy tools used for environmental protection purposes. Finally, it is conceivable that some of the CEE states will themselves join the group of environmental policy innovators in the EU. This may be especially true in relation to the development and use of new policy instruments.

Financing Environmental Catch-Up

In addition to EU institutional reforms that have helped strengthen the system of checks and balances related to decision making, several mechanisms were established over the course of the 1990s to deal with the disparities that exist within the EU15 economically and environmentally. Critical to the philosophy of the European Union has been the idea that redistributing wealth to less developed regions is critical for the smooth functioning of an integrated economy. Similarly, harmonisation of policies across the Community requires that less developed regions and states be aided by those with greater wealth and more experience with a given policy area. The EU spends as much as one-third of its budget for Structural Funds, which have been established for social and economic restructuring of less developed areas.

In the environmental field, several tools have been developed to aid less developed regions meet the environmental goals of the EU. One important tool is the Cohesion Fund, which was set up in 1994. The Cohesion Fund was established to provide transport and environment aid to the EU's four poorest member states: Ireland, Portugal, Spain and Greece. It had a total budget of 13.6 billion ECU for the period from 1994 to 1999. Approximately half of the funds distributed to these states in the 1993–97 period were for environmental projects, the other half going to the transport infrastructure sector. The focus of the environmental funds was on improving the supply of drinking water, wastewater treatment and sewage treatment. There were also measures for coastal protection, reforestation, addressing desertification and nature conservation [*European Report, 1998*].

In 1992 the LIFE programme was established as an environmental policy-financing instrument. Under LIFE's nature programme EU member states and some accession countries are receiving funding for nature conservation projects. Greece, for example, has received support for its efforts to protect species threatened with extinction. Sweden has received assistance to restore large areas of coastal meadows and wetlands on islands in the Baltic Sea. Italy has received financial assistance for its work to protect wolves, bears, bats and other species. The LIFE environment programme co-finances development projects that employ new technologies or procedures for waste management, land use development and planning, the reduction of the environmental impact of economic activities, and water management. For the period 2000–2004, the EU budgeted 300 million Euro for these initiatives [*Europa, 2003*]. Between 1992 and 2000 LIFE dispersed 850 million ECU [*European Parliament, 1998*].

Another programme, the Short and Medium-Term Priority Environmental Action Programme, which was adopted in November 1997, supports projects addressing environmental protection in the Mediterranean Sea region. The Environment DG also provides financial support for environmental organisations. The Energy DG, through its SAVE programme, supports proposals aimed at enhancing energy efficiency [*Europa, 2003*]. Overall, the EU allocated approximately 160 million Euro for internal policies related to the environment in 2000 [*European Communities, 2000*].

In order to assist the accession states to come into compliance with the environmental *acquis*, several other programmes were established or revised. The most important of these, the PHARE (Poland/Hungary Aid for the Reconstruction of the Economy) programme, which was originally set up to assist with the restructuring of economies and the development of democratic structures, spent approximately 7.3 billion ECU for environmental sector improvements in Central and Eastern Europe during the 1990s [*European Parliament, 1998*]. The LIFE programme also began providing financing to the accession states after 1995.

There is no doubt, however, that the cost of bringing CEE states up to EU standards will strain EU coffers and is of a magnitude not previously experienced by the EU. This is likely to be one of the biggest challenges for the EU in its bid to assist new accession states with coming into compliance with EU regulations. It is also likely to be one of the factors that will most strain public sentiment for expansion. One indicator of the substantial costs associated with bringing a former East bloc state into compliance with EU regulations is the case of the former East Germany. *Business Week* [*1999a*] put the total cost of German reunification at $560 billion, of which a large portion went to the clean-up of toxic waste sites and improving environmental infrastructure. In the former East Germany, which was among the wealthier of the former East bloc states, many municipalities lacked sewerage treatment facilities and chemical plants and coal mining operations turned huge swaths of land into brown fields. The costs of supporting clean-up in East Germany have strained public will and in some instances even support for German unification. Still, while the clean-up has come with a big price tag, it is also the case that for the same period, per capita GDP in Germany increased from $14,362 in 1989 to $19,776 in 1999. In comparison, in the EU as a whole, GDP per capita went from $11,494 in 1989 to $16,802 in 1999 [*Business Week, 1999b*]. Moreover, a decade later, environmental conditions have improved substantially in the east.

In September 1996 the European Commission calculated that the CEE economies will have to spend between 198.4 and 121.5 billion ECU to come into compliance with the environmental *acquis*. The European Parliament considers these statistics optimistic [*European Parliament, 1998*]. Undoubtedly, one of the biggest challenges for the EU with the accession of the CEE10 will be helping to finance clean-up and capacity-building initiatives. The German case suggests that the short-term, admittedly heavy costs are worth the long-term gains to be achieved.

The EU, Enlargement, and International Environmental Leadership: The Case of Climate Change

The preceding sections suggest that despite the accession of several poorer member states to the EU, over time the EU has continued to strengthen its environmental capacity and leadership role. It also suggests that while implementation problems have been serious in the periphery states, they have also plagued the richer states. The EU's ability to maintain a leadership role internationally in the environmental field may have as much to do with the ability of the EU15 to meet goals they already have established for themselves and committed to internationally as it does with the expansion of the EU to the east. It may even be the case that in some instances, EU expansion will provide Europe with new possibilities for meeting international commitments in innovative ways. The case of climate change is an interesting one through which to examine these possibilities.

Beginning in the mid- to late 1980s, the EC became an increasingly important international environmental actor. The United States initially opposed giving the Community the right to negotiate on behalf of its member states, during the negotiations over the Montreal Protocol. Eventually, however, the desire of the member states to cede this authority to the Community prevailed. The Commission now typically negotiates on behalf of the EU in areas where it has competence, but receives its mandate from a unanimously agreed decision of the Council. However, members also typically participate in the negotiations [*Sbragia, 2002*]. The EU has played a particularly visible, albeit complex role in relation to the climate change negotiations.

Throughout much of the 1990s, the EU was a firm voice calling upon industrialised states to take measures to reduce their domestic greenhouse gas emissions. At the same time, however, the EU developed a complicated internal burden-sharing agreement whereby those member states that are economically behind are permitted to increase their greenhouse gas emissions. Others with low emission levels to begin with were exempted from making cuts (Finland and France), while the others (Austria, Denmark, Germany, Great Britain and The Netherlands) were expected to make sharp cuts in their emissions. The EU as a whole agreed, under the auspices of the 1997 Kyoto protocol, to cut their total greenhouse gas emissions by eight per cent of 1990 levels by 2012. Under pressure from the United States, and later Japan, EU officials eventually agreed to the use of emissions trading and joint implementation – a system whereby the industrialised states can choose to assist a transition economy to reduce its emissions and get credit towards its own emissions reduction for doing so – as long as the industrialised states were still required to make some of their emissions cuts domestically.

In the spring of 2001, President George Bush pulled the United States out of the Kyoto Protocol. Without United States participation, the future of the Kyoto Protocol has been on the line. In order for the Protocol to go into force, at least 55 countries, representing 55 per cent of industrialised states' carbon dioxide emissions must ratify the agreement. Because the US is such a large emitter, this requires that almost all other major producers of carbon dioxide, a major greenhouse gas, ratify the agreement. EU officials reacted forcefully to the US withdrawal, calling the US move irresponsible. Although the EU has failed to change the US administration's position, it has been successful in helping to convince over one hundred nations to ratify the agreement. If Russia ratifies the Protocol, it will go into force even without the United States [*Schreurs, 2002, 2004*]. EU and member state officials continued to pressure Russia to do so through 2002 and 2003.

The EU's leadership role in the climate change negotiations and in reaction to the US pull-out from Kyoto means that there is now considerable pressure on the EU to meet its Kyoto Protocol reduction goal of eight per cent. While the European Environment Agency [*2002*] found that the EU had met its goal of stabilising emissions at 1990 levels by 2000, there are concerns that only six member states were on track to meet their emission targets (Finland, France, Germany, Luxembourg, Sweden and the United Kingdom). Other member states, including such 'green' states as The

Netherlands, Austria, Denmark, Finland and Belgium will have trouble meeting their reduction commitments under the EU bubble.

This has put pressure on the EU to consider new approaches to policy implementation. The EU will begin the world's first international carbon dioxide emissions trading system in 2005 among power plants in the EU15 plus the accession states. The EU, furthermore, has introduced a goal of sharply increasing energy supply from renewable energies. There will be increased use of joint implementation projects and the accession countries are prime targets. If Russia ratifies the Kyoto Protocol, it is expected to benefit from such projects. There are already joint implementation projects among EU15 member states and the accession countries. Sweden, for example, has been active in the Baltic States and The Netherlands in the Czech Republic, Romania and Poland [*Michaelowa and Betz, 2001*]. Because reduction of emissions in the accession states in many cases will be more cost effective than reducing an equal level of emissions in an EU15 member state, there are strong possibilities for expanding joint implementation schemes with the accession states [*Arnentevos and Michaelowa, 2002*].

The ten candidate countries for 2004 have all established emission reduction targets under the Kyoto Protocol. As countries in transition, they were allowed to establish baselines other than 1990. Thus, Bulgaria and Poland use a baseline for emissions at 1988 levels; Estonia, Latvia, Lithuania and Slovakia use 1990; and Hungary the average for 1985–87. Most have emission reduction goals relative to their respective baselines of six to eight per cent. In total the ten countries have already reduced their greenhouse gas emissions by 33.7 per cent between their base year and 1999. Poland, the largest emitter of the group, has reduced its emissions by 29.1 per cent between 1988 and 1999 [*European Environment Agency, 2002*]. Much as was the case with the former East Germany, the sharp reduction in emissions in these countries is largely attributable to the shutting down or insolvency of highly polluting firms shortly after the transition process began [see *Pavlínek and Pickles, this volume; Archibald et al., this volume*]. This provides the EU15 with some real possibilities.

The EU's ability to maintain regional and global environmental leadership in this and other environmental policy areas will be tested by how well it implements the bold commitments it has made. Whether the CEE accession states prove a drag on this process or aid the EU15 in their efforts to fulfil international goals will depend greatly on how well member states and accession states learn to work together and develop win-win policy scenarios, as may be provided with joint implementation efforts.

Finally, the EU has made environment a 'pillar' of its foreign policy. Since the EU has been criticised both internally and internationally for its weak foreign policy presence (especially in the security realm), it is critical that the EU maintain strength in areas where it does exhibit considerable leadership. This suggests that even with an expanded EU, maintaining a progressive environmental foreign policy will be important for the EU.

Conclusion

While many problems remain with implementation of environmental regulations in the EU15, there is no doubt that the EU has done much to help European countries think about environmental protection as a matter that demands cooperation among nations. The EU also helps foster greater environmental concern among nation states where national attentions remained heavily focused on economic development, without resulting in a back peddling of environmental policies among leader states. There are still capacity problems among the poorer economies of the EU15 and the will to enforce regulations is not always strong. Yet it is not only the poorer states of the EU that have implementation problems to address. In many instances, the richer states are also having problems complying with EU mandates. Still, the EU's experiences in environmental regulation and innovation, especially in the period since 1987, suggest that nations may be able to do more for environmental protection working together than working alone. The expansion of the EU to a body of 25 – and likely more – member states will no doubt place a great strain on EU environmental institutions and finances. The challenges that lie ahead are far greater than existed with any previous accession group. Nevertheless, in the long run, the benefits for the environment are likely to be far higher than the likely high costs in the initial post-accession period of harmonising CEE environmental laws and programmes with those of the EU.

The case of an expanded EU, moreover, could provide important lessons for the international community as it looks to a future where global environmental protection will require greater interaction and cooperation among the first and third world nations of the world. Expansion to the east is really in many ways only a small first step towards a much larger challenge facing Europe and the rest of the global north: determining how to help improve rapidly the environmental performance of developing countries even as they reduce the size of their own ecological footprints. There is growing interest in Africa and East and Southeast Asia in learning from European experiences with environmental harmonisation and multinational problem solving. The experiences of the CEE10 with integrating the body of EU environmental law into their own national programmes will provide many relevant lessons for future accession countries that must bring their environmental protection systems up to first world standards in a very short period of time. Similarly, there are many industrialising states, particularly in East and Southeast Asia but also in Latin American and Africa that could learn much from the environmental transformations occurring in Central and Eastern Europe.

ACKNOWLEDGEMENTS

I wish to express my thanks to Sonja Walti for her helpful comments on an earlier draft of this study along with the anonymous reviewers for *Environmental Politics* for their insightful comments and suggestions.

REFERENCES

Arnentevos, Mercedes Fernández and Axel Michaelowa (2002), 'Joint Implementation and EU Accession Countries', Hamburgisches Welt-Wirtschafts-Archiv (HWWA), Discussion Paper 173. Available at http://www.hwwa.de/Publikationen/Discussion_Paper /2002/173.pdf.

Axelrod, Regina, Norman Vig and Miranda Schreurs (2004), 'The European Union as an Environmental Governance System', in Norman Vig and Regina S. Axelrod, *The Global Environment: Institutions, Law and Policy*, Washington, D.C.: Congressional Quarterly Press, 2004.

Baukloh, Anja and Jochen Roose (2002), 'The Environmental Movement and Environmental Concern in Contemporary Germany', in Axel Goodbody (ed.), *The Culture of German Environmentalism: Anxieties, Visions, Realities*, New York: Bergham Books, pp.81–101.

Börzel, Tanja A. (2003), *Environmental Leaders and Laggards in Europe: Why there is (Not) a 'Southern Problem'*, Aldershot: Ashgate Publishing Co.

Business Week (1999a), 'Europe Ten Years After the Wall', *Business Week*, 8 Nov.

Business Week (1999b), 'Before and After: Europe, 1989–1999, By Almost Every Measure, Life for Europeans Has Improved', *Business Week*, 8 Nov.

De Clercq, Marc (2002), *Negotiating Environmental Agreements in Europe*, Cheltenham: Edward Elgar.

Delbeke, Jos and Hans Bergman (1998), 'Environmental Taxes and Charges in the EU', in Jonathan Golub (ed.), *New Instruments for Environmental Policy in the EU*, London: Routledge, pp.242–60.

Deketelaere, Kurt (1998), 'New Environmental Policy Instruments in Belgium', in Jonathan Golub (ed.), *New Instruments for Environmental Policy in the EU*, London: Routledge, pp.86–106.

Dubal, Veena, Jung Lah, Ian Monroe and Martha Roberts, 'Why are some Trade Agreements "Greener" Than Others? Nafta Versus the European Union,' Earth Island Journal, Vol.16, No.4, Winter 2001–2002, http://www.earthisland.org/eijournal/new_articles.cfm?articleID= 2958journalID=49

Eder, Klaus and Maria Kousis (2001), *Environmental Politics in Southern Europe: Actors, Institutions and Discourses in a Europeanizing Society*, Dordrecht: Kluwer.

Eiderström, Eva (1998), 'Ecolabels in EU Environmental Policy', in Jonathan Golub (ed.), *New Instruments for Environmental Policy in the EU*, London: Routledge, pp.190–214.

Europa (2003), http://europa.eu.int/comm/environment.

European Commission (1986), 'Final Act and Declarations of The Single European Act', Luxembourg, 17 Feb. 1986, p.17. Available at www.eurotreaties.com/singleeuropeanact.pdf.

European Commission (2000), 'Implementing and Monitoring of the Application of Community Legislation on the Environment', working document SEC(2000)1219. Available at http://europa.eu.int/scadplus/leg/en/lvb/128111.htm.

European Commission (2002), 'Third Annual Survey on the Implementation and Enforcement of Community Environmental Law: January 2000 to December 2001', Luxembourg: Office for Official Publications of the European Communities, 10 Jan.

European Communities (2000), 'The Budget of the European Union: How is Your Money Spent', Luxembourg: Office for Official Publications of the European Communities.

European Environment Agency (2002), 'Greenhouse Gas Emission Trends in Europe, 1999–2000', Topic Report 7/2002, Mar. Available at http://reportseea.eu.in/topic_report_2002.7/en/tab_abstract_RLR.

European Parliament (1998), 'Environmental Policy and Enlargement', Briefing No.17, Luxembourg, 23 Mar., PE 167.402. Or. DE. Available at http://www.europarl.eu.int/enlargement/briefings/17a3_en.htm#6.

European Report (1998), 'Cohesion Fund: More Environment Projects', 14 Oct. Available at http://www.poptel.org.uk/aries/members/funds/archive/msg00349.html.

Fernandez, Susana Aguilar (1998), 'New Environmental Policy Instruments in Spain', in Jonathan Golub (ed.), *New Instruments for Environmental Policy in the EU*, London: Routledge, pp.125–41.

Fruchart, Vincent (2002), 'Subsidiarity Versus Sovereignty: An Institutionalist Analysis of the Treaty of Maastricht', PhD dissertation, University of Maryland.

Getimis, Panagiotis and Georgia Giannakourou (2001), 'The Development of Environmental Policy in Greece', in Hubert Heinelt, Tanja Malek, Randall Smith and Annette E. Töller (eds.), *European Union Environment Policy and New Forms of Governance*, Aldershot: Ashgate, pp.289–307.

Glachant, Matthieu (2001), *Implementing European Environmental Policy: The Impacts of Directives in the Member States*, Cheltenham: Edward Elgar.

Hildebrand, Philipp M. (2002), 'The European Community's Environmental Policy, 1957 to "1992": From Incidental Measures to an International Regime', in Andrew Jordan (ed.), *Environmental Policy in the European Union*, London: Earthscan, pp.13–36.

Hix, Simon (1999), *The Political System of the European Union*, New York: Palgrave.

Jiménez, Manuel (2001), 'National Policies and Local Struggles in Spain: Environmental Politics Over Industrial Waste Policy in the 1990s,' paper presented at the workshop, 'Environmental Politics at the Local Level,' ECPR Joint Sessions, Grenoble, 6–11 Apr.

Jordan, Andrew (ed.) (2002), *Environmental Policy in the European Union*, London: Earthscan.

Jordan, Andrew, Duncan Liefferink and Jenny Fairbrass (2004), 'The Europeanization of National Environmental Policy: A Comparative Analysis', in J. Barry, B. Baxter and R. Dunphy (eds.), *Europe, Globalisation and Sustainable Development*, London: Routledge.

Jordan, Andrew and R. Wurzel (2003), *New Instruments of Environmental Governance*, London: Frank Cass.

Koutalakis, Charalampos (2003), 'The Impact of EU Environmental Policies on Patterns of Social Mobilization in Southern Europe', paper presented at the 1st PhD Symposium on Modern Greece, Current Social Science Research on Greece, London School of Economics, London, 21 June.

Lenschow, Andrea (2002), *Environmental Policy Integration: Greening Sectoral Policies in Europe*, London: Earthscan.

Liberatore, Angela (1997), 'The European Union: Bridging Domestic and International Environmental Policy Making', in Miranda A. Schreurs and Elizabeth Economy (eds.), *The Internationalization of Environmental Protection*, Cambridge: Cambridge University Press.

Liefferink, Duncan (1998), 'New Environmental Policy Instruments in the Netherlands', in Jonathan Golub (ed.), *New Instruments for Environmental Policy in the EU*, London: Routledge, pp.86–106.

Liefferink, Duncan and Mikael Skou Andersen (2002), 'Strategies of the "Green" Member States in EU Environmental Policy-making', in Andrew Jordan (ed.), *Environmental Policy in the European Union*, London: Earthscan, pp.63–81.

Michaelowa, Axel and Regina Betz (2001), 'Implications of EU Enlargement on the EU Greenhouse Gas "Bubble" and Internal Burden Sharing', *International Environmental Agreements: Politics, Law and Environment*, Vol.1, pp.267–79.

Moravcsik, Andrew (1991), 'Negotiating the Single European Act: National Interests and Conventional Statecraft in the European Community', *International Organization*, Vol.45, No.1, pp.19–56.

National Center for the Environment and Sustainable Development (2001), 'Greece – the State of the Environment – A Concise Report', Oct. 2001. Available at http://www.epa.gr/documents/NCESD-EN-State_of_Environment.pdf.

Nugent, Neill (2003), *The Government and Politics of the European Union*, Durham: Duke University Press.

Pridham, Geoffrey (2002), 'National Environmental Policy-making in the European Framework: Spain, Greece and Italy in Comparison', in Andrew Jordan (ed.), *Environmental Policy in the European Union,* London: Earthscan, pp.81–99.

Sbragia, Alberta (2002), 'Institution-building from Below and Above: The European Community in Global Environmental Politics', in Andrew Jordan (ed.), *Environmental Policy in the European Union*, London: Earthscan, pp.275–98.

Schreurs, Miranda A. (2002), *Environmental Politics in Japan, Germany, and the United States*, Cambridge: Cambridge University Press.

Schreurs, Miranda A. (2004), 'The Climate Change Divide: The European Union, the United States, and the Future of the Kyoto Protocol', in Norman J. Vig and Michael Faure, *Green Giants? Environmental Policy of the United States and the European Union*, Cambridge: MIT Press.

Taschner, Karola (1998), 'Environmental Management Systems: The European Regulation', in Jonathan Golub (ed.), *New Instruments for Environmental Policy in the EU*, London: Routledge, pp.215–41.

Theis, Fons, Phillippe Bautier and Anette Simes (2001a), 'Regional GDP Per Capita in the EU in 1998, A Difference from One to Six Between Lowest and Highest Level', Eurostat Press Office, No.22, Feb.

Theis, Fons, Phillippe Bautier and Anette Simes (2001b), 'Central European Candidate Countries Per Capita GDP in 41 out of 53 Regions Below 50% of the EU Average in 1998', Eurostat Press Office, No.31, 15 Mar.

Wallström, Margot (2000), 'Speech to the Environment Committee of the European Parliament', Public Hearing on Enlargement, 2000. P-reference: p.1.41.33. Briefing request D 220034, 20 June.

Walti, Sonja (2002), 'Competing for the Environment? Evidence from the European Union', paper presented at the Midwest Political Science Conference, Chicago, 25–28 Apr.

Walti, Sonja (2003), 'Interstate Competition in Europe', paper presented at the 44th International Studies Association Convention, Portland, OR.

Wätzold, Frank and Alexandra Bültmann (2001), 'The Implementation of EMAS in Europe: A Case of Competition between Standards for Environmental Management Systems', in Matthieu Glachant (ed.), *Implementing European Environmental Policy*, Cheltenham: Edward Elgar, pp.134–77.

Weale, Albert, Geoffrey Pridham, Michelle Cini, Dimitrios Konstadakopulos, Martin Porter and Brendan Flynn (2000), *Environmental Governance in Europe: An Ever Closer Ecological Union?*, Oxford: Oxford University Press.

Wilkinson, David (2002), 'Maastricht and the Environment: The Implications for the EC's Environment Policy of the Treaty on the European Union', in Andrew Jordan (ed.), *Environmental Policy in the European Union*, London: Earthscan, pp.37–52.

Differential Effects of Enlargement on EU Environmental Governance

INGMAR VON HOMEYER

Since the second half of the 1990s, post-communist transition processes in Central and Eastern European (CEE) countries have become increasingly bound up with European integration. The wish of many CEE countries to join the European Union (EU) and the conditions that the EU formulated for accession – the so-called Copenhagen criteria and the adoption of existing EU legislation by these countries – have played an important role in shaping transition processes. Conversely, it has become increasingly clear in recent years that eastern enlargement also poses a major challenge for the EU. This is demonstrated, for example, by the institutional changes proposed by the 2003 European Convention. Following the failure of EU member governments at the 2000 Nice Summit to agree on sufficiently radical institutional reforms to prepare the EU for enlargement, the Convention drafted a proposal for a European constitution that envisages major additional reforms.

The following analysis looks at interrelationships between the transition processes in the eight 'vanguard' CEE countries – the Czech Republic, Estonia, Hungary, Latvia, Lithuania, Poland, Slovakia and Slovenia – which are set to join the EU in 2004 and the EU at the sectoral level, specifically in the field of environmental policy. However, rather than focusing directly on environmental transition processes in these countries, the discussion primarily considers the potential impact of eastern enlargement on EU environmental governance. Two closely related considerations suggest that an understanding of this factor is a crucial precondition for analysing the future of environmental transition processes in CEE states. First, EU environmental policy generally has a very strong, if not overriding influence on the environmental policies of EU member states – a group of countries to which the accession countries will soon belong. Second, as a result of the accession process, EU environmental legislation already largely determines the environmental policies of the CEE countries [cf. *Jehlička, 2002*].

The analysis of the potential impact of enlargement on EU environmental governance proceeds in five parts. Against the background of relevant debates among policymakers and analysts, the first part presents the main hypotheses. The second part introduces the historical– institutionalist perspective on which the subsequent analysis draws. Part three identifies and describes three EU environmental policymaking regimes which together constitute EU environmental governance. This is followed by a discussion of the main features of environmental policy in the ten CEE accession countries that are set to join the EU in 2004. The last part discusses how enlargement could affect each of the three EU environmental governance regimes.

Decline, Extension or Transformation of EU Environmental Governance?

European environmental policy has developed in a surprisingly dynamic way since the 1970s. In particular, it has often defied those who predicted a lowering of environmental standards and a 'race to the bottom' as a result of the creation of the EU's internal market and the associated increasing regulatory competition among the member states [*Sbragia, 1996; Héritier, 1995; Eichener and Voelzkow, 1994; Holzinger, 1994*]. But opinions are divided as to whether EU environmental policy will continue to develop at such a pace once the CEE accession countries have joined the EU [see also, *Schreurs, this volume; Jehlička and Tickle, this volume*]. Many observers and political actors tend to see enlargement as a threat to EU environmental governance. For instance, there is considerable concern that enlargement will have a negative effect on decision-making efficiency at the European level, lead to the adoption of lower EU standards of environmental protection, and further increase the implementation deficit of EU environmental legislation [cf. *Bär et al., 2001: 215–17; Baker, 2000; Holzinger and Knoepfel, 2000; Homeyer et al., 2000*]. Frequently these expectations are based on a view of EU environmental governance that seems to have much in common with liberal intergovernmental theories of European integration [cf. *Homeyer, et al. 2000; Moravcsik, 1993*]. From this perspective, EU environmental governance essentially reflects the interests of member state governments which, in turn, are shaped primarily by domestic conditions, in particular economic variables. Because the economies of most CEE accession countries are less developed than those of the present EU member states, it is expected that these countries will oppose a high level of environmental protection. Among other things, CEE accession countries are assumed to be unwilling to spend their relatively scarce resources on achieving a high level of environmental protection. They may also be afraid of losing comparative trade advantages resulting from lax environmental standards. All of this implies that enlargement is likely to significantly weaken political support for EU environmental policy.

Other analysts expect enlargement to have a less dramatic impact. Frequently their views seem to be related to neo-functionalist theories of the EU that emphasise the positive role of policy experts and are more optimistic with respect to the integrative capacities of European policies and institutions [cf. *Caporaso and Keeler, 1995; Haas, 1964*]. Relatively optimistic assessments are based on a number of observations. For example, throughout the 1990s environmentally progressive countries such as Denmark, The Netherlands and Sweden have been particularly active in providing environmental assistance to CEE accession countries. Consequently, many environmental experts and policymakers in the recipient countries now adhere to models of environmental policy that are similar to those promoted by the donor countries. The diffusion of progressive environmental ideas among experts in CEE accession countries suggests that, despite relatively unfavourable general economic and political conditions, these countries may not simply align themselves with present member states, such as Spain, which tend to oppose high-level EU environmental standards. Rather, they might often even be supportive of relatively strict legislation [*Jehlička, 2002*]. In addition, EU environmental governance may to some extent be able to accommodate the difficult political and economic conditions in many CEE accession countries. Since the late 1980s, the EU has put considerable efforts into developing more flexible, cost-efficient environmental regulations, for example framework laws and information-based instruments. This approach should make it easier to address the increased diversity among member states following enlargement and the more severe financial constraints in many CEE countries [*Homeyer et al., 2000*].

The argument developed here draws on a more differentiated framework for assessing the potential impact of eastern enlargement on EU environmental governance which allows recognition of the potentially serious negative implications of enlargement, but also identifies opportunities for policymakers to intensify reform efforts already under way to reduce the implementation deficit of European environmental legislation and increase the effectiveness and legitimacy of EU environmental governance. This approach is based on three main premises. First, enlargement is likely to have a differential effect on the various mechanisms which have led to the adoption of high-level EU environmental standards in the past. Second, while enlargement poses a threat to the future functioning of some of these mechanisms, awareness of this threat is already producing reactions that may reduce the impact of enlargement and may eventually even allow these mechanisms to evolve and continue to operate in the enlarged EU. Third, enlargement may reinforce emerging changes in the relative importance of the different mechanisms driving EU environmental policy. These shifts are produced by the broader challenge of sustainable development.

A Historical–Institutionalist Perspective

It is possible to analyse the potential medium- to long-term implications of enlargement for EU environmental governance from a historical–institutionalist point of view. First, such a perspective allows for consideration of the time dimension. It not only focuses on the factors that stabilise institutions in changing contexts, but also specifies ways in which institutions may break down or evolve. Second, a historical–institutionalist perspective can draw on a broad understanding of political institutions. Arguably, this is essential for analysing governance. A broad understanding goes far beyond formal rules and effective sanctioning mechanisms. In a historical–institutionalist perspective, institutions can be formal or informal; they are comprised of legitimising ideas in the form of normative and cognitive scripts, decision-making procedures, legislative outputs, and entire policies [cf. *Peters, 1999; Pollack, 1996; Hall and Taylor, 1995*].

A historical–institutionalist perspective suggests that most of the time, EU environmental governance will be relatively stable. If anything, gradual shifts in underlying conditions, such as economic and technical developments or environmental trends, will have a mediated and delayed effect. The reason for this is that institutions are conditioned by the past. More specifically, institutional reproduction tends to be path-dependent [*Pierson and Skocpol, 2000; Hall and Taylor, 1996*]. Path dependence has two main features. First, earlier events have a much larger influence on institutional development than later events. The critical formative period of an institution arises from a 'critical juncture' and establishes the essential mechanisms of institutional reproduction. It transforms a situation in which there appear to be at least two plausible alternatives into a much more predictable process of path-dependent institutional development [*Mahoney, 2001, 2000; Pierson, 2000; Hall and Taylor, 1996*]. Second, in terms of reproductive mechanisms, path dependence may result from either positive feedback or take the form of a more dialectical 'reactive sequence'. In the first case, institutions persist even if new circumstances suggest that alternative institutional arrangements may be more effective [*Mahoney; 2000*]. The positive feedback processes underlying this persistence of 'suboptimal' institutions are based on mechanisms that render it increasingly difficult to switch from a given institutional arrangement to an alternative one as time progresses [*Mahoney, 2001; Pierson, 2000*]. In the second case of path-dependent institutional reproduction, 'chains of temporally ordered and causally connected events' [*Mahoney, 2001;*

Pierson, 2000a] produce predictable reactive sequences. Whereas positive feedback reinforces initial institutional arrangements, 'reactive sequences are marked by backlash processes that transform and perhaps reverse early events' [*Mahoney, 2000*].

Despite the conservative effects of self-reinforcing mechanisms of institutional reproduction, institutional change remains possible. It can take two forms. First, there is the possibility of a sudden institutional breakdown. With path-dependent processes, institutional breakdown is caused by shocks that disable the causal mechanisms underlying institutional reproduction [cf. *Schwartz, 2001; Pierson, 2000, 2000a; Thelen, 2000*]. Second, institutions may either evolve and adapt to more limited and gradual changes or face a process of slow decline *if* such changes significantly reduce the effectiveness of the mechanisms of institutional reproduction. Adaptation may occur in at least three ways. First, in a process of 'institutional conversion', existing institutions may come to serve new purposes. Second, new features may be added to an institution without modification of the pre-existing rules. This may be analysed as a process of 'institutional layering' [*Thelen, 2000*]. Finally, adaptation may also result from shifts in the relative importance of different parts of an institution causing what might be termed an 'internal drift'. If, however, institutions fail to adapt to changes that negatively affect the mechanisms of institutional reproduction, this may lead to institutional decline [cf. *Deeg, 2001; Schwartz, 2001*].

As summarised in Table 1, Mahoney [*2000*] has identified a typology of four mechanisms that are crucial for institutional reproduction and change. First, utilitarian mechanisms may allow an institution to persist as long as it makes a positive contribution in terms of actors' cost-benefit calculations. Second, institutional reproduction may depend on whether the institution continues to serve a function for a larger system. Third, an institution may be reproduced if it helps an elite group to consolidate its power. Finally, reproduction may be the result of an institution's conformity with prevailing values. As a consequence of the operation of these mechanisms, institutions may persist in circumstances in which alternative institutional arrangements would appear to be more effective generators of utility, system output, power or legitimacy.

Superficially, a historical–institutionalist perspective may suggest that EU enlargement – which is often referred to in phrases such as the 'big bang' of the near-doubling of the number of member states' [*Secretariat of the European Convention, 2002: 3*] – may lead to a relatively sudden, dramatic change of the conditions under which the mechanisms sustaining EU environmental governance operate. However, on closer inspection, it seems unlikely that enlargement will lead to the breakdown of EU

TABLE 1

TYPES OF PATH-DEPENDENT INSTITUTIONAL REPRODUCTION

	Utilitarian	Functional	Power	Legitimation
Mechanism of Reproduction	Actors make a rational cost-benefit assessment of the institution	Institution serves a function for a larger system	Institution is supported by an elite group	Prevailing belief that the institution is morally just or appropriate
Potential Institutional Deficiencies/ Characteristics	Institution is less efficient than previously available alternatives	Institution is less functional than previously available alternatives	Institution provides elite status to a previously subordinate group	Institution conforms less to prevailing beliefs than previously available alternatives
Mechanism of Change	Increasing competition, learning	Exogenous shock that transforms system requirements	Weaker elites and stronger subordinate groups	Changes in values or subjective beliefs

Source: Adapted from Mahoney [*2000: 517*]. Reproduced with permission of Kluwer Academic Publishers

environmental governance. As argued below, EU environmental governance is best viewed not as a single institution but as a set of several institutions or 'governance regimes' [*Bulmer, 1997*]. These regimes are based on different types of reproductive mechanisms. Therefore, the mechanisms underlying some governance regimes may be more affected by enlargement than others. While there may be potential for a breakdown of one of these regimes, others may be much more resilient.

EU Environmental Governance Regimes

There are at least three distinct policymaking regimes which together constitute EU environmental governance: the 'internal market regime', the 'environmental regime' and the 'sustainability regime'. Table 2 sums up the main characteristics of each regime in terms of its aims, legal base in the EU Treaties, general approach to policymaking, typical instruments and the prevailing mechanisms of institutional reproduction. The following sections describe each regime in more detail.

The Internal Market Regime

The internal market regime primarily applies to environmental product standards. In this area, EU legislation is usually based on Article 95 TEC

TABLE 2

EU ENVIRONMENTAL GOVERNANCE REGIMES

Regime:	**Internal Market**	**Environmental Policy**	**Sustainability**
Aim	Free circulation of goods and environmental protection and sustainable resource consumption	Environmental protection	Decoupling of economic growth from environmental degradation and resource consumption
Legal base	Article 95 TEC	Articles 174–176 TEC	Articles 2, 6 TEC
Approach	Legislation	Legislation/subsidies	Environmental Policy Integration (EPI); Open Method of Co-ordination (OMC)
Instrument	Product regulation	Process regulation	Objectives/indicators
Mechanism	Functional/utilitarian	Power/utilitarian	Legitimacy

that provides for the adoption of harmonisation measures to create the internal market. The environmental legislation in question simultaneously pursues the goals of creating the internal market and of securing a high level of environmental protection. Some of the most important EU environmental laws have been adopted on this basis. Examples include EU legislation limiting the exhaust and noise emissions of various kinds of vehicles and machinery, a directive regulating the release of genetically modified organisms into the environment, environmental fuel quality standards, and a directive on packaging waste. Various provisions of Article 95 have contributed to the adoption of progressive environmental standards. For example, under Article 95, EU legislation must be adopted using the Co-decision Procedure. Besides requiring qualified majority voting in the Council, this procedure allows the European Parliament, which tends to be more supportive of progressive environmental legislation than either the European Commission or the Council [*Judge, Earnshaw and Cowan, 1994; Judge and Earnshaw, 1992*], to exert relatively strong influence. Article 95 also stipulates that the Commission must ensure that its legislative proposals reflect a high level of environmental protection. In addition, the member states may under certain conditions retain stricter environmental standards than those agreed by the Community.

The principal mechanisms underlying the internal market regime are functional and utilitarian. Functional reproduction is linked to the fact that the internal market performs essential functions for the European project and its success may be crucial for the survival of the Community. For example, the aim of completing the internal market was a critical factor in the relaunching of the Community in the 1980s through the adoption of the Single European Act. As regards utilitarian mechanisms, relatively strict European environmental product standards primarily result from the fact that highly regulated member states have a significantly weaker economic interest in harmonisation than countries with lower product standards. The latter group of countries needs harmonisation to gain access to markets in the highly regulated member states. Consequently, highly regulated member states can put considerable pressure on the countries with lower standards to accept relatively stringent EU regulations in exchange for harmonisation [*Scharpf, 1994*]. The Co-decision Procedure and the fact that the member states may, under certain conditions, maintain environmental standards that exceed Community legislation, further increase the leverage of highly regulated countries.

The Environmental Regime

The second regime is the environmental one. It generally applies to environmental process standards relating to the production process rather than the final product. Process standards are less relevant to the internal market because they do not directly affect trade. The environmental regime is based on Title XIX TEC on the Environment. This Title is particularly important for EU environmental governance because it creates original EU environmental competences. Key pieces of EU environmental legislation have been adopted on the basis of these provisions. The Integrated Pollution Prevention and Control (IPPC) Directive, which lays down comprehensive environmental licensing requirements for industrial facilities, is a good example. Other important examples include EU legislation on nature protection (the Habitats Directive) and the Water Framework Directive, the latter of which brings together various pieces of EU legislation aimed at improving water quality. So-called 'new' instruments, such as a directive on access to environmental information and a regulation on environmental management and auditing (EMAS Regulation) have also been adopted on the basis of the provisions of the environmental Title.

Although Title XIX is the default legal basis for environmental legislation, some of its provisions are somewhat less favourable from an environmental point of view than those of Article 95. Depending on the substantive nature of a legislative measure, the provisions of Title XIX either prescribe the Co-decision Procedure or the Consultation Procedure. The Consultation Procedure empowers member state governments *vis-à-vis* the Commission and the Parliament in the legislative process. In particular, each member state has a right to veto legislation. In addition Title XIX calls on the Community to pursue a high level of environmental protection, but this requirement appears to be less strict than the similar provision of Article 95. While Title XIX is more liberal than Article 95 as to the possibility for member states to exceed Community environmental standards, it also contains provisions that allow for temporary derogations from compliance, or for compensation if member state authorities face 'disproportionately high costs' as a result of EU environmental measures [*Ehlermann, 1995; Beck, 1995*].

The environmental regime is based on utilitarian and power mechanisms. The adoption of relatively tough minimum environmental standards on the basis of Title XIX is largely driven by the interest of highly regulated member states in avoiding competitive disadvantages for their industries and in minimising the administrative and economic costs that arise if established national practices have to be adapted to new EU environmental process standards. These factors create incentives for the environmentally progressive countries to use their influence on EU policymaking to 'export' their standards and practices to the remaining member states [*Héritier, 1995*]. Although the environmental pace-setters among the member states do not command enough votes in the Council to form a qualified majority, they exert disproportionate influence on EU environmental policy because of the institutional insulation of environmental policymaking from rival interests in the European Commission, and in particular in the Council, where the Environment Ministers are usually responsible for the adoption of environmental legislation. In addition, highly regulated countries use their superior technical and economic resources and their extensive regulatory experience to set the EU environmental agenda and exploit first mover advantages [*Homeyer, 2001; Sbragia, 1996*].

The Sustainability Regime

The emerging 'sustainability regime' is the most recent of the three regimes. Although sustainable development is commonly understood as having an environmental, an economic and a social dimension, in practice the EU sustainable development regime focuses primarily on the environmental and economic dimensions. In particular, the regime aims at decoupling economic growth from environmental degradation and resource consumption. Although the instruments employed by the internal market and the environmental regimes may assist in achieving this aim, implementation of the EU sustainability regime crucially relies on the integration of environmental concerns into other policies (Environmental Policy Integration – EPI) [cf. *European Commission, 2001*].

The sustainability regime dates back to the 1997 Amsterdam Treaty. It derives from Article 2 TEC calling for 'a harmonious, balanced and sustainable development of economic activities'. To implement this commitment, Article 6 TEC requires the integration of environmental concerns into all Community policies. The regime initially relied mainly on the Cardiff Process, which was initiated in 1997/98 by the Luxembourg and Cardiff European Councils to put environmental policy integration into practice. In the framework of this process, several sectoral Council formations – such as Agriculture, Transport and Economic and Financial Affairs ('Ecofin') – are working on strategies to integrate environmental concerns into their activities [*Kraemer, 2001; Lenschow, 2001*]. The Cardiff Process has recently, these efforts have been complemented by the adoption of the EU Sustainable Development Strategy and the related decision to add an environmental dimension to the so-called Lisbon Process. Initiated by the Lisbon European Council in 2000, this ten-year strategy originally focused only on economic and certain social measures intended to turn the EU into 'the most competitive and dynamic knowledge-based economy in the world ...', but the 2001 Gothenburg European Council added the environmental dimension. Like the other aspects of the Lisbon Process, the implementation of the environmental dimension relies on the Open Method of Coordination (OMC) that was codified at the Lisbon Summit. The OMC is a non-legislative approach to policymaking based on informational instruments and social pressure, for example through disseminating 'best practice' and 'naming and shaming'. It includes setting short, medium- and long-term policy guidelines, establishing performance indicators and benchmarks, translating targets from the European to national and regional levels, and periodic monitoring, peer review and evaluation [cf. *De la Porte et al., 2001*]. So far, the OMC has mainly been applied to policy areas where the EU has weak competencies, such as economic and social policy coordination.

The EU Sustainable Development Strategy, which was also adopted by the 2001 Gothenburg European Council, identified basic objectives and measures concerning four environmental priority issues: climate change and clean energy, public health, natural resource management and transport, and land use [cf. *European Commission, 2001*]. Like the economic and social objectives of the Lisbon Process, progress towards reaching the environmental goals was to be assessed annually at the spring European Council meeting. The Council and the Commission subsequently agreed on several structural indicators, such as greenhouse gas emissions and the amount of municipal waste collected, to measure progress.

The principal mechanism underlying the sustainability regime is legitimacy. The regime depends on the acceptance and internalisation by sectoral actors of sustainable development as a 'leading idea' [*Kohler-Koch, 2000*]. The dissemination of 'best practice', benchmarking, peer review, etc. is used to stimulate social learning and practically implement relevant measures on the ground [cf. *De la Porte et al., 2001; Lenschow, 2001*]. More specifically, in the framework of the Lisbon and Cardiff Processes, certain highly regulated member states, parts of the Commission, such as DG Environment, and the European Environment Agency (EEA) act as political entrepreneurs who try to diffuse and operationalise the concept of sustainable development at European and national levels [*Lenschow, 2001*]. However, while sustainable development has been established as a norm that is rarely openly opposed, its practical implications remain contested [cf. *Lafferty, 2002*]. This appears to be one of the reasons why relevant instruments do not (yet) sufficiently pervade sectoral policies and measures, such as the integration strategies that have been produced in the framework of the Cardiff Process [cf. *Kraemer, 2001*].

Environmental Policy in the Accession Countries

What are the likely impacts of enlargement on the institutional foundations of EU environmental governance outlined above? An answer to this question requires a closer look at environmental problems and policy in the CEE accession countries that are set to join the EU in 2004. In the present context, it is not possible to look at each country individually. Thus, the analysis is restricted to those impacts that can be attributed to characteristics that most or all accession countries share. However, even using this limited approach [cf. *Smith and Pickles, 1998*] it seems possible to identify some of the most significant potential impacts because there are many environment-related similarities among the CEE accession countries which are rooted in their common 'post-socialist condition'. As Forsyth and Herrschel [*2001a: 570*] put it: there is a 'degree of commonality among formerly socialist states with individual modulation of this common theme'. (In addition, some impacts stem from factors that are not country specific by definition, for example the sheer number of new member states and the 'homogenising' influence of the preparations for EU accession.) In this sense, the factors listed below reflect a set of highly generalised conditions important for environmental policymaking in all or most of the eight CEE accession countries.

Perhaps most importantly, the case for sustainable development is particularly persuasive in CEE countries. Environmental quality in CEE states is still characterised by a sharp contrast between, on the one hand, heavily polluted areas such as the Polish Katowice region, abandoned military and industrial sites, unsafe nuclear and industrial facilities and, on the other hand, large unspoiled areas possessing rich biodiversity, a relatively favourable modal split between alternative means of transport, and a significant share of more traditional farming. At the same time, CEE countries are in the midst of a process of economic and social restructuring that is accelerated by EU accession. While restructuring may further reduce industrial pollution in the heavily polluted areas, the transition to a more market-based economy, the expansion of the service sector, changing investment, consumption, settlement and mobility patterns, as well as increasing integration into European and global markets and unsustainable policies – for example the EU agricultural and transport policies – pose a threat to the preservation of the expansive natural reserves, the relatively favourable split between alternative modes of transport, and traditional farming practices [*Hager, 2000; EEA, 1999*]. Preserving these environmentally favourable conditions in CEE countries would require a much more sustainable restructuring of the various sectors of economic activity than is currently the case [*Stritih, 2003*].

Second, despite the wave of environmental reforms in the early 1990s, environmental protection now ranks relatively low on the political agenda in many CEE accession countries. When conditions in some of the most notorious environmental 'hot spots' – for example the 'Black Triangle' region between the Czech Republic, Germany and Poland – began to improve, the economic and social problems associated with transition started to dominate politics. In addition, environmental concerns are only weakly rooted in party systems and civil society. Strong green parties are rare in CEE countries and the influence of environmental NGOs on policymaking is often quite limited. First, NGOs still suffer from a certain lack of staff and financial resources as well as institutionalised access to decision-makers [*Fagin and Tickle, 2001*]. Their weakness also stems from other factors. Perhaps most importantly, environmental NGOs tend to be apolitical, professionalised, and small in terms of membership. Their unwillingness to link environmental concerns to broader political and ideological issues reduces their ability to enter into strong alliances with political parties. At the same time, due to low membership, a certain reluctance to stage public protests and direct actions [cf. *JehliŹka, Sarre and Poboda, 2002; Herrschel and Forsyth, 2001*], and widespread attitudes among citizens, which encourage political passivity [*Tickle and Welsh, 1998*], environmental NGOs also have difficulties in generating sufficient mobilisation and political pressure on their own. Consequently, even if they manage to gain access to decision-makers – which is most frequently the case with environment ministries – the political influence of environmental NGOs often remains limited. In addition to the weak position of environmental NGOs, so far non-environmental civic and business associations have rarely integrated environmental concerns into their agendas [*OECD, 1999; Baker and Jehlička, 1998*].

Third, these problems are compounded by weak administrative capacities and inefficient, bureaucratic structures and decision-making procedures. For example, there is often rivalry and overlap of competencies between the environment ministries and the sectoral ministries responsible for, among other things, water management, transport, energy and agriculture, as well as conflicts between the different divisions of the environment ministries themselves. Administrative and technical monitoring and enforcement capacities are weak. Although this is often also true for the central environmental inspectorates, in many CEE accession countries the problem appears to be most acute at the regional and local levels [cf. *European Commission 2002a, 2002b, 2002c*; *OECD, 1999; EEA, 1999*].

Finally, the process of accession to the EU amounts to a major exercise in legal and administrative restructuring, administrative and technical capacity building, and investment in clean-up technology, such as wastewater treatment plants. This is because the EU expects the accession countries to practically apply most of the *acquis communautaire* of EU environmental legislation by the date of accession. To support the necessary upgrading and streamlining of administrative structures and the requisite investment in the environmental infrastructure, the EU has established various funds and programmes (PHARE, ISPA, SAPARD) with an annual budget exceeding 3,000 million Euro. The EU has also granted long transitional periods for certain provisions of EU environmental legislation that are particularly difficult to implement, owing to heavy investment requirements. This applies, for example, to the area of wastewater treatment [cf. *European Commission, 2002a, 2002b, 2002c; Stritih, 2003; Homeyer et al., 2000*].

As a result of these developments, the accession process has already significantly affected environmental policy in the CEE accession countries in a number of ways. For example, external pressure by the EU has partly compensated for the lack of domestic incentives for an active environmental policy [cf. *Jehlička, 2002*]. Environmental legislation has been modernised and administrative structures have been reformed and streamlined. However, there is still a need for further improvements to administrative capacities and more investment in environmental infrastructure to meet the requirements of the EU [*European Commission, 2002a, 2002b,2002 c*]. The total investment in environmental protection in most of the eight CEE accession countries between 1996 and 2000 only represents a minor share of the estimated total costs. Even in the two countries – the Czech Republic and Poland – which spent significantly more, the total amount of investment reached no more than 40 per cent and one-third, respectively, of the estimated expenditure necessary to achieve full compliance with EU environmental legislation [*Eurostat, 2002*]. These shortfalls, which illustrate the challenge that the investment requirements pose for the CEE accession countries, exist despite the fact that in recent years environmental expenditure as a share of GDP has tended to be higher in the CEE accession countries than in the present member states [*Eurostat, 2002*].

On the whole, it seems unlikely that the positive effects of joining the EU in terms of the modernisation of legislation, better administrative capacity, and more investment in environmental protection will suffice to compensate for past neglect and, in particular, for the potential negative medium- and long-term environmental effects of economic restructuring and accession which, among other things, threaten the extensive natural reserves, relatively favourable modal split among alternative means of transport, and more traditional farming practices which can still be found in CEE accession countries. After all, traditional EU environmental policy as constituted by the internal market and the environmental regimes has largely failed to halt similar trends in the existing member states [cf. *EEA, 1999*].

Potential Impact of Enlargement

Having elaborated on the main independent variables – the basic institutional structure of EU environmental governance and the general conditions for environmental policymaking in CEE accession countries – it is now possible to consider the potential impact of enlargement on EU environmental governance. The following scenario assumes that the eight CEE accession countries will join the Community in 2004. It seems useful to assess each of the environmental regimes separately because the three regimes are characterised not only by different aims, legal bases and instruments, but also by different mechanisms of institutional reproduction. It also seems helpful to distinguish between impacts on EU environmental policy formulation and implementation.

The Internal Market Regime

As pointed out above, the internal market regime is based on the functional and utilitarian mechanisms underlying the maintenance of the internal market. It seems unlikely that this logic will be undermined by enlargement. First, as the internal market performs core functions for the Community as a political institution promoting European integration, the EU has been particularly keen on ensuring that it will not be weakened by enlargement.

The special importance placed on maintaining the internal market was demonstrated, for example, by the fact that the EU pressed the accession countries at an early stage of the 'pre-accession' process to give priority to the legal transposition and practical implementation of internal market legislation [*Sedelmeier and Wallace, 2000*]. This included a number of important EU environmental laws. Similarly, at the beginning of the actual accession negotiations the EU declared that it would not grant transitional periods for product standards [cf. *European Union, 1999*], including environmental product standards.

Second, in the enlarged EU the new member states will have a considerable economic interest in harmonising environmental product standards. This is likely to lead to the adoption of relatively strict EU legislation. More specifically, the new member states will probably have much more interest in harmonising product standards than the (old) highly regulated member states, because exports from the new member states will be threatened by the standards of these highly regulated countries, which may ban imports not conforming to their requirements. Therefore, the highly regulated member states will be able to put pressure on the new member states to accept strict product standards in exchange for harmonisation. A similar logic applies to the practical implementation of EU environmental product standards: the general implementation difficulties of EU environmental legislation in the accession countries are unlikely to seriously undermine the internal market regime, because highly regulated member states may ban products that do not conform to EU environmental standards from entering their markets. Consequently, at least the major export-oriented producers in the accession countries have an economic interest in facilitating the necessary environmental investments and ensuring compliance.

The Environmental Regime

In contrast to the internal market regime, it seems possible that, in the absence of reforms, enlargement might critically weaken the power and utilitarian mechanisms underlying the environmental regime. There are several reasons for this. First, enlargement will dramatically increase the number of member states represented in the EU Council of Ministers. This is generally expected to decrease decision-making efficiency and raise the potential for deadlock. Fewer, or weaker, Community regulations would probably be agreed to under these conditions [cf. *Bär et al., 2001; Baldwin et al., 2001*]. Second, most of the new member states are likely to prefer a low level of environmental regulation once they have joined the Union and the external pressure resulting from the accession process has disappeared.

This seems to be largely due to the structural conditions in many accession countries, such as limited administrative and financial capacities, lack of domestic political incentives for a more proactive environmental policy, and the preoccupation with the economic and social implications of transition and EU accession – all factors addressed by various contributors to this volume. In the absence of a broader domestic support base for more progressive environmental policies, countervailing trends, such as the opinion of CEE environmental experts who may have adopted views similar to those held by their colleagues in the highly regulated member states [*Jehlička, 2002*], are unlikely to have a strong impact, in particular with respect to investment heavy process standards.

This scenario suggests that enlargement will significantly weaken the position of the highly regulated member states in the Council. Under these circumstances, it seems questionable whether factors that have so far compensated for the low number of votes of these countries in the Council – such as the political insulation of EU environmental policymaking and the superior resources of the highly regulated member states – will continue to secure the adoption and implementation of strict environmental process standards in an enlarged EU [cf. *Homeyer, 2001*].

Enlargement may also negatively affect the implementation of environmental process standards. First, in contrast to the internal market regime, producers have no direct economic self-interest in complying with environmental process standards. On the contrary, at least in the short term, non-compliance may render their products more competitive [cf. *Scharpf, 1994*]. Because non-compliance with process standards may be rational for many producers, the lack of administrative capacities for monitoring and enforcement in the accession countries is likely to have a particularly strong negative impact on the effectiveness of the environmental regime.

Second, many process standards require considerable public and private investment. The EU Urban Wastewater Treatment Directive is a case in point. In the course of the accession negotiations, the accession countries were granted long transitional periods for the full, practical implementation of this directive. In addition, implementation is supported by special Community funds. But the long duration of the transitional periods and the example of previous enlargements raise doubts as to the prospects for full implementation of those environmental process standards requiring substantial investment. For instance, Spain joined the EU in 1986 but is still far from having achieved full implementation of the 1991 Urban Wastewater Treatment Directive, despite massive EU financial support. Other member states also lag behind in the practical implementation of the directive [*European Commission, 2002*]. More generally, as pointed out above, CEE accession countries are likely to continue to face considerable difficulties in funding the installations needed for the practical implementation of expensive EU legislation. In fact, in the period 1996–2000 investment in environmental machinery and equipment even decreased in several CEE accession countries [*Eurostat, 2002*].

In recent years the EU has increasingly used procedural and flexible regulations, such as the EMAS Regulation and the IPPC Directive, to increase the efficiency of environmental measures and to reduce the implementation deficit of EU environmental legislation. However, effective implementation of these measures is particularly demanding with respect to administrative capacities and the mobilisation of civil society. Because of the weak structures in CEE countries in both of these respects, it seems unlikely that procedural and flexible regulations will significantly reduce implementation problems in the accession countries. On the contrary, the lack of clear, substantive targets may create additional opportunities to avoid taking effective measures [cf. *Börzel, 2000; Scott, 2000; Knill and Lenschow, 1999*].

In sum, weak political support, economic incentives, financial resources and administrative capacities are likely to negatively affect 'on the ground' implementation of both 'traditional' and more flexible EU environmental process standards. This implies the possibility of a further deterioration of the implementation record of EU environmental legislation after enlargement. Against the background of this short-term and medium-term scenario, it is possible to identify two very different potential long-term trends. First, there is the possibility of a breakdown of the environmental regime. Considering that the interests of highly regulated member states in effective environmental protection *and* in minimising economic competitive disadvantages and adjustment costs are at the root of the environmental regime, it seems questionable whether these countries would continue their efforts to maintain and extend the environmental regime in the face of increasingly more relaxed EU process standards and ever growing implementation problems in an enlarged Union. In such a situation, the highly regulated countries might decide to abandon the environmental regime and 'renationalise' environmental process standards.

Second, there is the possibility of adaptation of the environmental regime. For example, the provisions on Enhanced Cooperation introduced by the Amsterdam and Nice Treaties may offer opportunities for adaptation. These rules give a minimum of eight member states the possibility of using the EU institutions to agree on standards that exceed those adopted by the Community as a whole. Highly regulated member states may choose to use Enhanced Cooperation to renew their leadership role. More specifically, the use, or even the mere threat, of Enhanced Cooperation in the environmental field may generate economic, administrative and political advantages for the participants. For example, the participants in a particular instance of Enhanced Cooperation could assume the role of technological and institutional pace-setters who unilaterally determine the standards and procedures with which the non-members would have to comply once they decided to increase their level of environmental protection. Similarly, by threatening to take recourse to Enhanced Cooperation highly regulated countries could put pressure on member states with a lower level of regulation to accept stricter EU process standards than they would otherwise have been willing to agree to [*Bär et al., 2002; Homeyer, 2001*]. These dynamics could preserve the power and utilitarian mechanisms underlying the environmental regime.

The Sustainability Regime

Factors such as the low domestic political priority of environmental protection in many CEE accession countries, bureaucratic administrative structures, and the lack of both mobilisation and awareness of civil society in environmental matters, as well as the legally non-binding character of the sustainability regime, all seem to suggest that enlargement would negatively affect this regime. But this need not be the case. In fact, if the existing member states gave sufficient support to the emerging sustainability regime, it could probably make a particularly important contribution to environmental policymaking in an enlarged EU.

As pointed out above, given present environmental conditions and the pace and scale of economic and social restructuring in the CEE accession countries, a shift towards sustainable development seems particularly promising for these countries. In particular, it might help to avert a situation in which CEE countries would repeat those choices made by the existing member states, which are now widely recognised as environmentally unsustainable. But the sustainability regime also offers additional opportunities to address the challenge of enlargement. Since the regime is still relatively new, it will be easier for the accession countries to actively adapt the future development of the regime to their specific conditions and needs than to effect the older, more established regimes. In addition to more effective and efficient concrete measures, this could create a sense of 'ownership' and increase the commitment of CEE accession countries to the sustainability regime. Furthermore, the regime focuses on sustainable development which, in addition to the environmental dimension, also includes the economic and social dimensions. This broader approach seems to be particularly suitable for addressing some of the key structural conditions for environmental policymaking in CEE countries. More specifically, the accession countries are in a situation in which they must deal with the social implications of transition, for example the consequences of the increasing income disparities between urban and rural areas [*Fiedler and Janiak, 2003*]. At the same time they need to catch up economically with the existing EU member states. Against the background of these specific political conditions [cf. *Herrschel and Forsyth, 2001*], pursuing environmental objectives in the framework of the broader sustainable development approach might help to gradually increase and widen the domestic support base for environmental protection in the CEE accession countries.

The sustainability regime also offers advantages for the CEE accession countries with respect to the resources needed for implementation. The regime is primarily driven by 'soft' reproductive mechanisms based on information, learning, peer review, etc. It therefore requires relatively few financial resources and administrative capacities, which are particularly scarce in the accession countries. This advantage is reinforced by the fact that a significant share of the investment that will nevertheless be needed tends to be multi-purpose in that it has an environmental, but also an economic and social rationale. Given these multiple benefits, it should be relatively unproblematic to raise money for corresponding projects in the private sector. The potential for such multi-purpose investment appears to be particularly large in the accession countries, where environmental problems are often closely linked to economic inefficiencies [cf. *Hager, 2000*]. Investment to reduce the high energy intensity of the economies of the CEE accession countries is an outstanding example; Slovenia, the least energy intensive economy of the CEE accession countries, is still twice as energy intensive as the average of existing EU member states [*Fiedler and Janiak, 2003*].

Perhaps CEE accession countries could also use some of the experience and organisational structures created to facilitate EU accession to promote sustainable development through environmental policy integration. The accession process has led to the formation of entirely new intra- and inter-sectoral coordination structures. For example, Slovenia has established an Office for European Integration which coordinates and monitors the transposition and implementation of EU legislation by the various ministries [*ECE, 1999*]. Similar structures have been created in other accession countries. In addition to a central coordinating body, they usually include a European integration unit in the ministries that are most affected by the accession process. It may be possible for the accession countries to use their experiences in coordinating the accession process to design and implement measures to improve EPI. There may even be a possibility to use the existing coordination structures for this purpose during the accession process and, in particular, beyond the date of accession, when they have lost their original function.

However, owing to the fact that the sustainability regime is relatively young, its significance with respect to enlargement still depends to a large extent on whether the existing member states and the EU institutions will firmly establish the regime. Compared to the adoption and implementation of legislation, which is at the centre of the internal market and environmental regimes, the internalisation of the norms and practices of sustainable development into the administrative practices and decision-making routines of non-environmental sectors and actors is a more subtle and time-consuming process. It is therefore unlikely to be achieved in the absence of stabilisation of the sustainability regime over time by sufficient institutionalisation. This must be ensured primarily by the EU institutions and the existing member states.

Conclusions

Contrary to some expectations, the 2004 EU enlargement is unlikely to lead to a major disruption of EU environmental governance as such because it will probably affect the three environmental policymaking regimes within EU environmental governance in very different ways. The internal market regime is based on functional and utilitarian reproductive mechanisms rooted in structures – the institutionalisation of the internal market at the core of the Community and the economic interest of countries with low environmental product standards in EU harmonisation – that are unlikely to be strongly affected by enlargement. In contrast, it seems possible that enlargement will considerably weaken the environmental regime. If the accession of CEE countries shifts the balance of power significantly in favour of countries supporting a low level of environmental regulation, this will seriously affect the power and utilitarian mechanisms on which the environmental regime is based. As a result, the highly regulated countries will no longer be able to use the environmental regime as an instrument to impose their regulations on the remaining member states. This will deprive the environmental regime of its most important support base among the highly regulated countries. However, emerging instruments, such as Enhanced Cooperation, may offer ways for the highly regulated member states to maintain their superior position and adapt the environmental regime to enlargement.

Although the conditions for the implementation of Environmental Policy Integration may be even worse in the CEE accession countries than in many present member states in the short term, this may change in the longer run as a result of the good opportunities and potentially large benefits of a shift to sustainable development in these countries. Whether or not this shift will take place may depend less on the scarce financial resources and administrative capacities of the accession countries than on further consolidation of the sustainability regime at the Community level. Ideally, the highly regulated member states, which have so far promoted the emergent sustainability regime, should achieve some consolidation under relatively favourable conditions in the months before enlargement takes place.

As the important role of the existing member states with respect to a consolidation of the sustainability regime indicates, the potential trends outlined above cannot be taken for granted. In particular two factors seem worth mentioning that have the potential to lead to outcomes which differ to some extent from the ones outlined above. Perhaps most importantly the analysis focuses on the impact of the 2004 enlargement. Yet, additional countries are likely to join the EU in the next 15 or so years. More specifically, the accession of Bulgaria and Romania is currently scheduled for 2007. In addition, the EU has granted official candidate country status to Turkey. Croatia has also applied for EU membership and more south-eastern European countries are likely to follow suit. However, the assumption made in this study that it is possible to explore some of the most significant potential impacts of enlargement on EU environmental governance by focusing on the commonalties among the CEE accession countries is more difficult to justify if extended to additional countries, in particular Turkey, the former Yugoslavian states and Albania. These countries either do not share the 'post-socialist condition' (Turkey) or this condition has been considerably overshadowed, especially by the Balkan ethnic conflicts and wars. Despite these limitations, the analysis provides a stating point for thinking about the potential implications for EU environmental governance of post-2004 enlargements.

Weak institutionalisation of the sustainability regime is another factor that could lead to a modification of the effects of enlargement on EU environmental governance. As pointed out above, the sustainability regime may not be sufficiently stable to lead to an internalisation of new administrative practices and decision-making procedures that better reflect the requirements of sustainable development. But the lack of institutionalisation also has a substantive dimension relating to the contents of these practices and norms. For example, there is concern in some quarters that the concept of sustainable development may be redefined by an undue emphasis on the economic dimension. If this happens, a strengthening of the sustainability regime at the EU level could weaken the environmental regime as some measures under the environmental regime might be seen as too burdensome from an economic point of view. So far these concerns are not supported by the emerging EU sustainability regime which primarily relies on environmental policy integration. Because environmental policy integration focuses on non-environmental sectors, it is unlikely to have negative implications for the environmental regime. Nevertheless, given that the EU sustainability regime is not yet fully established, a future refocusing on the economic dimension cannot be excluded.

The dynamic development of EU environmental policy has so far defied those who expected a lowering of European environmental standards as a result of regulatory competition among the member states in the internal market or because of political resistance, in particular by the UK government in the 1980s and early 1990s. Given the domestic political structures and priorities of most CEE accession countries as well as their limited financial and administrative resources and capacities, the 2004 enlargement certainly poses a serious challenge to the future of effective European environmental policies. However, as argued here, the differentiated structure of EU environmental governance offers a good chance of accommodating the new member states. The consolidation of the EU sustainability regime is likely to be of crucial importance in this respect – not least because the most serious environmental challenges are increasingly similar in the new and the existing member states. Addressing problems such as climate change, waste generation, urban sprawl, soil sealing and habitats protection requires the integration of environmental concerns into other policies to achieve a significant decoupling of economic development on the one hand, and environmental degradation and resource consumption on the other.

ACKNOWLEDGEMENTS

The author would like to thank JoAnn Carmin, Stacy D. VanDeveer, an anonymous reviewer, and R. Andreas Kraemer for helpful comments on previous versions of this contribution.

REFERENCES

Baker, Susan (2000), 'Between the Devil and the Deep Blue See: International Obligations, Enlargement and the Promotion of Sustainable Development in the European Union', *Journal of Environment and Planning*, No.2, pp.149–66.

Baker, Susan and Petr Jehlička (1998), 'Dilemmas of Transition: The Environment, Democracy and Economic Reform in East Central Europe: An Introduction', in S. Baker and P. Jehlička (eds.), *Dilemmas of Transition: The Environment, Democracy and Economic Reform in East Central Europe*, Ilford: Frank Cass, pp.1–26.

Baldwin, Richard, Erik Berglof, Francesco Giavazzi and Mika Widgren (2001), 'Nice Try – Should the Treaty of Nice be Ratified?', *Monitoring European Integration 11*, Brussels: The Centre for Economic Policy Research.

Bär, Stefani, Ingmar von Homeyer and Anneke Klasing (2001), 'Fit for Enlargement? Environmental Policy After Nice', *European Environmental Law Review*, Vol.10. No.7, pp.212–20.

Bär, Stefani, Ingmar von Homeyer and Anneke Klasing (2002), 'Overcoming Deadlock? Enhanced Co-operation and European Environmental Policy After Nice', in H. Somsen (ed.), *Yearbook of European Environmental Law*, II, Oxford: Oxford University Press, pp.241–70.

Beck, Heiko (1995), *Abgestufte Integration im Europäischen Gemeinschaftsrecht unter besonderer Berücksichtigung des Umweltrechts*, Europäische Hochschulschriften, Frankfurt a.M.: Peter Lang.

Börzel, Tanja A. (2000), 'Best Practice Solution or Problem for the Effectiveness of European Environmental Policy?', *EUI Review*, Spring 2000, Florence: European University Institute, pp.32–6.

Bulmer, Simon J. (1997), 'New Institutionalism, The Single Market and EU Governance', *ARENA Working Papers*, Vol.97, No.25, University of Manchester.

Caporaso, James A. and John T.S. Keeler (1995), 'The European Union and Regional Integration Theory', in Carolyn Rhodes and Sonia Mazey (eds.), *The State of the European Union: Building a European Polity?*, Harlow: Longman, pp.29–61.

Deeg, Richard (2001), 'Path Dependence and National Models of Capitalism: Are Germany and Italy on New Paths?', paper prepared for delivery at the Annual Meeting of the American Political Science Association, 30 August–02 Sept 2001 in San Francisco, CA.

De la Porte, Caroline, Philippe Pochet and Graham Room (2001), 'Social Benchmarking, Policy-Making and the Instruments of New Governance in the EU', *Journal of Social Policy*, Vol.11, No.4., pp 291–307.

ECE (Economic Commission for Europe) (1999), *EPR of Slovenia: Report on Follow-Up*, Environmental Performance Reviews, Geneva: United Nations.

EEA (European Environment Agency) (1999), *Environment in the European Union at the Turn of the Century*, Luxembourg: Office for Official Publications of the European Communities.

Ehlermann, Claus-Dieter (1995), 'Increased Differentiation or Stronger Uniformity', *EUI Working Papers of the Robert Schuman Centre* (RSC), Vol.95, No.21, Florence: European University Institute.

Eichener, Volker and Helmut Voelzkow (1994), 'Ko-Evolution politisch-administrativer und verbandlicher Strukturen: Am Beispiel der technischen Harmonisierung des Europäischen Arbeits-, Verbraucher und Umweltschutzes', *Politische Vierteljahresschrift*, Special Issue No.25, pp.256–90.

European Commission (2001), 'A Sustainable Europe for a Better World: A European Union Strategy for Sustainable Development', COM(2001)264 final, Brussels, 15 May 2001.

European Commission (2002), 'Commission Acts against Portugal, Spain, Italy, Sweden, Belgium, Luxembourg, The Netherlands, France and Greece for Not Complying with Water Quality', Press Release, Brussels, 2 July 2002.

European Commission (2002a), '2002 Report on Czech Republic's Progress Towards Accession', COM(2002)700 final, Brussels, 9 Oct. 2002.

European Commission (2002b), '2002 Report on Hungary's Progress Towards Accession', COM(2002)700 final, Brussels, 9 Oct. 2002.

European Commission (2002c), '2002 Report on Poland's Progress Towards Accession', COM(2002)700 final, Brussels, 9 Oct. 2002.

European Union (1999), 'Conference on Accession to the European Union – Slovenia – European Union Common Position, Chapter 22: Environment', CONF-SI 60/99, Brussels, 19 Nov. 1999.

Eurostat (Statistical Office of the European Communities) (2002), *Environmental Protection Expenditure in Accession Countries, Data 1996–2000*, Luxembourg: Office for Official Publications of the European Communities.

Fagin, Adam and Andrew Tickle (2001), 'Globalisation and the Building of Civil Society in Central and Eastern Europe: Environmental Mobilisations as a Case Study', paper for the European Consortium for Political Research (ECPR) Conference, University of Kent, Canterbury, 2–8 Sept. 2001.

Fiedler, Joanna and Paulina Janiak (2003), *Environmental Financing in Central and Eastern Europe 1996–2001*, Szentendre: The Regional Environmental Center for Central and Eastern Europe.

Haas, Ernst (1964), 'Technocracy, Pluralism and the New Europe', in Stephen R. Graubard (ed.), *A New Europe*, Boston: Houghton Mifflin, pp.62–88.

Hager, Wolfgang (2000), *The Environment in European Enlargement*, Brussels: Centre for European Policy Studies.

Hall, Peter A. and Rosemary C.R. Taylor (1996), 'Political Science and the Three New Institutionalisms', *MPIFG Discussion Papers*, Vol.96, No.6, Köln: Max-Planck-Institut für Gesellschaftsforschung.

Héritier, Adrienne (1995), 'Leaders and Laggards in European Clean Air Policy', in F. van Waarden and B. Unger (eds.), *Convergence or Diversity? The Pressure of Internationalization on Economic Governance Institutions and Policy Outcomes*, Aldershot: Ashgate, pp.278–305.

Herrschel, Tassilo and Timothy Forsyth (2001), 'Constructing a New Understanding of the Environment Under Postsocialism', *Environment and Planning A*, Vol.33, pp.573–87.

Herrschel, Tassilo (2001a), 'Environment and the Postsocialist "Condition"', *Environment and Planning A*, Vol.33, pp.569–72.

Holzinger, Katharina (1994), *Politik des kleinsten gemeinsamen Nenners? Umweltpolitische Entscheidungsprozesse in der EG am Beispiel der Einführung des Katalysatorautos*, Berlin: Sigma.

Holzinger, Katharina and Peter Knoepfel (2000), 'The Need for Flexibility: European Environmental Policy on the Brink of Enlargement', in K. Holzinger and P. Knoepfel (eds.), *Environmental Policy in a European Union of Variable Geometry? The Challenge of the Next Enlargement*, Basel: Helbing & Lichtenhahn, pp.3–35.

Homeyer, Ingmar von (2001), 'EU Environmental Policy on the Eve of Enlargement', *EUI Working Papers*, RSC, No.2001/35, Florence: European University Institute.

Homeyer, Ingmar von, Alexander Carius and Stefanie Bär (2000), 'Flexibility or Renationalization: Effects of Enlargement on EC Environmental Policy', in M.G. Cowles and M. Smith (eds.), *The State of the European Union: Risks, Reform, Resistance and Revival*, Oxford: Oxford University Press, pp.347–68.

Jehlička, Petr (2002), 'Environmental Implications of Eastern Enlargement of the EU: The End of Progressive Environmental Policy?', *EUI Working Papers*, RSC, No.2002/23, Florence: European University Institute.

Jehlička, Petr, Phillip Sarre and Juraj Poboda (2002), 'Czech Environmental Discourse After a Decade of Western Influence: Transformation Beyond Recognition or Continuity of the Pre-1989 Perspectives?', *EUI Working Papers*, RSC, No.2002/24, Florence: European University Institute.

Judge, David and David Earnshaw (1992), 'Predestined to Save the Earth: The Environment Committee of the European Parliament', *Environmental Politics*, Vol.1, No.4, pp.186–212.

Judge, David, David Earnshaw and N. Cowan (1994), 'Ripples or Waves: The European Parliament in the European Community Policy Process', *Journal of European Public Policy*, Vol.1, No.1, pp.27–52.

Knill, Christoph and A. Lenschow (1999), 'Neue Konzepte – alte Probleme? Die institutionellen Grenzen effektiver Implementation', *Politische Vierteljahresschrift*, Vol.40, No.4, pp.591–617.

Kohler-Koch, Beate (2000), 'Framing: The Bottleneck of Constructing Legitimate Institutions', *Journal of European Public Policy*, Vol.7, No.4, pp.513–31.

Kraemer, R. Andreas (2001), 'Results of the "Cardiff-Processes" – Assessing the State of Development and Charting the Way Ahead', Report to the German Environmental Agency and the German Federal Ministry for the Environment, Nature Conservation and Nuclear Safety, Research report no.290 19 120 (UFOPLAN), abridged and translated version from the German original version. Available at http://europa.eu.int/comm/environment/enveco/integration/german_study.pdf (7 Nov. 2002).

Lafferty, William M. (2002), 'Adapting Government Practice to the Goal of Sustainable Development', Program for Research and Documentation of Sustainable Society, Center for Development and the Environment, Working Paper No.1/02, Oslo: Oslo University.

Lenschow, Andrea (2001), 'New Regulatory Approaches in Greening EU Policies', LANAGE (Law and New Approaches to Governance in Europe) workshop, 29–30 May 2001, European Union Center, University of Wisconsin–Madison.

Mahoney, James (2000), 'Path Dependence in Historical Sociology' *Theory and Society*, Vol.29, No.4, pp.507–48, Kluwer Academic Publishers.

Mahoney, James (2001), 'Path-Dependent Explanations of Regime Change: Central America in Comparative Perspective', *Studies in Comparative International Development*, Vol.36, No.1, pp.111–14.

Moravcsik, Andrew (1993), 'Preferences and Power in the European Community: A Liberal Intergovernmentalist Approach', *Journal of Common Market Studies*, Vol.31, No.4, pp.473–524.

OECD (Organisation for Economic Co-operation and Development) (1999), *Environment in the Transition to a Market Economy. Progress in Central and Eastern Europe and the New Independent States*, Paris: OECD.

Peters, B. Guy (1999), *Institutional Theory in Political Science: The 'New Institutionalism'*, London, New York: Pinter.

Pierson, Paul (2000), 'Increasing Returns, Path Dependence, and the Study of Politics', *American Political Science Review*, Vol.94, No.2, pp.251–67.

Pierson, Paul (2000a), 'Not Just What, but *When*: Timing and Sequence in Political Processes', *Studies in American Political Development*, Vol.2000, No.14, pp.72–92.

Pierson and Skopcol, 'Historical Institutionalism in Contemporary Political Science.' Paper presented at the 2000 Annual Meeting of the American Political Science Association (APSA) in Washington DC, 31 Aug–03 Sept 2000.

Pollack, Mark A. (1996), 'The New Institutionalism and EC Governance: The Promise and Limits of Institutional Analysis', *Governance*, Vol.9, No.4, pp.429–58.

Sbragia, A. (1996), 'Environmental Policy: The Push-Pull of Policy-Making', in H. Wallace and W. Wallace (eds.), *Policy-Making in the European Union*, 3rd edition, Oxford: Oxford University Press, pp.235–55.

Scharpf, Fritz W. (1994), 'Community and Autonomy: Multi-Level Policy-Making in the European Union', *Journal of European Public Policy*, Vol.1, No.2, pp.219–42.

Schwartz, Herman (2001), 'Down the Wrong Path: Path Dependence, Markets, and Increasing Returns', unpublished paper, on file with the author.

Scott, Joanne (2000), 'Flexibility in the Implementation of EC Environmental Law', in H. Somsen (ed.), *Yearbook of European Environmental Law*, Vol.1, Oxford: Oxford University Press, pp.37–60.

Secretariat of the European Convention (2002), 'New Institutions for a New Europe', CONV 452/02, CONTRIB 166, Brussels: Contribution Submitted by Mr Alain Lamassoure, member of the Convention.

Sedelmeier, Ulrich and Helen Wallace (2000), 'Eastern Enlargement – Strategy or Second Thoughts?', in H. Wallace and W. Wallace (eds.), *Policy-Making in the European Union*, 4th edition, Oxford: Oxford University Press, pp.427–60.

Smith, Adrian and John Pickles (1998), 'Introduction: Theorising Transition and the Political Economy of Transformation', in A. Smith and J. Pickles (eds.), *Theorising Transition: The Political Economy of Transition in Post-Communist Countries*, London: Routledge, pp.1–22.

Stritih, Jernej (2003), 'Central and Eastern Europe Face More Environmental Work After EU Accession', *The Bulletin. Quarterly Magazine of the Regional Environmental Center for Central and Eastern Europe*, Vol.11, No.4, pp.14–24.

Tickle, Andrew and Ian Welsh (1998), 'Environmental Politics, Civil Society, and Post-Communism', in A. Tickle and I. Welsh (eds.), *Environment and Society in Eastern Europe*, Harlow: Longman, pp.156–85.

Thelen, Kathleen (2000), 'Timing and Temporality in the Analysis of Institutional Evolution and Change', *Studies in American Political Development*, No.14, pp.101–8.

Environmental Implications of Eastern Enlargement: The End of Progressive EU Environmental Policy?

PETR JEHLIČKA AND ANDREW TICKLE

In the years immediately following the fall of the communist regime, Central and Eastern European (CEE) governments paid significant attention to environmental policy developments, both at home and abroad, as a consequence of domestic environmental mobilisation. However, after 1992 the nature of environmental policy reform changed. Legal and institutional change could no longer rely on its domestic base and instead policy development became a function of the 'structural imperative' of reform in CEE countries [*Slocock, 1996*]. In other words, reform was an unwelcome, but necessary part of EU environmental harmonisation. By the late 1990s, an overriding concern in CEE states was to minimise the impact of these EU reforms on economic growth and competitiveness. Thus, in contrast to the initial proactive approach to environmental policy, CEE countries have become passively compliant with EU requirements and a strictly national perspective in this policy arena has been eclipsed by EU hegemony.

The 'Europeanisation' perspective of CEE national environmental policy has generally focused on the one-way process of CEE adaptation to EU requirements and on the management of this process by EU institutions, predominantly the European Commission. Most scholars, regardless of their analytical approach, conclude that enlargement will have an adverse effect on progressive EU environmental policy. In these considerations, the impact of eastern enlargement on the EU is seen as a function of EU-related variables such as its administrative and financial resources and strategies of their deployment in the process of CEE integration with the Union. A top-down mode of analysis is thus essentially maintained. However, little attention has so far been paid to the interests, preferences and priorities of CEE countries themselves.

Using the four Visegrád countries (V4)[1] as a case study, this study offers some preliminary observations on the possible implications of eastern enlargement for EU environmental policy from the applicant states' perspective. To that end, 29 in-depth interviews were carried out with environmental policy experts in the V4 countries in 2000,[2] supplemented by five interviews conducted at the end of 2000 and at the beginning of 2001 with experts from EU countries who have substantial experience with environmental policymaking in CEE candidate countries.

We begin with a discussion of insights from the existing academic literature on the implications of eastern enlargement for future EU environmental policy. We then outline some of the theoretical difficulties in exploring future environmental policy implications of CEE accession to the EU. Primarily following Andersen's and Liefferink's [*1997*] approach to studying the influence of individual states on EU environmental policy, this study then presents the domestic and international policy activities of the V4 countries in the 1990s that may be indicative of their attitude towards future EU environmental policymaking. The ensuing sections provide an overview of the V4 states' involvement in international environmental politics and present expert ideas of future strategies for the V4 countries in an enlarged EU. We conclude by discussing the extent to which the patterns revealed in this study challenge commonly held views of how eastern enlargement will impact on EU environmental policy.

The Need for an Applicant States-Centred Approach

Analyses of the environmental policy dimensions of eastern enlargement, by both West [e.g. *Connolly et al., 1996; Slocock, 1999*] and East European authors [e.g. *Stehlúk, 1998; Kerekes and Kiss, 1998; Zylicz and Holzinger, 2000*] mainly focus on adaptation to the environmental *acquis* in a narrow 'technical' sense (namely, the effectiveness of the transfer of these norms to the domestic context). Other authors are more critical of the power relations between the EU and applicant states. Caddy [*1997*], for example, describes EU–CEE relationships as 'hierarchical imposition' while Baker and Welsh [*2000*] emphasise the undemocratic character of the harmonisation process in contrast to previously positive impacts of environmental policy on legitimation and governance, both in the EU and in Central and Eastern Europe.

Authors who discuss the implications of eastern enlargement for EU environmental policy commonly assume that the accession of CEE applicant countries will lead to a 'downward pressure on environmental policy' [*Baker, 2000*]. Pellegrom [*1997: 55*] expresses this view when she suggests that, 'within only a limited number of years, environmental policy will be subject to many more conservative positions than the progressive ones'. In a similar vein, Holzinger and Knoepfel [*2000: 15*] more recently have argued that:

> ... CEECs do not have a tradition of strong environmental policy and in the future they will probably give economic development priority

over stringent environmental policy. Hence, most of them will presumably join the group of environmental 'laggards' within the European Council of environment ministers.

Much of the existing literature on enlargement is dominated by past and present debates on the Europeanisation of policy in candidate states (that is, the 'top-down' approach) and only rarely adopts an alternative perspective that focuses on the ways that enlargement will affect EU environmental policy. Those authors who do discuss the future implication of enlargement [*Homeyer et al., 2000*; *Carius et al., 2000*] pay only cursory attention to CEE-related variables other than the number and length of requested transition periods. CEE country strategies, priorities, resources and interests in the post-accession period are accorded little attention within these analyses. In other words, most analysts consider the applicant countries as passive subjects of the EU's governance in the post-accession period. This implies that their domestic base of environmental policy will be weak and incapable of transferring domestic environmental policy concepts to the EU level.

Concurrently it is also assumed that the new member states will maintain a negative approach to progressive EU environmental policy after they join the Union. In practice, this would mean that they would either try to block the adoption of new legislation or press for lower standards. However, as Aguilar Fernández [*1997*] points out, passivity in EU environmental policymaking can be ultimately disadvantageous for a given member state because if 'leader' states succeed with their policy proposal, passive states will still have to adopt new legislation in whose formulation they had little or no influence. Further the expectation that the accession of CEE countries will lead to the downward pressure on EU environmental policy is based on the assumption that the CEE states will coordinate their conservative stance not only among themselves but also with the current group of 'laggards'.

Other arguments qualify such conclusions and suggest instead the possibility of CEE countries taking a more progressive and proactive approach to EU environmental policy. As Homeyer [*2001*] suggests, despite unfavourable economic, administrative and political factors, there are incentives (for example, reducing EU-sourced transboundary pollutants, geographical and cultural proximity to 'leader' countries) for CEE countries to take an active part in EU environmental policymaking. Furthermore, there is the historical lesson of the UN-based 'Environment for Europe' process that was instigated by CEE countries in 1991. There are also several specific pieces of legislation (for example, the Czech law on strategic environmental impact assessment or SEIA) adopted in the early 1990s that were more progressive than extant EU legislation. In addition, Thorhallsson [*2000*] argues that, despite their limited resources, smaller member states can be effective in pursuing their interests at the EU level owing to special features of their administration and the necessity to prioritise between sectors.

Nonetheless, it is true that it is almost impossible to identify EU integration-related interests of CEE countries other than full membership in the Union. As Ágh [*1999*] observes in the Hungarian case, this means there are still many non-articulated interests currently subsumed within the accession process with some potential for unpredictable outcomes. This lack of articulation clearly distorts the policymaking process. Our research provides abundant evidence that the environment is a prime example of such a policy area of non-articulated interests.

This study searches for evidence supporting an alternative view of the approach of V4 countries to EU environmental policy, compared to the model of reactive and passive adaptation. We make the idealised assumption that once the current process of V4 countries environmental adaptation is concluded and these countries become full members, they will have an equal opportunity to follow their interests as other member states. We also assume that the European Commission's high degree of influence on the candidates and its insistence on their full adoption of environmental *acquis* with only a limited number of transition periods will lead to a relatively high degree of harmonisation when CEE countries join the Union, thus considerably reducing the threat of re-nationalisation of environmental policy in the future.

On this basis we formulate two sets of questions. The first set is: what is the domestic base of environmental policy in the V4 states? Are there signs indicating that in some areas of domestic environmental policy, the strategy of passive adaptation to the EU has an alternative, more proactive approach, either now or in the future? The second set of questions is: what capacity do the V4 states have to shape EU environmental policy? Are V4 countries likely to pursue – at least in some areas – proactive environmental policy at the EU level? What is the likelihood that they will coordinate their efforts to slow down the development of EU environmental policy among themselves and with the group of 'laggards'?

The Lack of Theory of National Integration

When analysing the implications of eastern enlargement on EU environmental policy from the applicant states' perspective, one is confronted with a paucity of suitable theoretical frameworks, *viz.*:

> Integration theory has focused on describing and explaining integration processes and the role of supra-national actors such as the

> Commission and the European Parliament. On the other hand, the role of the policies, interests and actions of its most important actors, the nation-states, has been neglected by theory; existing efforts are mainly empirical and a-theoretical, concentrating on national peculiarities rather than on establishing a theory of national integration policy [*Petersen, 1998b: 87*].

Thorhallsson [*2000*] also notes the neglect of smaller states and their impacts in international relations. He observes that the highest priority has been given to the study of the adaptive policy of small states in regard to the power politics of superpowers and not to the participation of small states in integration processes [*Thorhallsson, 2000*].

Elsewhere Petersen [*1998a*] formulates a general theory of national integration in the EU that is based on the premise of adaptation theory. This assumes that foreign policy consists of policymakers' actions to manipulate the balance between their society and their external environment in order to secure an adequate functioning of societal structures in a situation of growing interdependence [*Petersen, 1998a*]. Depending on the balance between the degree of control over the external environment (influence capacity) and the degree of sensitivity to it (stress sensitivity) a state can pursue four types of integration strategies. The first is dominance (high influence capacity, low stress sensitivity), under which the state is able to make demands on partner states in the integration process without giving concessions in return. The second is policy of balance (high influence capacity and high stress sensitivity) that describes an ideal form of national integration strategy. The third is a policy of acquiescence (low influence capacity, high stress sensitivity), essentially a subordination of domestic priorities to external pressures and the fourth category is a policy of quiescence (both low influence capacity and stress sensitivity), which is typical for low-influence countries.

Petersen's typology enables us to make an initial assessment of the approach of V4 countries to policymaking in an enlarged EU. In line with general expectations, it seems likely that, owing to their lack of both tangible resources (such as a strong economy and military power) and intangible resources (such as diplomatic skills, policy expertise and willpower) for influence capabilities, after accession, V4 countries will oscillate between acquiescence and quiescence. The former, which presupposes a limited degree of influence capability and high stress sensitivity, is typical for applicant states that make numerous concessions as they adapt their policies to membership. The latter indicates a preference for a low-participation strategy aimed at limiting concessions in the integration process. It is also a strategy of reduced commitment such as having loose ties to the integration process, perhaps with a concentration on particular goals or aspects.

Thorhallsson's [*2000*] analysis of smaller member states also suggests that in certain areas, even those with limited resources can exert an influence on the EU. Despite its different sectoral perspective (CAP and regional policy), Thorhallsson's [*2000*] work on the behaviour of smaller states in EU integration is pertinent to this study as it provides a framework for consideration of the behaviour of CEE countries in the area of environmental policy. With the exception of Poland, the other V4 countries (and all other CEE countries invited to join the EU) fall into the category of smaller states. Thorhallsson's [*2000*] argument is that small and large state integration behaviours differ owing to the size of their administrations, with small states not having sufficient capacity to address all negotiations owing to their lack of staff, expertise and other resources. As a consequence, while they behave reactively in most sectors they adopt proactive behaviours in the most important sectors. This is enabled by certain features of their administration such as informality, flexibility and greater room for their officials to manoeuvre. Owing to their smaller range of interests they tend, unlike the larger states, to prioritise between sectors.

One of the few attempts to develop a more theoretical approach to the question of influence of member states on the EU, specifically in the area of environmental policy, is that of Andersen and Liefferink [*1997*]. Recognising the reciprocal nature of EU policymaking, they examine the domestic politics of environmental policymaking of different member states and analyse how links are made to Brussels' politics. However, the applicability of this approach to CEE countries is subject to several limitations. First, their work is partly concerned with the 'domestication' of EU environmental policy in connection with countries that are already member states, although in the cases of Sweden, Austria and Finland they also analyse the pre-accession period of environmental policy. Second, the countries that were analysed are usually environmental policy 'pioneers' with highly developed domestic policies that they then seek to transfer to the EU level. In contrast, literature on the adaptation strategies of 'laggard' countries' to European environmental policy and their attempts to influence EU level is much rarer (but see Aguilar Fernández [*1997*]).

However, the concepts from which Andersen and Liefferink [*1997*] derived their analytical tool were developed neither specifically for the group of 'pioneer' countries, nor in fact for the purpose of EU studies. Thus, in the absence of a theoretical framework that would better match the dynamic discussed in this contribution, we loosely utilise their approach by analysing the past developments of V4 countries' domestic environmental policymaking that may be indicative of their attitude to future EU environmental policy as well as the past record of V4 states' international (extra-EU) environmental strategies. As the EU dimension of V4 states' environmental policy is overwhelmingly defined by the goals of harmonisation and as instances of their own strategies are almost invariably ruled out, we therefore analyse areas of environmental foreign policy that were not constrained by the EU integration process. The future implications of V4 accession are then discussed on the basis of the Regional Environmental Center's document summarising the suggestions of CEE countries for the 6th Environmental Action Programme [*REC, 2000*] and expert interviews conducted for this study.

Domestic Factors Influencing V4 Countries' Approach to EU Environmental Policy

The enthusiasm of the immediate post-1989 environmental reform in V4 countries was partly a consequence of the role of environmental protests in overthrowing the previous regimes. In some cases this led to adoption of progressive environmental legislation that went further than the existing EU legislation, such as the SEIA legislation in the former Czechoslovakia. It is symptomatic that such legislation is now subject to 'downward' harmonisation. This is also a reflection of the view commonly held in V4 states that many environmental problems were resolved by the end of the 1990s. For instance, there is a considerable degree of complacency in the Czech Republic, particularly in local government and industry, with regard to reductions in air and water pollution during the last decade. The prevailing feeling is that 'we have done too much for the environment'.

The demise of the socialist system in CEE countries broadly coincided with what some describe as the culmination of the most significant shift in Western environmental governance over the last 30 years, characterised by Bernstein [*2000*] as the convergence of environmental and economic norms towards 'liberal environmentalism'. This corresponded with pre-1989 domestic environmental (oppositional) discourse that stressed free markets and democracy as key conditions for successful environmental reform in V4 countries. 'Liberal environmentalism' thus found a ready niche in V4 states shortly after 1989.

Environmental Policy Actors

Despite the significant role played by environmental mobilisation in the 1989 revolutions, for most of the 1990s V4 countries lacked powerful domestic actors in environmental politics. Within two years of the fall of the old regime, relatively strong green parties disappeared almost without trace [see *Jehlička and Kostelecký, 1995*]. Thus in V4 countries, environmental groups are now the main source of political and social communication about the environment. However, after more than a decade of activity, the existence of these groups is still critically dependent on external (that is, Western) financial assistance.[3] Despite substantial foreign support aimed at the development of civil society, combined membership of environmental groups in V4 countries was, at the end of the 1990s, still lower than in the late 1980s. While a significant part of their funding is EU-based, environmental groups in V4 countries have not showed much interest in EU environmental policy, mainly because the funding aimed to strengthen domestic capacity. Only recently has the availability of EU pre-accession funds to V4 states started to generate some interest in EU environmental policy, although still related mainly to implementation issues. The scale of these activities is still national at best:

> We organised training for NGOs about EU integration, but our knowledge of integration is very weak … We organised it, but we also (were) listeners, as the others … Only (a) few speakers had the knowledge that we wanted. Their knowledge was also very narrow (Polish sustainability expert, interview 5 July 2000).

Policy Structures and Networks

According to experts interviewed in both V4 and EU countries, the most important (and often the only) environmental policy institution relevant to EU accession were the ministries of the environment and more specifically their departments of EU integration. The only exception seems to be Hungary, where a powerful prime minister's office concentrates large competencies in the area of EU integration, including environmental policy. Elsewhere, departments of EU integration have a relatively short history and, compared to the scale of their task, an inadequately small staff.[4] For instance, in 1997 only one person at the Czech Ministry of the Environment worked on environmental integration with the EU. Young people with knowledge of foreign languages, who in some cases have a background in environmental NGO activities, often staff these departments. Since their inception, their activity has quite understandably been limited to a passive adoption of the environmental *acquis*. None has developed a proactive policy agenda *vis-à-vis* the EU.

This weak institutional base is clear evidence of state and EU failure to use the process of accession as a stimulus for strengthening indigenous policy structures. For example, a recent report blames the Czech central state authorities for the poor capacity within regional and local public authorities to harmonise legislation [*GA&C and UK, 2001*]. Homeyer [*2001*] also emphasises the European Commission's overriding concern in the process of approximation with the formal requirements of transposing and legally implementing the environmental *acquis* and the relative neglect of administrative reforms needed for effective implementation and enforcement on the ground [see also *Kružíková, this volume*].

Given the weak position of environmental ministries within V4 governments, it comes as little surprise that ministries often used the process of approximation as a power enhancing tool by which they seek 'to out manoeuvre rival ministries in a "two-level" game' [*Homeyer, 2001*]. Ágh [*1999*] argues in a similar way when he maintains that 'it is particularly true for Hungary that numerous interests are better represented in Brussels than in national capitals, first of all in the field of environmental protection'. This may also explain the rather uncritical acceptance of the existing environmental *acquis* by environmental policy communities in V4 states.

It seems likely that after accession, the ability of V4 countries to participate in Brussels' environmental politics will be undermined by a lack of experts with appropriate training and experience. This prediction is based on three factors. The first is the past and current educational structure of V4 societies, which mainly emphasise narrow technical and scientific disciplines. This has serious consequences for the way in which environmental issues are understood by experts in V4 states:

> The root of the problem lies in [the] educational system. We lack people with an interdisciplinary background. We have specialists, but at the same time lack technically educated people and economists who would be concerned with the environment (interview, Polish sustainability expert, 5 July 2000).

The second factor expected to limit post-accession participation in environmental politics is the scant attention paid in the 1990s by social and political elites to the environmental dimension of the approximation process. The third factor is the fact that EU assistance programmes aimed at strengthening V4 states' environmental capacity have also neglected this area. Recent efforts by the EU to enhance the environmental policy capacity of the candidate states through various assistance programmes [see *Carius et al., 2000*] are invariably aimed at strengthening implementation *per se*, rather than wider policy thinking. As noted by one official, '[A] lot of the capacity building is being done on implementation issues and this is not the same thing' (interview, British government official, 9 January 2001).

EU environmental policy communities in V4 countries were described by respondents as small and closed groups of experts that developed on the basis of expertise applicable at the sub-national or national levels. This community is usually centred on a single personality who has a strong influence on the way in which the discourse on EU environmental policy develops. Thus, it seems that in Hungary and the Czech Republic, for instance, only one school of thought on EU environmental policy exists. This echoes recent suggestions that countries with centralised state structures and weak civil society tend to promote clientelistic relations that do not enhance social capital and thereby inhibit the capacity of networks for policy learning [*Paraskevopoulos, 2001*].

It does not seem likely that this situation could change significantly in the foreseeable future. At present, few institutions, whether research institutes, think-tanks or study and research programmes exist to address environmental policy research in V4 countries. Apart from the environmental component of a major Hungarian Academy of Sciences' project researching the effects of EU integration on Hungary, no policy research programmes in V4 countries have been initiated. Existing environmental research institutes, such as those affiliated with the Czech Ministry of Environment, have an almost exclusively scientific and technical orientation.

Policy Content

Environmental policy in V4 countries did not start from scratch in 1990. For example, systems based on fees paid by polluters, national environmental quality standards and pollution permits were – to varying extents – in place in V4 states in the 1980s. In the 1990s however, virtually all newly introduced policy concepts and instruments were imported from the West. The majority of respondents believed that the environmental *acquis* meets the needs of their countries in terms of the most pressing environmental problems, particularly in the areas of water and air pollution and waste management. Most of them also consider implementation of the *acquis* by their countries as a major innovation in environmental policy. Horizontal legislation aimed at public participation and access to information is regarded as an important means of opening the political system to a wider spectrum of actors. There are only a few areas that are regarded as being more developed in candidate states than in the EU. These include the system of nature protection, and land use planning and SEIA procedures, although this varies, depending on which country is examined.

By the mid-1990s, Caddy [*1997*] noted that the majority of CEE policymakers rejected wholesale the idea that pre-1989 policy was relevant within the framework of the EU–CEE policy dialogue. The interviews conducted for this study also confirmed little demand for an indigenous approach to environmental policy or for alternatives to EU concepts. In fact, most respondents see the current EU environmental policy model as optimal. Respondents were aware of some environmentally positive practices of V4 countries such as lower production of household waste per capita, recycling, wider use of public transport and, in some cases, less intensive forms of agriculture. These positive features were inherited from the socialist period. As such, they have mainly negative connotations, both for society at large and for most decision makers. Respondents were not aware of any domestic efforts aimed at their retention or even expansion. Furthermore, these features do not correspond with the traditional perception of environmental issues such as industrial pollution endangering human health that can be relatively easily resolved by clean-up programmes. As a consequence, these positive practices (such as glass bottle recycling) have been marginalised and replaced by other products (aluminium cans and plastic cartons) that are usually environmentally less beneficial [*Gille, 2000; Gille, this volume*]. However, international institutions shaping the development of environmental policy in the region have also neglected these features. Sometimes the consequences of their involvement in V4 countries are also environmentally questionable, such as EU infrastructure support for road building. Such issues were seldom discussed by V4 respondents.

V4 International Environmental Strategies

Despite the neglect of international environmental policy initiatives during the socialist period, at the beginning of the 1990s the V4 countries developed a relatively ambitious foreign environmental policy agenda, at both the European and global levels. The development of a foreign policy agenda signalled a significant break from the perception of environmental issues commonly held during the socialist period:

> [At the beginning of the 1990s] we realised how little we knew about global environmental issues, which we simply did not discuss during the socialist period, since until the end of the communist regime we were more interested in environmental issues at the local or national level (interview, Czech government official, 13 April 2000).

Partly as a response to this perspective, the initiative that led to the first ever set of pan-European environmental strategies came from within the region itself [*Vavroušek, 1993*]. But this initially strong, proactive approach to global environmental politics largely came to an end with the ratification of the basket of global conventions signed at the Rio 'Earth Summit' in 1992. After the signing of association agreements with the EU (between 1991 and 1996), the focus of V4 countries' international strategies shifted almost exclusively to the goal of EU harmonisation. In all four V4 countries this is underlined by the paucity of official documents on regional or global goals for the post-accession period. The dominant focus on the fulfilment of EU requirements has therefore led to the virtual abandonment of most other spheres of international environmental policy:

> I estimate that 95 per cent of the Czech Republic's activity in foreign environmental policy is oriented to the EU, the remaining five per cent covers all the rest, including UNEP and UN conventions (interview, Czech government official, 6 January 2000).

It is an irony then that even in the 'Environment for Europe' process, V4 states are now perceived as merely passive participants as described by a British government official:

> But there's a bit of a problem in the Environment for Europe process which is meant to be a pan-European, west–east cooperation, as not enough of the initiative comes from the east and they don't seem to set enough of the agenda as to what it is that they need (interview, British government official, 9 January 2001).

Since the mid-1990s V4 states' approach to global environmental regimes, most importantly to the climate change regime, has become fully dependent on the position of the EU. V4 countries do not have defined goals for their activities in the field of foreign environmental policy other than membership in the EU. For instance, the Czech Ministry of the Environment has an annual plan of action at the international level, but it resembles a list of forthcoming events rather than a programmatic document setting out short-, mid- or long-term goals. Hungarian and Slovak respondents in particular emphasised that in areas of policy unrelated to the EU approximation, it is the personality and field of expertise of ministers of the environment that define the country's activity at the international level. However, consistent with domestic policy, V4 environmental diplomacy is based in weak institutions and networks and relies instead on certain key individuals:

> There are occasional individual personalities [in CEE] who do have an influence in other international conventions, climate change, and sustainable development … but that is not the same as having a clearly defined foreign environmental policy … that is just the case of an individual who is having an effect in some fora or another (interview, British government official, 9 January 2001).

Despite their similar history, common environmental problems, and the shared goal of EU membership, these have not prompted V4 countries to engage in systematic cooperation either in the area of global environmental agreements or in the process of approximation with the EU. The lack of mutual information about the process of harmonisation within the V4 group and minimal contact between their experts is striking. With the growing distance from the fall of the socialist system and with the successes of some clean-up programmes, it seems that previously common problems are becoming less important. More diverse definitions of environmental problems are now emerging including, for example, differing attitudes within V4 countries to nuclear energy. Another example is Hungary's preference for framing environmental problems in terms of water policy, for which the optimal unit of management is the Danube basin. Several Western experts also noticed a certain degree of rivalry among CEE countries during the approximation process.

The pre-accession process was staged by the European Commission as a contest that promoted rivalry among the candidates. For instance, annual assessment reports on each candidate country's progress in adopting and implementing the *acquis* and the publication of subsequent tables ranking the countries according to the number of concluded thematic 'chapters' of negotiations fostered competition rather than coordination and cooperation among applicants. The purpose of the European Commission's assistance was to enhance the compliance of individual candidates with EU demands, not to encourage them to take joint positions towards EU requirements or even develop joint proposals.

Future Impact on EU Environmental Policy

In the opinion of the interviewed experts from V4 states, the ideal future EU environmental policy should build on its current trends. In their view, the ultimate goal of an enlarged EU environmental policy should be sustainable development. The key mechanism for achieving this goal is integration of environmental and other public policies [see *Homeyer, this volume*]. The area in which this is most urgently desired is the interface between the environment and transport. EIA and strategic environmental assessment (SEA) are perceived as the most promising means of effective integration. Rather than command-and-control legislation, the type of policy preferred by the V4 countries relies on new policy instruments including market-based instruments such as green taxes and horizontal legislation such as access to information and participation of civil society and economic actors.

As Knill and Lenschow [*2000*] argue, new policy instruments assume a certain level of societal responsiveness and organisational mobilisation (supported by an appropriate resource level). Given the deeply unfavourable context for such policy styles in V4 states, including over-centralised state administrations and under-developed civil society, the seemingly unreserved acceptance of new modes of environmental governance by V4 states' experts appears striking. We argue that this can be explained by two self-reinforcing factors. The first is the connotation of concepts such as flexibility, freedom of information and market-based instruments as a symbolic break from the oft-criticised socialist model of bureaucratic environmental regulation. The second is the hegemonic power of Western institutions over environmental transition in V4 states as virtually the sole source of environmental policy innovation.

An analysis of the document synthesising the contribution of CEE countries to the 6th Environmental Action Programme [*REC, 2000*] corresponded closely with the views of the experts interviewed for this study, notably in the parallel advocacy of key concepts such as subsidiarity, adaptable policymaking, framework legislation and stakeholder involvement in response to the expected growing diversity of environmental problems. Despite this emphasis on a more flexible style of policy the document does not seem to confirm fears of the re-nationalisation of EU environmental policy resulting from eastern enlargement [*Homeyer et al., 2000*]. Rather, it is flexibility at the regional or sub-national level, rather than at the national level, which the document sees as crucial for effective implementation of EU legislation. Second, apparently disregarding the accession states' failure in the past decade to take advantage of environmentally favourable features of CEE countries (mostly inherited from the pre-1989 period), the REC document suggests that applicant states' accession to the EU offers an opportunity to enhance sectoral integration. Third, the strengthening of the effectiveness of current legislation should be given priority over the development of new legislation. Particular attention should be paid to the strengthening of institutions in CEE countries. Fourth, the experts who contributed to REC [*2000*] believe that actions at the local level, such as better planning and local action plans, are key policy concepts on the path to sustainable development.

According to the experts interviewed for this study, possible innovative policy contributions in an enlarged EU could emerge in the areas of nature and landscape conservation and land use planning. This supports suggestions that the historic strength of networks in such policy areas can act as a buffer against asymmetric EU–CEE relations [*Tickle, 2000*]. Some respondents expressed an idea that the experience of harmonisation of CEE countries with stringent EU environmental directives may lead to more cost-effective approaches. However, these ideas did not go beyond general proposals, as the experts were unable to specify strategies and mechanisms by which they could be developed and promoted by V4 countries with the prospect of future EU-wide application.

The same holds true for some environmentally positive features of societies in V4 countries. Theoretically, all these features could become stimuli for innovations in EU environmental policy. Furthermore, these environmentally beneficial features of V4 societies could be best preserved and expanded through sectoral integration. However, despite being strong advocates of integration, none of the interviewed experts had suggestions as to how these positive features of V4 countries could be transformed into policy proposals applicable at the EU level. Instead, the experts tended to defer to the European Commission's initiative in this respect.

Most respondents did not expect active and innovative participation of V4 countries in the development of future environmental policy of the EU. There is clearly an absence of ambition concerning the post-accession period. Exhaustion from the demanding harmonisation process was indicated as the main reason. Another possible explanation was related to the lack of experience and will:

> The barrier is little knowledge of environmental policy. In Hungary nobody expects Hungary to be able to do something important on its own (interview, Hungarian environmental policy consultant, 28 June 2000).

Among other reasons for the expected passivity and reactivity were the overriding priority ascribed to economic growth, a political culture that functions as a barrier for effective environmental integration, and also the ability of heavy industry to pursue vested interests.

The interviews also addressed the theme of V4 states' future alliance politics and revealed a fundamental discrepancy between the expectations of most Western commentators and the views of V4 country experts. First, contrary to Western analysts and apparently some southern European politicians [see *Viñas, 2000*], V4 experts unanimously ruled out alliances with south European EU countries in the Council of Environmental Ministers. Second, despite the declared similarities and shared interests among V4 countries, the experts did not reckon on coordination between them in the post-accession period. Third, all the respondents (V4 and Western) anticipated that a stable pattern of voting behaviour of V4 countries in the Council of Ministers was unlikely to emerge. In their view, individual V4 countries will behave in an *ad hoc* manner depending on specific opportunities and interests rather than on any systematic strategy.

The marginal attraction of south European countries as allies is explained by their negligible involvement in the CEE transformation process, including its economic and environmental dimension as well as by minimal historic contacts between these two regions. Instead, if any discernible alliance pattern occurs, the respondents expect it to be generally oriented to north-western Europe states (such as The Netherlands and the Baltic states) and primarily to neighbouring countries such as Germany and Austria. This would likely stem from cultural and geographical proximity, from the intensity of current economic relations, and also from the environmental assistance of these countries to the V4 group. For example, throughout the 1990s it was Dutch and Danish styles and concepts that were cited most in terms of policy influence.

Conclusion

The December 2002 invitation extended to eight CEE countries to join the EU has initiated a new phase of interaction between the EU and the V4 countries. Post-accession, the strong leverage that the EU exerted on candidate states (hierarchical imposition) should recede and an altered set of relations may emerge in which the new member states will have more resources and political opportunities for pursuing their own interests and priorities. Thus it is now appropriate to try to identify the interests and priorities that may shape V4 states' approach to EU environmental policy in the future. Most perspectives on eastern enlargement of the EU are based on a top-down perspective and an assumption that the current mode of asymmetrical relations will be maintained in the future. We have attempted to extend the scope of existing accounts by adding the applicant states' perspective.

We now examine whether this extension may lead to potential changes in the conclusions drawn by existing studies. For several reasons this is a complicated inquiry. First, the articulation of many domestic interests has been suppressed by the one-way process of approximation. This holds true for the environment, despite some limited evidence of proactive policy initiatives. Second, the future-oriented perspective inevitably renders our conclusions at least partly speculative. Third, a serious obstacle is a lack of an appropriate theoretical framework for such an inquiry. Integration theory has focused on describing and explaining integration processes from a top-down perspective focusing on the role of supranational actors. On the other hand, the role of the policies, interests and actions of its most important actors, the nation states, has been somewhat neglected.

Despite initial evidence of a proactive approach to international environmental policy in the V4 countries, this model became quickly subsumed by the 'hierarchical imposition' of EU requirements, which has since become the dominant framework for the development of their domestic environmental policy. As a consequence, the preferred environmental policy outcomes in V4 countries correspond closely with the current trend in the EU towards flexibility, economic instruments, stakeholder participation and sectoral integration.

Owing to the weak domestic base of environmental policy and the acceptance of EU environmental policy as a hegemonic model, it is highly unlikely that V4 states are, in the short term, capable of adopting a proactive approach to environmental policymaking at the EU level when they become full members. Based on our interview data, V4 policy experts neither expected nor required any major changes – based on indigenous experiences – to this model. The conditions of asymmetrical relations in which the transfer of the EU policy model took place have thus reduced the scope of policy considerations to the national and sub-national level. The strengthening of their environmental capacity – facilitated by various EU assistance programmes – has also centred on policy implementation at the domestic level, rather than enhancing V4 states' ability to influence the EU.

We also found that V4 states have not, and do not seem likely to coordinate their strategies – either among themselves or with environmentally 'laggard' member states. Instead, it appears that they would rather align themselves with the north-western 'pioneer' member states that have been most active in transferring environmental know-how and have made environmental policy discourse in V4 countries largely compatible with their policy models. Thus, we find that extant expectations about V4 states joining the current group of 'laggards' and putting a brake on development of EU progressive environmental policy may be premature and should be qualified. However, unknown variables and as yet unarticulated interests render such conclusions tentative.

NOTES

We wish to acknowledge the contribution of Ivan Rynda in helping design the questionnaires and Ivan Rynda and Robin Webster for conducting some of the interviews. This contribution is based on two separate projects, one funded by the Czech Ministry of the Environment (project No.EU/043/99) and the other by the Czech Ministry of Foreign Affairs (project No.RB 5/14/00). Petr Jehlička worked on an earlier draft of this text while holding a Jean Monnet Fellowship at the European University Institute, Florence.

1. The V4 comprise the Czech Republic, Hungary, Poland and Slovakia. The term Visegrád refers to the location where the first summit meeting of the loose political alliance met in 1991. In December 2002, all four countries were invited to join the European Union in 2004.
2. Fifteen interviews were conducted in the Czech Republic, five in Hungary, four in Poland and five in Slovakia. Among interviewed experts were members of parliaments, former ministers of the environment, academics, NGO activists, civil servants and consultants.
3. In 2000 85 per cent of the annual income of the most active Czech 'new' environmental group *Hnutú DUHA* came from foreign grant agencies or Czech foundations that distribute foreign funding. Membership fees made up only three per cent of the income [*Hnutú DUHA, 2000*].
4. According to Wajda [*2000*], the Polish Ministry of the Environment is seriously understaffed with only some 300 staff members. In particular, there is a consistent lack of EU specialists.

REFERENCES

Ágh, Attila (1999), 'Europeanization of Policy-Making in East Central Europe: The Hungarian Approach to EU accession', *Journal of European Public Policy*, Vol.6, No.5, pp.839–54.

Aguilar Fernández, Susana (1997), 'Abandoning a Laggard Role? New strategies in Spanish Environmental Policy', in D. Liefferink and M.S. Andersen (eds.), *The Innovation of EU Environmental Policy*, Oslo: Scandinavian University Press, pp.156–72.

Andersen, Mikael S. and Duncan Liefferink (1997), 'Introduction: The Impact of the Pioneers on EU Environmental Policy', in M.S. Andersen and D. Liefferink (eds.), *European Environmental Policy: The Pioneers*, Manchester: Manchester University Press, pp.1–39.

Baker, Susan (2000), 'Between the Devil and the Deep Blue Sea: International Obligations, Eastern Enlargement and the Promotion of Sustainable Development in the European Union', *Journal of Environmental Policy and Planning*, Vol.2, No.2, pp.149–66.

Baker, Susan and Ian Welsh (2000), 'Differentiating Western Influences on Transition Societies in Eastern Europe: A Preliminary Exploration', *Journal of European Area Studies*, Vol.8, No.1, pp.79–103.

Bernstein, Steven (2000), 'Ideas, Social Structure and the Compromise of Liberal Environmentalism', *European Journal of International Relations*, Vol.6, No.4, pp.464–512.

Caddy, Joanne (1997), 'Harmonisation and Asymmetry: Environmental Policy Co-ordination between the European Union and Central Europe', *Journal of European Public Policy*, Vol.4, No.3, pp.318–36.

Carius, Alexander, Ingmar von Homeyer and Stefani Bär, (2000), 'The Eastern Enlargement of the European Union and Environmental Policy: Challenges, Expectations, Multiple Speeds and Flexibility', in K. Holzinger and P. Knoepfel (eds.), *Environmental Policy in a European Union of Variable Geometry: The Challenge of the Next Enlargement*, Basel: Helbing & Lichtenhahn, pp.141–80.

Connolly, Barbara, Tamar Gutner and Hildegard Bedarff (1996), 'Organisational Inertia and Environmental Assistance to Eastern Europe', in R.O. Keohane and M.A. Levy (eds.), *Institutions for Environmental Aid: Pitfalls and Promise*, Cambridge, MA: The MIT Press, pp.281–323.

GA&C and UK (2001), *Posilování připravenosti ČR k implementaci norem EU v oblasti ochrany životního prostředí* (Strengthening the Czech Republic's capacity for implementation of EU environmental norms), Praha: Gabal, Analysis & Consulting; Univerzita Karlova v Praze, Centrum pro otázky životního prostředi.

Gille, Zsuzsa (2000), 'Legacy of Waste or Wasted Legacy? The End of Industrial Ecology in Post-Socialist Hungary', *Environmental Politics*, Vol.9, No.1, pp.203–30.

Hnutú DUHA (2000), *Výročni zpráva 2000* (Annual Report 2000), Brno: Hnutú DUHA.

Holzinger, Katharina and Peter Knoepfel (2000), 'The Need for Flexibility: European Environmental Policy on the Brink of Eastern Enlargement', in K. Holzinger and P. Knoepfel (eds.), *Environmental Policy in a European Union of Variable Geometry: The Challenge of the Next Enlargement*, Basel: Helbing & Lichtenhahn, pp.3–35.

Homeyer, Ingmar von (2001), 'Enlarging EU Environmental Policy', paper presented at Environmental Challenges of EU Eastern Enlargement workshop, Robert Schuman Centre, European University Institute, Florence, 25–26 May.

Homeyer, Ingmar von, Alexander Carius and Stefani Bär (2000) 'Flexibility or Renationalization: Effects of Enlargement on Environmental Policy', in M.G. Cowles and M. Smith (eds.), *State of the European Union: Risks, Reforms, Resistance and Revival*, Oxford: Oxford University Press, pp.347–68.

Jehlička, Petr and Tomáš Kostelecký (1995), 'Czechoslovakia: Greens in a Post-Communist Society', in D. Richardson and C. Rootes (eds.), *The Green Challenge: The Development of Green Parties in Europe*, London: Routledge, pp.208–31.

Kerekes, Sándor and Károly Kiss (1998), 'Hungary's Accession to the EU: Environmental Requirements and Strategies', *European Environment*, Vol.8, No.5, pp.161–70.

Knill, Christopher and Andrea Lenschow (2000), '"New" Environmental Policy Instruments as a Panacea? Their Limitations in Theory and Practice', in K. Holzinger and P. Knoepfel (eds.), *Environmental Policy in a European Union of Variable Geometry: The Challenge of the Next Enlargement*, Basel: Helbing & Lichtenhahn, pp.317–43.

Paraskevopoulos, Christos, J. (2001), *Interpreting Convergence in the European Union: Patterns of Collective Action, Social Learning and Europeanization*, Basingstoke: Palgrave.

Pellegrom, Sandra (1997) 'The Constraints of Daily Work in Brussels: How Relevant is the Input from National Capitals?', in D. Liefferink and M.S. Andersen (eds.), *The Innovation of EU Environmental Policy*, Oslo: Scandinavian University Press, pp.36–58.

Petersen, Nikolaj (1998a) 'National Strategies in the Integration Dilemma: An Adaptation Approach', *Journal of Common Market Studies*, Vol.36, No.1, pp.33–54.

Petersen, Nikolaj (1998b), 'Review Article: Small States in European Integration', *Scandinavian Political Studies*, Vol.21, No.1, pp.87–93.

REC (2000), *Applicant Countries' Contribution to the 6th Environmental Action Programme*, Budapest: Regional Environmental Center. Available at http://www.rec.org/REC/Programs/6thEAP/.

Slocock, Brian (1996), 'The Paradoxes of Environmental Policy in Eastern Europe: The Dynamics of Policy-Making in the Czech Republic', *Environmental Politics*, Vol.5, No.3, pp.501–21.

Slocock, Brian (1999), '"Whatever Happened to the Environment?": Environmental Issues in the Eastern Enlargement of the European Union', in K. Henderson (ed.), *Back to Europe: Central and Eastern Europe and the European Union*, London: UCL Press, pp.151–67.

Stehlík, Jiří (1998), 'The Environment in the Czech Republic and the Association Agreement', in B. Lippert and P. Becker (eds.), *Towards EU-Membership: Transformation and Integration in Poland and the Czech Republic*, Bonn: Europa Union Verlag, pp.271–92.

Thorhallsson, Baldur (2000), *The Role of Small States in the European Union*, Aldershot: Ashgate.

Tickle, Andrew (2000), 'Regulating Environmental Space in Socialist and Post-Socialist Systems: Nature and Landscape Conservation in the Czech Republic', *Journal of European Area Studies*, Vol.8, No.1, pp.57–78.

Vavroušek, Josef (1993), 'Institutions for Environmental Security', in G. Prins (ed.), *Threats Without Enemies*, London: Earthscan, pp.101–30.

Viñas, Angel (2000), 'The Enlargement of the European Union: Opportunities and Concerns for Spain', *Mediterranean Politics*, Vol.5, No.2, pp.60–75.

Wajda, Stanislaw (2000), 'Harmonisation – The Commitment to Change', paper presented at conference held at the Regional Environmental Center, Budapest, 12 June.

Zylicz, Tomasz and Katharina Holzinger, (2000), 'Environmental Policy in Poland and the Consequences of Approximation to the European Union', in K. Holzinger and P. Knoepfel (eds.), *Environmental Policy in a European Union of Variable Geometry: The Challenge of the Next Enlargement*, Basel: Helbing & Lichtenhahn, pp.215–48.

Part II

ENVIRONMENTAL POLICY CHALLENGES

EU Accession and Legal Change: Accomplishments and Challenges in the Czech Case

EVA KRUŽÍKOVÁ

In the environmental sector alone, the legal, regulatory and organisational changes engendered in Central and Eastern European candidate countries by the EU accession process are enormous. This contribution examines the accomplishments of, and the challenges to, the reform of environmental laws in the Czech Republic as driven by the EU. Ongoing candidate state harmonisation and implementation efforts have largely been framed in terms of transposition of EU law into domestic law within various candidate countries. In addition, as discussed in other contributions to this volume, some necessary types of public sector capacity have received a fair amount of attention from both EU and domestic officials. This contribution argues that many remaining barriers to the effective administration, implementation and enforcement of EU environmental policy are posed by the challenges of merging the existing legal cultures, expectations and practices of EU law with those of candidate countries.

The Czech Republic, like other Central European states, is scheduled to join the EU in 2004. Czech officials have engaged in an enormous and somewhat rushed effort to conform to all EU requirements for membership, including those concerning the environment. EU environmental protection law belongs to the group of sectors widely considered among the most difficult for candidate countries, including the Czech Republic, to comply with. The following section briefly presents three waves of environmental legal changes in the Czech Republic since 1990, changes which culminate in a massive 1999–2002 effort towards legal harmonisation and implementation of the Community environmental law. The study then reviews some accomplishments of this effort and catalogues a host of challenges to legal implementation beyond the achievement of EU membership. While the character of the Community law presents accession states with one set of challenges, the domestic legal cultures, practices and participant expectations present a second set. The study closes with a special focus on the role of the European Court of Justice (ECJ) in Community environmental law – the making, interpretation and enforcement of it. The role of the ECJ provides excellent illustrations of the many changes in the legal cultures and practices facing accession countries upon EU membership.

A Decade of Change in Czech Environmental Law

The Czech Republic launched systematic approximation efforts in 1999, nine years after crucial changes of the Czech legal system were initiated and eight years after the first post-revolution environmental act was passed. Since the early 1990s, Czech environmental legislation has undergone three waves of changes. During the first wave (1991–92), the main body of environmental legislation was approved and brought into effect. The main driving force of legislative 'storm' was the need to transform the communist system of law to a new one based on democratic grounds. By the end of the first wave, the legislation covered almost all aspects of environmental protection, with several exceptions: access to environmental information, chemicals, major industrial accidents, and genetically modified organisms (GMOs). It involved such major environmental issues as the protection of air, water, soil, nature and landscape, forest, waste, environmental impact assessment. Regarding water and forest management, new legislation was not enacted at that time, keeping the acts from the 1970s in force.[1]

As Table 1 illustrates, the wave of environmental legislation in the early 1990s included framework legislation on environmental protection and laws specific to air pollution, nature and landscape protection, agricultural soil protection, wastes and environmental impact assessment. In this period important acts concerning the institutional framework for environmental protection were also approved, including those that established the Czech Environmental Inspectorate and the National Environmental Fund. In the second wave of environmental law development (1995–98), acts were passed regarding access to environmental information, forests, wastes, ozone layer protection, international trade in endangered species, nuclear energy and technical product standards. The main impetus for this wave was partly the Czech Republic's international obligations that needed to be incorporated into the national law, partly the effort to replace the rest of the old communist-type environmental laws (in the case of forest legislation) and partly to improve the state of laws enacted during the first wave (in the case of the Waste Act).[2]

In 1999, following the establishment of a new government in 1998, a third wave of environmental law development was launched. At that time, the Czech environmental legislation needed substantial changes to comply with Community law. None of the EC directives were fully transposed to Czech law and a number of directives were not even partially transposed. Of particular concern were laws regarding water protection, waste

TABLE 1

THREE WAVES OF CZECH ENVIRONMENTAL LEGISLATION SINCE 1990

First Wave (1991–1992)
Act on the Air Protection (1991)
Act on Waste (1991)
Act on the Czech Environmental Fund (1991)
Act on the National Environmental Fund (1991)
Act on Environment (1992)
Nature and Landscape Protection Act (1992)
Agricultural Soil Protection Act (1992)
Act on Environmental Impact Assessment (1992)

Second Wave (1995–1998)
Act on the Right to Access to Environmental Information (1998)
Act on Forests (1995)
Act on Waste (1997)
Act on the Ozone Layer Protection (1995)
Act on Conditions of International Trade with Endangered Species of Wild Fauna and Flora and Other Measures of Protection of Such Species (1997)
Act on Peaceful Use of Nuclear Energy (1997)
Act on Technical Requirements for Products (1997)

Third Wave (1999–2002)
Act on Chemicals and Chemical Preparations (1998)
Act on the Prevention of Major Accidents caused by Certain Dangerous Chemical Substances and Preparations (1999)
Act on Handling with Genetically Modified Organisms and Products (2000)
Act on Indemnification of Damage Caused by Certain Protected Animals (2000)
Act on Hunting (2001)
Act on Environmental Impact assessment (2001)
Water Act (2001)
Act on Waste (2001)
Act on Air Protection (2002)
Act on Integrated Prevention and Pollution Control (2002)
Act amending Penal Code in the Field of the Environment (2002)
Act amending the Act on Peaceful Use of Nuclear Energy (2002)
Act on Conditions for Introduction on the Market of Biocide Preparations and Substances (2002)

management, chemicals, and integrated pollution prevention and control [*Miko, 2000; Ministry of the Environment, 1999*]. As such, the core environmental legislation had to be essentially rewritten or newly drafted. In addition, EU environmental law continued to develop, with new directives being approved throughout the accession negotiation process. For example, the Integrated Prevention and Pollution Control (IPPC) Directive, Water Framework Directive, and a new Environmental Impact Assessment (EIA) Directive were all issued at about this time.

The Czech environmental legislation enacted during the last three years of the accession negotiations (1999–2002) is entirely focused on compliance with EU legal requirements. During this period numerous acts were approved, either covering new issue areas or substantially amending previously enacted legislation, including laws regarding chemicals and chemical preparations, prevention of major accidents caused by certain dangerous substances, genetically modified organisms and products, indemnification of damage caused by certain protected animals, hunting, environmental impact assessment, water, wastes, air protection, integrated prevention and pollution control, nuclear energy, and biocide preparations and substances. Almost all the acts on this list have been followed by a large number of decrees or governmental regulations implementing the acts, particularly in the field of air and water protection, waste management, noise, civil protection and chemicals.

The Czech Republic was the first candidate country to close negotiations on the Environment chapter, on 1 June 2001. Only two transition periods were agreed by the European Commission for the Czech Republic: the first for packaging waste (Directive 94/62/EC) and the second for municipal wastewater (Directive 91/271/EEC). By comparison, the number of transition periods for other Central European states includes four for Hungary, nine for Poland, seven for Slovakia and two for Slovenia.

In many respects, the Czech Republic has been quite successful in the transposition of the major EU environmental directives. According to the 2002 European Commission Report [*Commission of the European Communities, 2002*], progress continues regarding the transposition and implementation of the environmental *acquis*. By 2002, the Czech Republic had largely eliminated transposition delays in the industrial pollution and nuclear safety sectors – a subject of criticism in the 2001 Report that states that in these fields 'no particular development has been noted in terms of transposition of legislation over the past year' [*Commission of the European Communities, 2001: 83*]. While the Czech Republic has achieved considerable alignment with the EU environmental *acquis*, a number of areas of legal action remain to be completed. These areas include so-called horizontal legislation (such as environmental impact assessment); waste management (regarding titanium dioxide and implementing legislation); industrial pollution and risk management (implementing legislation); water quality (alignment of the Public Health Act and the Water Act, implementing legislation); and nature protection (transposition of the Habitats And Birds Directives).

As accession negotiations drew to a close in the autumn of 2002, several acts were either being drafted, under discussion within the Cabinet, or pending in the Czech Parliament. These included necessary amendments to laws on environmental impact assessment, the right to environmental information, water quality, endangered species, chemicals management and genetically modified organisms. Similarly, a substantial amount of legislation regarding implementation remains in various stages of development, including that aimed at air and water protection, IPPC, nature conservation and chemicals.

Analysing Results: Accomplishments and Challenges

The post-1999 process of harmonisation and implementation of EU environmental legislation offered chances to consider many overall concepts of environmental legislation. During this time, the opportunity was present to establish a comprehensive, transparent and consistent system of legal norms coordinated within the environmental legal system itself as well as with other parts of the Czech legal order. Three years into these implementation efforts, it is now appropriate to evaluate what has been achieved and what challenges remain.

Regarding accomplishments, the Czech Republic is relatively well prepared for the accession as far as the environmental chapter of the *aquis* is concerned. By late 2002, many Czech officials and environmental advocates considered harmonisation and legal implementation efforts a tremendous success. Without a doubt, the implementation of the Community environmental legislation has produced increased stringency and specificity of Czech environmental legislation. This is particularly clear in the fields of chemical management, waste management (including packaging waste, batteries and end-of-life vehicles), GMOs, and the prevention of major industrial accidents. In fact, a number of acts, especially those laying out new approaches and policy instruments, would not have been enacted without the need to comply with the EU requirements. Examples of such legislation include acts on IPPC and the restoration of planning instruments to environmental policy and legislation in areas of waste and water management and air protection. Therefore, in terms of implementation, the post-1999 legal changes have been positive for the Czech legal system, as well as for the environment and for environmental policy. Clearly, Czech law has been drawn closer to the legal systems of the democratic societies of Western Europe.

Implementation Challenges Stemming from the Community Law

Implementation of Community law in the Czech Republic constitutes a large set of difficult and complex tasks [*Miko, 2000; Auer and Legro, forthcoming*]. Their answering and working out represents a major challenge for a country endeavouring to become a standard European democratic country based on rule of law. One set of reasons explaining why this remains a demanding and challenging exercise stems from the Community law itself. First, the Community law, as founded by the

European Court of Justice (ECJ) is a new, distinctive legal order.[3] Its legal basis is the Treaty of Rome, which represents a primary source of law.[4] Community legislation is adopted by EU bodies and it is binding on member states. The Community has its own system of enforcement and its own judicial body – the ECJ. In fact, ECJ decisions can impose fines on those member states that do not comply with obligations established by the Community law. The Community bodies carry out their power in the fields listed in Article 3 of the Treaty of Rome (such as market policy, policies in the field agriculture, fisheries, transportation, environment, social affairs and research), aiming to achieve the goals laid down in Article 2.

The EU system of law is based on the legal culture of West European democratic countries that has been developing (at least) since the end of World War II. During this time, Central and Eastern European (CEE) candidate countries experienced a 40-year breach of legal continuity. The socialist legal order established a very different system of principles and mechanisms, which were frequently unable to reflect the needs of a modern society or to protect the environment. Candidate countries have been attempting to catch a train that left the station long ago and has now travelled a long way down the tracks of the post-World War II era.

A second challenge for the merger of EU and candidate country legal systems lies in the relationship between the Community and national legal systems. This relationship is governed by the principle of supremacy of the former. As argued below, candidate countries remain unfamiliar and unprepared for this substantial change in legal systems and cultures. Thus, national law is subordinated to the Community rules. By delegating certain powers to the Community, the member states give up traditional forms of state autonomy to decide in what ways they will comply with treaty obligations within their national legal order. Community directives take priority over provisions of national law, even when the latter are contained in subsequent statutes or even in a national constitution.[5] The aim is to ensure a uniform effect of the relevant rules of Community law in every member state.[6] Candidate countries are not accustomed to this principle of the supremacy of Community law.

An additional example of the supremacy of Community law over national law can be seen in the important role played by the European Court of Justice (ECJ). Its judgments are a key source of interpretation of Community law and are crucial to its development over time. Also, so-called preliminary rulings of the ECJ are of substantial importance for ensuring the unified interpretation of Community law. As discussed below, Central European candidate countries are not used to such roles for the court.

Third, Community environmental law is constituted by a somewhat scattered set of legal norms that continue to change even while candidate countries attempt to transpose and implement them. Community environmental law has not developed systematically. Rather, it consists of approximately 200 particular directives and regulations mostly covering single-issue areas in a very detailed fashion. Only recently has the EU begun to enact more comprehensive framework directives to cover broad areas of environmental policy (such as water management or air pollution). The fact that Community law remains a moving target further complicates the implementation tasks faced by the candidate countries [*Miko, 2000*]. Community law has been experiencing comparatively rapid development throughout the period of accession negotiations and candidate harmonisation efforts. Put simply, every year new directives and regulations are approved. This fact alone makes candidate countries' dual tasks of implementing the 'old' legislation and transposing the newest provisions extremely difficult. Further complicating this task is the fact that EU legal developments continue apace across many of the non-environmental legislative areas as well. Until EU candidates sign final EU accession agreements they cannot participate in discussions accompanying drafting of particular pieces of EU environmental legislation. Thus, while being strongly urged to implement the legislation during the approximation period, CEE officials do not have a word on its drafting.

Implementation of Community legislation is also a challenge for current EU member states [see *Schreurs, this volume; Knill and Lenschow, 2000*]. Many of these challenges result from their different legal cultures and systems, varying division of powers among different levels of public authorities, varying economic and social conditions, and so on. Furthermore, and partially as a result of these differences, the wording of Community law provisions is not always clear and unambiguous [*Auer and Legro, forthcoming*]. The difficulty of implementing Community legislation is evidenced by problems encountered by the current member states, including those with highly developed environmental policy. These states face difficulties realising accurate and full implementation of particular Community directives even though their legal systems have been absorbing Community provisions for decades and they participate in the drafting and approval processes of new Community legislation. As an example, Germany had problems with the implementation of Directive 90/313/EEC on freedom of access to environmental information. Similarly practically all the member states had problems implementing Directive 92/43/EEC on the conservation of natural habitats and of wild fauna and flora and Directives 85/337/EEC and 97/11/EC on the assessment of the effects of certain public and private projects on environment.

Fourth, certain directives, particularly recent ones, set out new, innovative (or unusual) policy and legal instruments and approaches [*Knill and*

Lenschow, 2000; Holzinger and Knoepfel, 2000]. For example, the IPPC Directive aims at the protection of the environment as a whole, by integrating existing multi-media permitting processes into a single process covering all aspects of environmental protection against pollution. The Directive does not prescribe obligatory standards for relevant facilities. It prescribes the use of the so-called best available techniques (BAT) in facilities. The BAT are also considered the basis for setting emission limits in the integrated permits for each facility. Legislation such as the IPPC Directive requires substantial intervention in the existing domestic legal order and in functioning of domestic institutions. Their implementation affects a large number of legal norms, and not only those that are environmental.

The IPPC requires coordination and integration of different administrative/permitting procedures. In the Czech Republic they were regulated by sectoral environmental legislation and also by the Construction Act and the Act on Administrative Procedure. All the procedures, according to the legislation in force before the implementation of the IPPC Directive, were carried out by different competent authorities. These authorities were obliged to take into account the opinion of public authorities dealing with special concerns such as air, water, soil or nature protection.

IPPC calls for one procedure linking together all the relevant aspects of the environmental protection, carried out by a single competent authority as a priority. When designing a new IPPC Act the Czech drafters had to keep in mind that the integrated permit requires a high professional level of those who issue the permit because of a broad range of issues covered. It was also necessary to change the concept of the permit which, in the case of the IPPC, does not stipulate a particular technology to be used by a facility. It lays down emission limits or other measures to reach a high level of environmental protection by the way of BAT. Also, competent authorities need to carefully follow technical developments in each field.

The need to ensure the professional competence of decision makers gave rise to frequent discussions about which authority should be charged with the authority to grant the integrated permit. A suggestion to create an Environmental Protection Agency that would be entrusted with this responsibility was rejected. Instead, the approved Act on IPPC delegated this power to regions. The crucial reason for this decision was an application of the subsidiarity principle and the reluctance to establish another administrative authority at the national level.

It was a particular challenge to avoid the introduction of an additional administrative procedure and thus to further multiply the number of procedures required by Czech environmental legislation. In this context it was inevitable that there were issues that had to be addressed regarding the relationship between the new integrated permits and procedures according to the Act on land use planning and construction rules. Only implementation and enforcement of the new Act will demonstrate the extent to which the approved solution is efficient and effective.

Implementation Challenges from within the Czech Republic

A second set of challenges to implementing Community environmental law are related to attitudes, traditions and practices within the Czech Republic. While this study addresses the Czech case, there is little reason to doubt that most of the domestic challenges outlined here are applicable to other Central and Eastern European candidate countries. In these countries, implementation requires more than a transposition of legal provisions. It requires their application and enforcement, both of which require building and maintaining appropriate institutional capacities [*Crisen and Carmin, 2002; Holzinger and Knoepfel, 2000*]. The institutional capacity in the Czech Republic is not yet developed to the extent necessary to ensure full and correct implementation of the Community legislation [see also *Auer and Legro, forthcoming*].

Institutional arrangements for environmental protection and enforcement are in place and competencies for the main requirements of the *acquis* have been identified. The Czech institutional framework for environmental protection and enforcement was established at the beginning of the 1990s. However, the existence of a large number of institutions dealing with environmental issues does not contribute to efficient allocation of resources or clear administrative responsibilities. Competencies set up by Community environmental legislation do not necessarily fall under the jurisdiction of the Czech Ministry of Environment. Instead, many fall under other ministries, especially the Ministries of Agriculture, Industry and Trade, Public Health, and Transportation and Communication, as well as the National Agency for Nuclear Safety. The Czech environmental administration employs a very large number of staff with strong technical knowledge. In contrast, the staffing at regional and local levels seems quite low [*Commission of the European Communities, 2002*].

During the EU accession negotiations, the Czech Republic also has been engaged in broad-based public administration restructuring and reform. Regions, or geographic areas comprised of a number of districts were abolished in 1991 and then re-established in 2001. In 2003, the districts will disappear, resulting in national, regional and municipal levels of administration. The administrative reform process is intended to transfer numerous powers from the national to the local levels (and vice versa). By removing layers of administration, government is supposed to become more efficient and to establish closer ties between the local and national levels.

The transfers are not entirely well conceived and coordinated. Nonetheless, it will influence the efficiency of environmental enforcement, potentially weakening the goals of Community environmental policy. The last Report from the European Commission states:

> While the setting up of the new three tier administrative structures has started well, the lack of clear allocation of competencies and overlapping seems to continue since the reform has not been based on a holistic approach, considering only the division of competencies among the different governmental levels. There is a need to establish decision-making procedures, co-operation and co-ordination among different governmental bodies at all levels in particular in the water sector. At the regional and local level, enforcement and application of environmental legislation need to be further improved by additional staff, financial support, equipment and well-defined division of competencies as well as the guidance provided by central administration [*Commission of the European Communities, 2002*].

Pre-existing attitudes and expectations of policymakers, legal practitioners and other administrators also pose challenges to the implementation of EU environmental law in accession countries [*Miko, 2000*]. Environmental experts, including legislators, have been accustomed to a certain established stereotype of legal approaches and enforcement practices. They are not always willing to change. Water management offers one example. The Water Framework Directive (WFD) has introduced an ecosystem approach, based on watersheds and the territory influencing the quality and quantity of water. The previous Czech water legislation worked only partly with watersheds. It distinguished special areas of water accumulation or buffer zones around waters used as sources of drinking water. However, it did not provide comprehensive protection of all watersheds. It did not require a systematic and strategic approach towards water resources management including, for example detailed management plans for whole watersheds.

The WFD also established a very different and sophisticated set of water quality goals (ecological and chemical water quality status) and related standards. This concept is also new for Czech environmental legislation. This system of goals and standards is so complex that even EU bodies and existing member states do not know how it will work in practice. Therefore, it is hard to say at present whether it is more efficient than the 'old' approach. Time will tell whether the water quality in the EU, including in the Czech Republic, is improved.

In addition, regulations – a secondary source of Community law – are directly applicable at the national level. In other words, they automatically become a part of member states' legal systems. They may not be transposed, modified or amended and they do not need to become part of national pieces of legislation. Yet, Czech legislation must comply with Community law including regulations. Czech law now has provisions identical with those of the regulations, such as those concerning ozone layer protection. Thus after becoming a member of the EU, Czech law will have two sets of provisions dealing with the same issues. Solutions of such accession state problems are not clear, and the approaches to dealing with them are often inconsistent. Such challenges extend beyond the environmental sector.

Likewise, while transposing provisions of directives, legislators have not taken into account judgments of the ECJ – because they are not used to doing so – and EU member states' experiences of Community environmental law implementation. In addition, high-quality translations of Community directives into the Czech language are not always available in a timely fashion. Thus, the future likely includes problems regarding legal terminology and concepts.

Lastly, a number of challenges are engendered by the rapid rush towards implementation. As in other candidate countries, implementation started relatively late in the Czech Republic and the majority of transposition has taken place in less than four years. Thus, candidate countries have transposed decades of EU law in a very short period of time. The transposition process was carried out under pressure resulting from a presumed EU accession date of 1 January 2004. Consequently, there has not been enough time or institutional capacity to establish a sufficiently conceptual and systematic approach towards the implementation of environmental law. In many respects, Czech officials have missed opportunities to improve the whole system of environmental law. Like EU law, Czech environmental legislation is now a complicated set of acts and other regulations that are relatively scattered, compared to other legal sectors. In the Czech Republic there are currently about 40 environmental acts, more than 30 Cabinet regulations and about 90 ministerial decrees – and these numbers can change monthly.

The rush to implement the new laws has a number of additional consequences. For example, the complete reform of Czech law has gone largely unplanned and uncoordinated, with legislators proceeding in a fairly unsystematic fashion. One possibility might have been to proceed from the general legal instruments to the special or particular ones. For example, it would be much more reasonable to approve the IPPC Act first and then sectoral or multi-media permit rules in the field of water or air protection and waste management. This could avoid the need to amend recently approved provisions after enacting the IPPC Act.

The rush towards implementation has left overlapping, and potentially contradictory, legislation and administrative procedures to be carried out under the law. This is likely to result in unclear interpretations of law. Also, many administrative procedures, such as permit procedures, have become more complicated than before and more complicated compared to practices and approaches in EU member states. This will leave a host of challenges for public administration, private sector actors and the general public. Furthermore, insufficient knowledge of Community requirements and particularly of interpretation of legal provisions has led to repeated amendments or rewriting of acts immediately following their approval. Waste management and chemical legislation are good examples of this. Still it is likely that many issues remain omitted, or are not properly understood or interpreted, during the legislative process. These omissions and contradictions are likely to produce numerous administrative and implementation problems in the years to come, some of which are likely to end up before the European Court of Justice.

The ECJ as a Potential 'Surprise'

The ECJ has the authority to influence environmental protection in several types of jurisdictions entrusted to it by the 1956 Treaty of Rome. First, Article 226 of the Treaty empowers the ECJ to hear cases brought against member states by the European Commission or by other member states. This is the basis of the so-called infringement procedure, the purpose of which is to ensure that member states comply with obligations set out by Community law.[7] Thus, the ECJ forces member states to take actions they have been unwilling and/or unable to take. In areas within ECJ jurisdiction, the court has succeeded in enforcing environmental legislation many times [*Koppen, 1993; McCormick, 2001*]. Its judgments have led to substantial changes in current practices, the end of activities carried out in breach of Community provisions, and the removal of facilities that were sited without respect of Community law.[8] This might be surprising and difficult to accept in countries where enforcing environmental concerns is often framed in opposition to development concerns. Furthermore, in such cases the judgment of the court 'overrules' a decision by public authorities considered more important in the Czech Republic. Courts are not always perceived as independent, unbiased bodies. Their decisions are still often challenged by even the highest politicians of the country who do not accept their independent character and who try to find ways and mechanisms to influence courts. In part, such practices are legacies of the old system, which lacked both the rule of law and an independent judiciary, and in which all the bodies, including courts, were obliged to follow Communist Party guidance.

The power of the ECJ also has been strengthened by the introduction of penalties in the Treaty of Rome. When a member state does not comply with the ECJ's judgments, the Court – after another action of the Commission –

may impose penalty payments. These sums are calculated according to Commission guidelines,[9] based on a flat-rate, a coefficient of seriousness, a coefficient of duration, and a member state's economic conditions (ability to pay). Penalties are to be paid for each day that Community law is violated.

A second 'surprise' for which candidate country legal systems may be unprepared lies in Article 234 of the Treaty of Rome. Accordingly, the ECJ interprets Community environmental law with preliminary rulings.[10] Preliminary rulings are initiated by national courts asking for ECJ interpretation, in particular cases, of Community provisions *vis-à-vis* national rules. Preliminary rulings contribute to the uniformity of interpretation and application of Community environmental law. They are crucial for development of the Community legal order and its concepts and principles. Yet, in candidate countries such as the Czech Republic, courts lack expertise on the ECJ and its powers. They are not used to asking higher courts for an opinion concerning the interpretation of legal norms.

A third potential ECJ surprise for candidate country legal practitioners lies in the ECJ's role in contributing to the progressive, participatory democratic nature of environmental law and decision making. Some environmental law, on both international and national levels, opens policymaking processes to the public, even giving the public *locus standi* (right for standing) in environmental matters.[11] Article 230 of the Treaty of Rome reads: 'Any natural or legal person may … institute proceedings against a decision addressed to that person or against a decision which, although in the form of a regulation or a decision addressed to another person, is of direct and individual concern to the former.' The ECJ law is not unambiguous in this respect. It shows certain limits to openness of its procedures to the public [*Jans, 2000; Winter, 1999*].[12] However, even its views are developing under the influence of the EU's international obligations (particularly under the Aarhus Convention on Access to Information, Public Participation in Decision-making and Access to Justice in Environmental Matters), public awareness and increasing understanding of environmental problems [*Ward, 2000*].

In sum, all three of the ECJ roles noted above offer significant opportunities to influence the enforcement of Community environmental legislation and demonstrate the importance of the Court judgments. Such ECJ influence may surprise, confound and/or frustrate many legislators, judges, attorneys and plaintiffs in accession countries following accession to full membership. Although the judgments are not precedents (as a source of law in the common law system), they are respected by EU institutions and by member state authorities. The Czech Republic and the other candidate countries will have to accept this significant change in domestic legal systems upon EU membership.

Conclusion

To date, the domestic implementation process has been perceived mainly as a process for transposing EU environmental legislation and, to some extent, to establish the necessary institutional frameworks [*Holzinger and Knoepfel, 2000*]. A strong influence of the Court jurisdiction on legislative and administrative decisions is an issue to which the relevant institutions and their staffs must get accustomed. In a country where legal culture and legal awareness was heavily influenced by decades of the communist regime, enormous tasks lie ahead. Intensive training of civil servants at all levels of public administration, as well as of judges and other lawyers, will be necessary. To date, little of this training has been done.

In 2004 the EU will accept ten more countries. The process preceding this important step is challenging both for the EU itself and for individual candidate countries in particular. The field of environmental protection requires substantial changes not only in legislation, but in the usual ways of legal enforcement, legal thinking and legal practices. Therefore, the changes taking place concern public authorities as well as the courts. The Czech example provides insight into how difficult these processes are and it identifies obstacles and challenges faced by accession countries. As the Czech case suggests, it is essential to acknowledge that challenges stem from legal thinking and administrative practices of individual applicant states as well as from specific features of EU legislation. It also illustrates that many implications of EU membership for domestic institutions and practices in CEE states and societies will be experienced well beyond the formal date of EU enlargement.

NOTES

1. The classification of the environmental legislation used in the European Communities and in the EU member states at the beginning of the 1990s was used to make these assumptions.
2. For comparison of the Czech situation with several other candidate countries see Homeyer, Kempmann and Klasing [*1999*].
3. Case C-26/62 *Van Gend en Loos* [1963] ECR 1 at 2; case C-6/64 *Costa v. ENEL* [1964] ECR 585 at 593.
4. Treaty Establishing the European Community.
5. Case C-11/70 *Internationale handelsgesellschaft mbH v. Einfuhr- und Vorratsstelle für Getreide und Futtermittel* [1970] ECR 1125 at 1134.
6. Case C-6/64 *Costa v. ENEL* [1964] ECR 585 at 593.
7. Article 226 of the Treaty reads as follows: 'If the Commission considers that a Member State has failed to fulfil an obligation under this Treaty, it shall deliver a reasoned opinion on the matter after giving the Member State concerned the opportunity to submit its observations. If the State concerned does not comply with the opinion within the period laid down by the Commission, the latter may bring the matter before the Court of Justice.'
8. Case C-355/90 *Commission v. Spain* [1993] ECR I-4221.
9. Commission Communication on the method of calculating the penalty payments pursuant to Article 171 EC Treaty [1997] OJ C63/2.

10. Article 234 reads: 'The Court of Justice shall have jurisdiction to give preliminary rulings concerning: a) the interpretation of this Treaty; b) the validity and interpretation of acts of the institutions of the Community; c) the interpretation of the statutes of bodies established by an act of the Council, where those statutes so provide. Where such a question is raised before any court or tribunal of a Member State, that court or tribunal may, it if considers that a decision on the question is necessary to enable it to give judgment, request the Court of Justice to give a ruling thereon.'
11. See particularly the UN ECE 'Convention on Access to Information, Public Participation in Decision-making and Access to Justice in Environmental Matters'.
12. Case T-585/93 *Stichting Greenpeace Council (Greenpeace International) and 18 other applicants v Commission of the European Communities, supported by Spain* [1995] ECR II-2205; Case C-321/95P, *Appeal by Stichting Council and Others v. Commission of the European Communities, supported by Spain* [1998] ECR I-1651.

REFERENCES

Auer, Mathew R. and Susan Legro (forthcoming), 'Environmental Reform in the Czech Republic: Uneven Progress after 1989', in M. Auer (ed.), *Restoring Cursed Earth: Appraising Environmental Policy Reforms in Central and Eastern Europe and Russia*, Boulder, CO: Rowman & Littlefield Press.

Commission of the European Communities (2001), 'Regular Report on the Czech Republic's Progress Towards Accession', SEC(2001)1746, Brussels.

Commission of the European Communities (2002), 'Regular Report on the Czech Republic's Progress Towards Accession', COM(2002)700 final, Brussels.

Crisen, Sabina and JoAnn Carmin (2002), *EU Enlargement and Environmental Quality: Central and Eastern Europe & Beyond*, Washington, DC: Woodrow Wilson International Center for Scholars.

Holzinger, Katharina and Peter Knoepfel (2000), *Environmental Policy in a European Union of Variable Geometry?: The Challenge of the Next Enlargement*, Basel: Helbing & Lichtenhahn.

Homeyer, Ingmar von, L. Kempmann and A. Klasing (1999), 'EU Enlargement: Screening Results in the Environmental Sector', *ELNI Newsletter*, No.2, pp.43–7.

Jans, Jan H. (2000), *European Environmental Law*, Groningen: Europa Law Publishing.

Knill, Christopher and Andrea Lenschow (2000), *Implementing EU Environmental Policy: New Directions and Old Problems*, Manchester: Manchester University Press.

Koppen, Ida J. (1993), 'The Role of the European Court of Justice', in J.D. Liefferink, P.D. Lowe and A.P.J. Mol (eds.), *European Integration and Environmental Policy*, London: Belhaven Press, pp.126–49.

McCormick, John (2001), *Environmental Policy in the European Union*, New York: Palgrave.

Miko, Ladislav (2000), 'The Czech Republic on the Way to Accession: Problems in the Environmental Field', in K. Holzinger and P. Knoepfel (eds.), *Environmental Policy in a European Union of Variable Geometry?: The Challenge of the Next Enlargement*, Basel: Helbing & Lichtenhahn, pp.183–214.

Ministry of Environment (1999), *National Environmental Policy: National Programme of Preparation of the Czech Republic for the EU Membership*, chapter 'Environment', Prague: Ministry of Environment.

Ward, Angela (2000), 'Judicial Review of Environmental Misconduct in the European Community: Problems, Prospects, and Strategies, in H. Somsen (ed.), *Yearbook of European Environmental Law*, Oxford: Oxford University Press, pp.137–59.

Winter, Gerd (1999), 'Individualrechtsschutz im deutschen Umweltrecht unter dem Einfluss des Gemeinschaftsrechts', *NVwZ*, Vol.5.

Europeanising Hungarian Waste Policies: Progress or Regression?

ZSUZSA GILLE

The hope that prevailed immediately after the collapse of state socialism was that Eastern Europe's environmental pollution would be 'swept away by democracy and economic rationality' [*Solomon, 1990: A14*]. While with time such expectations have become more modest [*Andersen, 2002; Andrews, 1993; Bochniarz, Kerekes and Kindler, 1994; Gille, 2000a, 2000b, 2001; Kaderják and Powell, 1997; Kindler, 1994; Manser, 1993*], some of the same hopes are now resurfacing as the accession of former socialist countries to the European Union (EU) becomes imminent. Most environmentalists and policy experts anticipate an improvement in regulatory standards, in law enforcement, and in the availability of funding for environmental purposes. The purpose of this contribution is to evaluate whether and how such expectations are being met in one area of environmental policies in Hungary, a country among the first wave of candidates to be admitted to the EU.

The area I examine here is waste. While according to Western European reports, the environmental accession requirements grew increasingly less significant in the enlargement negotiations [*Environment Daily, 1999a, 1999b*] – much to the disappointment of some industrial lobbies and environmental organisations in the West [*Friends of the Earth, 2002; Stirith, 2000*] – the Hungarian newsreader was compelled to form an entirely different impression. Reports in the Hungarian media about the progress made in meeting the accession requirements frequently mentioned waste legislation on the 'to-do-list' of several candidate countries, as well as of Hungary. In the latter, it was primarily the delay in ratifying the Act on Waste Management and its enforcement regulations that kept the Commission admonishing the country for not fulfilling the accession requirements quickly enough. Given that in the Hungarian adoption of the *acquis communautaire*, it was the environmental chapters that were lagging behind the others as well as behind the agreed-upon deadlines, waste legislation appeared as the single greatest obstacle to joining Europe [*Environment Daily, 1999b, 2000*]. Instead of concluding from this that Hungary's waste practices and policies, or the domestic expertise in this area were backward and that the country's waste situation carries the symbolic burden of non-Europeanness, as it is often the case in representations of the region, we must ask why waste legislation was held up.

I will demonstrate that a key reason for this delay is the fact that the EU itself has sent mixed messages to Hungary. Officially, the EU stands for preventative policies, primarily waste reduction and secondarily reuse. In practice, however, its economic constituents as well as its aid encourage remedial end-of-pipe technologies, such as waste dumps and incinerators. Not only does this create confusion in legislation and institution building in Hungary, and thus delay, it also establishes a practice that may lock in a certain path of development [*Stoczkiewicz, 2001*] that will be increasingly difficult to steer away from later. This is made all the more ironic by Hungary's waste history, which from 1949 to the late 1980s favoured preventative waste policies, rendering the present trend more a return to the past of the West than a step forward.

In order to understand the changes and challenges of harmonising waste legislation, it is important to put the EU's and Hungary's waste issues in their proper historical contexts. Therefore, I first review past waste policies in the EU and Hungary and then analyse the challenges brought on by joining the EU. In the main part of this study, I will compare requirements for sustainable development and for satisfying requirements for EU accession. The conclusion evaluates how these sometimes contradictory requirements are met, and what the compromises made by Hungary imply for the country's future waste practices.

Ecological Modernisation in the European Union's Waste Policies

The European Union first started to elevate environmental concerns to communal legislative actions in 1973, when it passed what came to be known as its First Environmental Action Programme (EAP). EAPs (now numbering six) define the general direction and tasks of environmental legislation in the period to come (ranging from four to nine years). Most legislation is passed in the form of directives written and proposed by the European Commission, examined by the European Parliament and the Council of the European Union.

The first piece of legislation concentrating on wastes was the Framework Directive on waste (75/442/EEC) passed in 1975. The ways in which this directive has been amended and modified by numerous subsequent directives illustrates very clearly changes in ways of thinking about environmental problems, which in turn express the public's changing environmental concerns. Initially, most of the EU's waste policies remained in a remedial paradigm, tackling environmental problems *post facto* through end-of-pipe technologies. Such technologies with regards to waste primarily include landfilling and incineration. These regulations included the aforementioned first EU directive on waste, the Waste Oils Directive (75/439/EEC), the Hazardous Waste Directive 78/319/EEC (later replaced by Directive 91/689/EEC), and the Titanium Dioxide Directives (78/176/EEC, 82/883/EEC), the Sewage Sludge Used in Agriculture Directive (86/278/EEC), and the Municipal Waste Incineration Directives (89/369/EEC and 89/429/EEC). While in 1981 the Council issued a Recommendation concerning the reuse of waste paper, it was not until 1989, five years after the Brundtland Report coined the concept of sustainable development, that the Community Strategy for Waste Management established waste prevention as a top priority [*Gervais, n.d.*]. Since then, one can speak of an accepted 'waste management hierarchy' (in order of priority): (1) minimisation; (2) reuse (without chemical transformation of waste's material); (3) material recovery (recycling); (4) energy recovery (some forms of incineration); and (5) final disposal (landfill, incineration).

In 1993, the Fifth EAP, called *Towards Sustainability*, made several steps towards what came to be known as Integrated Product Policy, the title of a White Paper from 2002. The essence of these newer sets of regulations is to integrate environmental and economic policies, that is, to incorporate environmental concerns into economic planning and technological innovation from the beginning rather than saving them as add-on, expandable, features. In this spirit numerous waste-related pieces of legislation have been passed in the last ten years. Some established economic incentives such as eco-taxes (for example the compulsory minimum tax rate on mineral oil introduced in 1993). Others laid down the principles of environmental certification, including eco-labelling and audit systems such as EMAS (Council Regulation No. 1836/93). Yet another group of regulations focused on defining the duties of producers and consumers aiming for full (cradle-to-grave) responsibility for products, most recently in the areas of packaging waste (94/62/EC), the wastes of electrical and electronic equipment (2002/96/EC), and vehicles (2000/53/EC). As a result of these changes in environmental paradigms, the principles of present EU waste legislation are the following (often referred to as the five P's):

1. Prevention Principle: top priority should be given to waste prevention and minimisation.
2. Proximity Principle: waste should be disposed of as close as possible to where it is generated.
3. Producer Responsibility Principle: waste producers should bear cradle-

to-grave responsibility for any damage caused by the waste they generate.
4. Polluter Pays Principle: polluters should bear the costs of safe management and disposal.
5. Precautionary Principle: waste management strategies should not take risks even if the causal relation between waste and damage is not fully proved.

Such changes were initiated and supported by a new model of production management, collectively referred to as industrial ecology. In the words of Robert M. White, the president of the USA's National Academy of Engineering, 'the objective of industrial ecology is to understand better how we can integrate environmental concerns into our economic activities' [*White, 1994: v*]. In terms of waste practices, industrial ecology has meant planning for reuse of products (after their initial life span is over), establishing cradle-to-grave product responsibility, minimising toxic by-products and in general increasing eco-efficiency.

Many debate whether this trend is real or whether it is simply clever public relations on the part of corporations trying to limit the punitive effects of environmental regulations. Proponents of ecological modernisation theory, whose focus is broader than the individual firm, however also agree that a transition in environmental discourse and practice has begun in these countries, with The Netherlands and Japan leading the way [*Hajer, 1995; Mol, 1995; Spaargen and Mol, 1992*]. They argue that even though environmental protection previously tended to focus on how to 'safely' displace hazards from production, and environmental politics concentrated on the distribution of hazards, now the intention is to keep and solve the problem of emissions and wastes within the sphere of production. It is in production that emissions can be reduced or prevented, and it is in production that by-products can be reused or recycled. Such an internalisation of environmental externalities is now seen as consistent with efforts to increase efficiency and even as conducive to technological innovation. Furthermore, relying on the environmental Kuznets curve, it has been argued that environmental impacts may become decoupled from economic growth. In other words, changes in management and new innovations will lead to economic growth that no longer spurs commensurate increases in emissions or wastes.

Independent of the accuracy of these views, I also argue that a qualitative change has occurred in environmental discourse. Acknowledging the change, however, does not necessarily mean endorsing it. Environmental modernisation theory in itself does not say anything about absolute volumes of waste or emissions, power, social costs, and the role of the public. With this critical note, I suggest that ecological modernisation theory can offer us a vantage point from which to interpret the implications of Hungary's accession to the EU for its waste practices.

Hungary's Waste Past

Visual, textual and statistical representations all describe state socialism as a wasteful social order. Visual representations of state socialism, such as Antonin Kratovchil's photos in the 1990 special issue of the *New York Times Magazine*, invoke the image of the state socialist landscape most familiar in the West – a grey still life composed of shoddy goods, of people wearing poor, idiosyncratic clothes surrounded by houses that look like they could fall apart at any time, and of piled-up garbage and filth. The juxtaposition of images of poverty with images of debris, dirt, toxic wastes and degraded nature tells a story about state socialism that has been retold for many decades: megalomaniac, yet outdated, industrialisation that left a good portion of society in poverty, generated tremendous amounts of waste, and caused environmental destruction.

Textual representations in journalistic accounts and scholarly works blame these conditions on poor management of the economy and outdated technology. 'Open hearth steel manufacturing and other outdated, inefficient technologies are still widely used by East European and Soviet industries', says the study of the World Watch Institute in an explanation of state socialist wastefulness and pollution [*World Watch Paper 99, n.d: 11*]. In addition to backwardness, a 'faulty, mismanaged economic system' has been also invoked as a key cause of environmental degradation.

Statistical data usually applied by studies done in the international financial and aid hubs and by scholarly works like to point out that state socialist countries' emissions/GDP, emissions per capita and waste/GDP indexes, as well as material and energy intensities have been significantly higher – often multiples of – Western equivalents [*Hughes, 1990; World Watch Paper 99*]. The result, as a *New York Times* author put it, is 'mountains of garbage. Literally, garbage' [*Lewis, 1990: A21*].

The textual, visual and statistical representations all suggest that state socialism was wasteful, both in the sense of squandering resources and in the sense of being full of wastes: it produced too much waste and garbage, and too many rejects, outdated and unusable or useless goods. Despite the infamy of many aspects of central planning, many socialist states, including Hungary, also pursued rigorous waste reuse and recycling and to a lesser extent waste reduction. In Hungary, one of the first economic laws of Stalinist leadership aimed at organising the reuse of waste materials for industry. Between 1950 and 1959, 34 central regulations on the collection, storage, delivery and price of waste materials were issued. In the reform period of Hungarian state socialism, which began in the mid-1970s, emphasis shifted from reuse and recycling to reduction that was encouraged with financial incentives. In 1981, the state implemented a system of monitoring, then rare even in the West, that obliged companies to prepare material flow charts and material balances in order to facilitate the tracking of toxic by-products in the production process and to discover inefficiencies. These material flow charts and material balances can be seen as precursors to voluntary environmental standards, such as ISO and EMAS.

While not entirely successful even in their own terms, some of the state's policies made significant progress. As a result of the new reform waste regulations implemented in 1981, by 1987 it was claimed that more than half of the total waste generated was reused. Progress in the field of plastic wastes was especially impressive as their reuse increased by 200 per cent, and with this, 20 per cent of total plastic wastes were reused [*NÉTI, 1987: 6*]. While the programme did not achieve its stated goal of substantially increasing the portion of secondary raw materials among industrial inputs [*KSH, 1988*], it was quite successful in finding uses for potentially dangerous wastes. Before the implementation of the 1981 Waste and Secondary Raw Material Management Programme, only 17–18 per cent of hazardous wastes were reused or recycled [*Tudományos Ismeretterjesztő Társulat, 1980*], while in 1982 the figure was 21 per cent and by 1986 it was 29 per cent [*Árvai, 1990*].

In sum, during the 50 years of central planning in Hungary, the state established an extensive infrastructure of waste registration, collection, redistribution, reuse and recycling. Even though the emphases of Hungary's waste policies changed over time, there has been an ongoing public discourse on the amount and sources of waste in industry and agriculture. Furthermore, partially as a result of waste laws and a profound cultural propaganda on waste reuse, bordering on the cult of waste, the public in Hungary had a strong material conservationist attitude. This mentality could have been built on as an important cultural asset in reforming Hungary's environmental policies; instead, it has been all but eradicated along with old policies. The socialist waste infrastructure had numerous shortcomings including a tendency to encourage waste production to fulfil waste quotas, ignorance of pleas to facilitate safe waste dumping, and undemocratic enforcement of waste laws. While it is necessary to acknowledge these flaws, it is also necessary to debunk the one-sided picture that socialist Hungary did not care about waste.

Why are these 'achievements' ignored? One of the main reasons is that statistical data prevailed in the representation of centrally planned economies, but more importantly, that such statistics concentrated on ratios of waste/GDP, rather than on waste per capita or ratios of recycling of industrial waste – data that would have demonstrated faults with Western waste practices and would have highlighted the advantages of state socialism, or at least could have presented a more balanced picture. This way, perhaps, the functioning elements of socialist waste policies could have been preserved. Second, 'indigenous' economists mostly informed by neo-liberal paradigms themselves despised the state's intervention in the economy, including policies of waste reduction and reuse.[1] Third, these waste policies and practices, while not entirely unknown to Western academics and journalists, were not considered to be of an environmental nature, but rather merely a curious element of central planning, and thus their significance was left unexplored. Finally, we must not ignore the fact that the increasingly visible and the truly horrendous environmental record of state socialism in Eastern Europe and the Soviet Union simply dwarfed any positive achievements of state socialism's relationship to the environment.

Why insist on the importance of this other environmental legacy of central planning in Hungary now? First, arguably there is a remarkable similarity between the above-mentioned industrial ecology and ecological modernisation discourse observed in the West, on the one hand, and Hungary's past waste practices on the other. If that is the case, then, what are the reasons and consequences of the EU's treatment of Hungary as innocent of any kind of environmental regulation, especially any progressive ones? If Hungary is indeed lacking effective waste policies, should we look for the causes in the demands of a liberal world made on Hungary, rather than in the country's socialist past? That is, is it possible that in order to meet the EU's economic criteria – a liberal market economy purged from the state's 'interventions' – Hungary returned to the West's past: a free market unfettered by environmental and conservation principles? Raising these questions, even if we currently lack the answers, is unavoidable if we are interested in the meaning of the EU's presence in post-socialist Hungary and the environmental implications of that presence.

EU Accession and Hungary's Dilemmas

Hungary expressed a desire to become a member of the EU even before state socialism collapsed (in 1988), but membership and especially its ultimate ratification was not taken for granted until relatively recently.[2] From the beginning there were two pressures that Hungary and the other post-socialist countries in Europe faced: an economic transition (often referred to as marketisation, privatisation or liberalisation) and an environmental revolution. The latter referred to an expectation that the new regimes would not only disclose and clean up the pollution caused by central planning, but that they would also avoid the kinds of development that led to the degradation of nature both in the West and in the East.

Two models circulated in public discourse about how to achieve both ends simultaneously – the economic rationalism and the environmental modernisation paradigms. The initially strongest expectation has been that liberalisation will automatically improve the state of the natural environment. This is the assumption that informs this transition discourse, which, adopting Dryzek's [*1997*] terminology, might be called the 'economic rationalism discourse'. Many activists and experts, however, hoped that Eastern Europe, enjoying the late-comers' advantage, could draw lessons from the mistakes of Western capitalism and build an economic system in which environmental concerns were integrated from the beginning instead of being saved for later as add-on features. Taking this course, however, would have required a more active role for the state – at least a more direct relationship between industry and government than the economic rationalism discourse sees as necessary.

The Early 1990s

Until the mid-1990s, post-socialist legislative events (or non-events, as the case might be), and actual waste practices conformed more to the economic rationalism model than to the ecological modernisation paradigm [*Gille, 2000b*]. To the extent that liberalisation, especially in the shape and form dictated by the IMF and World Bank, demanded a much-reduced role for the state both as owner and as regulator, this 'Wild East' period of the post-socialist transition may be seen as inevitable.

Economic rationalism manifested itself in the following markers. Privatisation meant that the state lost control over not only the means of production but also waste materials, and, without simultaneous new waste legislation, had a radically reduced ground on which to influence the fate of production wastes. The privatisation of retail translated into an easy evasion of previous obligations of grocery stores to take back bottles and jars for a deposit, which, with the simultaneous explosion in disposable packaging, meant a sudden accumulation of packaging wastes that neither consumers nor the garbage collection companies were able to handle [*Kaderják, 1997*].[3] Soon, garbage collection ceased to be a gratis public service provided by municipal governments. Since real incomes kept declining through the first half of the 1990s, most people and businesses resorted to illegal dumping. For example, the sale of the almost 40-year-old state waste collection company, MÉH, to a French corporation imposed radical limits on the scope of its activity, including its range of waste materials and circle of suppliers. Since the company was a monopoly, this single privatisation had national consequences.

Under such circumstances, it would have been crucial to pass new legislation redefining the duties of waste producers and the scope of authority of the state. Instead, Hungary's comprehensive waste act, which got stuck several times in each phase of its making, did not take effect until January 2001. Moreover, existing regulations were withdrawn or made unenforceable owing to legal, economic and political uncertainties. NGOs have also started calling on the EU to retain positive legacies of state socialism, where it still exists [*No author, 2001*]. The aforementioned system of monitoring established in the 1980s that made evaluation of a firm's waste activities easy and transparent by obliging firms to prepare material flow charts and material balances was also dismantled (interviews with Attila Takáts, Budapest, 1996; Gábor Romhányi, Budapest, 1995).

Exactly at the time of a deep economic crisis and great uncertainty, state funds for the rationalisation of material and energy use were eliminated. Even though there were proposals to include eco-taxes in the new tax system to create a fund for environmental purposes, these were ignored or watered down to such an extent that the new system, as one former ministry official wrote, 'in the case of waste reuse, … function[ed] as an actual counter-incentive' [*Takáts, 1990*].[4] The first half of the 1990s, therefore, witnessed the victory of the free market, and more specifically that of the economic rationalism paradigm. The state retreated from the economy and was left without sufficient income to motivate producers in environmentally friendly directions. Further, the reduction of the state apparatus left it ill-equipped to enforce what regulations remained or slowly passed and, in general, too discredited to intervene on behalf of an environmentally conscious economic transition.

Waiting for Membership

By the mid-1990s a change in direction was noticeable. The new Law on Environmental Protection, passed in 1995, laid down producers' reporting obligations and it introduced a number of eco-taxes, such as environmental load fees, use contributions and product charges. The latter initially had been imposed on fuel, tires, refrigerators, refrigerants, packaging materials and batteries (Act LVI of 1995) introduced in 1996. The amount of the charge is reduced by half for those products that are environmentally friendly. The revenues from these are, in turn, earmarked for investments that abate pollution as well as for waste recuperation and recycling. The long-awaited hazardous waste act was also put in force in 1996. Both the latter and the Law on Environmental Protection were looked upon as transitional measures that would be revised once the EU-conform Act on Waste Management was ratified (which took many years longer than expected). What changes are now needed in order to meet EU requirements? In order to understand Hungary's tasks, we need to know what characterised the country's waste situation and how it differed from that of the EU.[5]

Hungary produces roughly 70 million tons of waste, with 28 million tons being biomass.[6] Out of this non-biomass waste, roughly 70 per cent is production waste and six per cent is municipal waste (with municipal liquid waste roughly 11 per cent).[7] Four per cent of all wastes (roughly 3.4 million tons, that is, eight per cent of non-biomass waste) is hazardous.[8] In comparison, the EU's municipal waste ratio without liquid waste is 14 per cent, which means that in Hungary, the share of municipal solid waste is considerably lower than that in the EU. This reflects not so much a 'lower developmental stage', as it is commonly implied, but a state socialist past, where consumption was shunned and economic growth was forced, even if it meant high waste-ratio production and thus high levels of industrial waste. High waste ratios did not constitute a problem for a social order that ruthlessly pursued 100 per cent recycleability. In sum, the difference between the EU and Hungary is not simply that of quantity, indicating a lower stage on the same developmental trajectory, but of quality, owing to a difference in developmental trajectories.

The rate of recycling for municipal solid waste in Hungary is three per cent compared to the EU's average 15 per cent. The existing rates of recycling in former socialist countries are similar to several EU members' rates (such as those of Ireland and Portugal), but in general are lower than average rates in the 1980s. In contrast to this difference in municipal waste data, the recycling rate for hazardous wastes is 20 per cent in Hungary as compared to 8 per cent in the EU. Given that dumping and incinerating hazardous waste is environmentally more risky then recycling, we should welcome this variation and study what allows Hungary to surpass the allegedly environmentally more progressive EU in this regard.

Out of the country's 3,000 municipalities, 2,700 have landfills. As is often pointed out, there is a dump in practically every village this includes 'the smallest ditch' (interview with Hilda Farkas, Budapest, 7 August 2003). There are 665 registered and municipally run landfills for municipal waste, of which only 15 per cent meet current technological standards. In addition, however, there are also 620 smaller dump sites, not registered and most likely not fulfilling safety requirements. There is one major hazardous waste incinerator in the country, in Dorog, with a capacity of 25,000 tons/year, and there are some minor incinerators, some recently built, and some older cement kilns, which add up to an overall capacity of 85,000 tons for hazardous waste incineration. There is also just one modern hazardous waste landfill, in Aszód, which only takes wastes for final (rather than temporary) disposal, 10,000 tons a year. Its overall capacity is 300,000 tons, which is not expected to be filled until about 2020. Therefore, assuming no import and export of hazardous wastes, Hungary cannot incinerate more than three per cent of its hazardous wastes. The overall amount of waste incineration cannot be ascertained from the data available because it is lumped together with other types of elimination, such as the desiccation of liquid municipal waste (and possibly others). The environmental ministry official in charge of the database was unable to clarify this. However, the ratio of incineration for all wastes is unlikely to be more than 11 per cent. Compared to the practice of many EU countries that incinerate as much as a quarter of their *total* wastes and that have dozens of incinerators, Hungary's rates seem very low.[9]

On average, accession countries have higher industrial waste per capita ratios than EU members, the reverse of the municipal waste ratio difference noted above. My estimates however do not allow extrapolating from these averages to Hungary. Its industrial waste per US$1,000 of GDP is 72 kg, which can be said to be in the middle range of existing EU countries, and well below Finland's 118 kg and the Czech Republic's staggering 288 kg. Similarly, and even more unambiguously, with 490 kg of municipal wastes per capita, Hungary seems to belong more to the West than to the East (should we insist on such characterisations). Its municipal waste per capita is higher than other candidate countries we have data for, and even higher than that of some EU members.

In sum, the areas in which Hungary's waste situation seems to be significantly and structurally different from the EU's are: (a) smaller proportion of municipal wastes among total wastes generated; (b) significantly smaller proportion of reuse and recycling, except in hazardous wastes; and (c) significantly smaller proportion of incineration than the EU average. These differences lead to a higher rate of landfilling and a lower capacity for hazardous waste treatment through modern landfilling and incineration techniques. It is getting more difficult to make the claim that Hungary seems under-developed based on its waste generation structure and 'waste-efficiency' indicators. Yet, another commonly made claim, here in the words of a study by regional experts, that 'the situation in the Central and East-European countries resembles to a high degree the situation in OECD country in the 1980s when landfilling was the main disposal technique', is not completely off the mark [*Eurowaste, 2000*].

There are two problems with this statement, however. One is related to the misinterpretation of Hungary's waste past. In the 1980s Hungary actually favoured recycling over landfilling much more than it does now – and possibly more than the OECD did. Also, Western European countries have not made much progress in developing alternatives to landfilling either, which still comprises close to two-thirds of the treatment options for municipal waste [*European Environment Agency, 1999*]. Furthermore, unlike Hungary, the EU has not yet managed to uncouple its waste generation from economic growth; in the last ten years economic growth averaged six per cent in the EU, while waste generation grew by ten per cent annually. In Hungary, according to latest governmental data, overall waste generation declined from 106 million tons in 1990 to 68.7 million tons in 2000, during which period economic growth while initially negative, in the second half of the 1990s was around five per cent. While an increase in waste generation has been registered since then, no increase is expected in the next few years according to the National Waste Management Plan (OHT).

The EU's Role in Hungarian Waste Policies and Practices

The data discussed above were not made available or used as context at the time the European Commission established the accession requirements for former state socialist countries. The EU departed from three assumptions: (1) Hungary, as with all former socialist countries, has no environmental waste legislation to speak of; (2) it lacks the technical and institutional infrastructure necessary for implementing modern waste management methods; and (3) without Western assistance, Hungary is incapable of implementing progressive change. As a market analysis from 2001 stated, 'Most of the current environmental difficulties arise from the fact that environmental policy was virtually non-existent under the communist regime' [*Trade Partners UK, 2001: 2*]. The inscrutability and the etatism of state socialist waste policies reinforced the conviction in the early 1990s that Hungary must start from a blank page, which first requires that the old be demolished. According to some sources, the collapse of the state apparatus tore apart existing scientific and policy collaboration and put promising projects on hold.[10] It would be worthwhile to investigate whether this sudden rupture and the idea that we must start anew made the adoption of the environmental (and especially the waste) chapter of the *acquis* slower than was necessary.

Based on the discrepancies between Hungarian and EU waste practices (if not principles), the EU had two major concerns. First, it wanted to make sure that by admitting Hungary (and other former socialist countries) it would not unduly add to environmental problems caused by wastes. That the EU has such fears is apparent from studies lamenting that its waste output would double, given present-day data, when the first wave of candidate countries joined. Not only will enlargement worsen the EU's waste statistics, it is also feared that candidate countries' loose environmental standards will offer an undesirable competitive advantage. It was exactly this looseness of enforcement, no regulation, that lured EU-constituent businesses to the region. Apparently, EU officials could not decide between the EU's environmental principles and its economic interests. In 2002, it granted exemptions to Hungary, along with most other candidate countries, from fulfilling EU requirements concerning waste policies. Hungary received a four-year exemption from implementing EU-conform legislation on hazardous wastes and on the recovery and recycling of packaging wastes, as well as a 15-year grace period in the area of urban wastewater treatment.

Indeed one cannot appreciate the full relevance of the enlargement unless we treat the EU not merely as a bundle of legal and institutional arrangements, as is usually done in studies of the eastern enlargement, but also as a powerful global actor that represents and is supported by specific (rather than universal) economic interests. From-within critiques of the Union often bring attention not merely to the commonplace of democratic deficit, but with a positive spin, to the corporate voices that have control over many decisions and resources. Similarly, catching up with the EU with regard to waste treatment capacities is not merely a humanitarian or an environmentalist goal. While the EU sees itself as the environmental conscience of the world and has played important leadership roles in numerous environmental issues in the international arena, actual legislation, policy and aid within its environmental activities reflect certain biases. In this respect, I address two different modes of EU presence in former socialist countries and examine the extent to which these different modes are complimentary or contradictory to the EU's presence as a set of laws. I examine the effects certain EU-constituent businesses and EU aid have on Hungary's future waste practices.

As no market analysis of Eastern Europe fails to mention, the big opportunity for environmental business in the region comes from former socialist countries' desire to join the EU. 'The goal of EU accession is the main driver for improvement of the environment in Hungary and the recent attention given to the latter (after a sluggish start) by the Hungarian Authorities looks set to continue', says a UK analysis [*Trade Partners UK, 2001: 2*]. This is good news for investors. 'The size of the Hungarian environmental market was valued at US$700 million in 1997 and was expected to rise about 40 per cent to the end of 2000. Further growth between 2000 and 2010 is expected' [*Trade Partners UK, 2001: 3*]. A study published by Frost and Sullivan also emphasises that, 'by far the most important driver of the (municipal waste management) market over the forecast period is expected to be EU expansion and harmonisation, leading to the increasingly urgent need to raise standards and improve infrastructures' (quoted in Davies [*2000: 3*]). The study commissioned by the US Department of Commerce also implies that with Western doors closed to Hungary, the country's environmental record is unlikely to improve without outside intervention. As noted by Svastics [*1999: 1*], 'The fact that Hungary became a member of the OECD in 1996 and subsequently made a commitment to joining the European Union (EU) has increased government attention to environmental issues.' The same study explicitly laments that 'municipal waste incineration is very scarce in Hungary' [*Svastics, 1999: 8*].

Indeed, the identified 'shortage' in waste treatment capacity has been seen as a magnet for Western firms, increasingly unable to sell their facilities in the EU. However, a weakened state unable to enforce environmental regulations, or the laxness of some of these regulations themselves, has been important in attracting waste investors to the region as the capacity shortage. The years immediately following 1989 saw a veritable 'waste-rush' of multinational incinerator corporations to the East [*Gille, 2000b*].[11] Unfortunately, EU accession is unlikely to deter such imports. The National Waste Management Plan (NWMP) for 2003–2008 is planning six new incinerators, even though in the parliamentary debate of the NWMP one reader considered the four or five mentioned in an earlier draft excessive [*Parliamentary Committee Minutes, 2001*]. Furthermore, the NWMP states that 'Hungary must have a nation-wide incineration capacity of 170 thousand t/year' – a doubling of its present capacity. The plan appears hardly more negotiable than five-year plans under state socialism [*National Waste Management Plan, 2002: 28*]:

> In certain sensitive regions of the country (e.g. where industrial activities are concentrated, where the geological conditions are special, where tourism represents a seasonal change, or where more of these factors are present simultaneously) construction of an incinerator can only be postponed for a period of time, but in the long term there is no alternative for the disposal of waste remaining.

Was it the pressure of such lobbies or was it the Western opinion that Hungary is backward in terms of its incinerator capacity that compelled the OHT authors to elevate the pumping-up of incinerator capacity to a national law? Or is it the synergy between the backwardness discourse and the economic interests that put their stamps so powerfully on the future waste practices of the country? Either way, one needs to examine how such interests prevail, how this synergy comes about and how it is reproduced.

Processes of Change

First, the EU tells Hungary what waste treatment capacities it needs in order to be accepted as truly European. Then, it provides the 'aid' to fulfil the requirements. Finally, most of this aid makes its way to back to EU constituent producers of environmental technologies. Three main funds were established for assisting former socialist countries in their transitions and in adopting the *acquis*. The earliest was PHARE (Poland/Hungary Aid for the Reconstruction of the Economy), followed by ISPA (Instrument for Structural Policies for Pre-Accession) for transportation and environmental objectives, and SAPARD (Special Accession Programme for Agriculture and Rural Development) for agricultural and regional development. Among Hungary's ISPA Funds, waste management projects received priority. Out of the total ISPA contribution between 2000 and 2002 (547 million Euro), 30 per cent was awarded to various waste management projects (not including wastewater treatment, another preference area).

I do not have access to data regarding whether technologies or firms from the EU are the most common beneficiaries of these projects (and most of the bidding is still open). Clearly, however, firms from Western Europe have an advantage in this market. After all they know best what will truly conform to EU standards. Market analyses also demonstrate that East Europeans prefer Western environmental technology to domestic alternatives [*Regional Environmental Center, 1997*]. Also non-EU firms are excluded from bidding on projects to be completed with EU funds (unless through their European subsidiaries), much to the annoyance of the US and Canada. North American market studies explicitly lament the small and decreasing share of US companies in the East European waste technologies market. With the recent departure of the US giant, Waste Management Inc., this seems to be irreversible. Such an exclusion makes it clear that the primary aim of 'environmental' aid is not for Eastern Europe to adopt the environmentally most beneficial solutions through completely open bidding, according to the principle of 'best available technology'. Rather the goal is to turn the region into a market for Western European goods[12].

The compromise of environmental principles becomes even clearer when we notice that the bidding announcements for these waste projects are about building new landfills. With the pretext that old, dangerous landfills have to be eliminated, the Hungarian National Waste Management Plan aims at establishing a network of regional dumps, so all settlements would have a landfill for municipal solid waste within a 30-kilometre radius. The defenders of the Hungarian National Waste Management Plan argue that, at this density, landfills become profitable [*Parliamentary Committee Minutes, 2001*]. But if the goal to reduce waste amounts is achieved, would not that change the profitability calculations? How does this plan get closer to the EU ideal of reducing the share of landfilled wastes?

Perhaps policymakers are motivated more to follow the Proximity Principle than by decreasing waste and dumping? The Proximity Principle, according to which waste should be disposed of as close as possible to where it is generated, works well in countries where there is already a network of legal and up-to-date dumps. However, in the candidate countries, where the existing network is insufficient either in safety standards or in capacity, it may act as an unintended inducement to build dumps, further marginalising waste prevention goals. More importantly, as long as waste dumping receives generous EU funding, and waste reduction receives only symbolic gestures in programmatic documents, the policy emphasis on landfill capacity-expansion will reinforce existing motivation for end-of-pipe technologies. In the long run, it may lock in future waste practices making a (hypothetical) move in a more preventative direction less and less feasible.

The practical consequences of both EU aid and EU business may be too early to see, but if Hungary's National Waste Management Plan is any indication, end-of-pipe technologies are still the favoured method of dealing with unwanted materials. In fact, the Plan aims to radically overhaul the existing economic incentives for recycling and reuse in order to relax restrictions and penalties waste producers presently face. While a new policy tool, a weak alternative to the existing system of product charges was passed at the end of December 2002. The Hungarian Parliament voted against the further relaxation of the producers' responsibility in the product charges system. It is clear, nevertheless, that the pressure from industry to loosen waste legislation and especially to forego economic incentives for waste minimisation and recycling remains.

In this often-fierce struggle between industrial lobbies and environmentalists, the meaning of Europeanness is also contested. Since Europe is a powerful symbol, most participants in the debate try to define their positions as closer to that of the EU. There are several difficulties in evaluating such claims. First, it is never clear whether the reference is EU law or EU practice. In that regard, it may be said that the National Waste Management Plan is conform not with EU ambitions, which as mentioned, place reuse and recycling above incineration and dumping in the waste management hierarchy, but rather with EU Practice, which still favours dumping.[13]

Second, in the case of waste policy tools, what policies conform to EU guidelines remains quite ambiguous. While the EU lays down the desired principles and goals of waste management in directives that the member countries adopt in the course of a few years, how a member country's state apparatus attempts to achieve those goals remains within the scope of national authority, and thus varies. As a result, there are diverse systems in place within the EU in the different areas of waste management. Not all are equally effective and presumably not all are equally practical in the case of Hungary. There is quite a variety, especially in methods of achieving EU objectives concerning packaging wastes and recycling. While environmental activists continue to demand the maintenance and strengthening of the system of product charges and the reinstitution of deposit systems, their opponents (occasionally Ministry officials, and the association representing manufacturing interests) point to various EU principles protecting the uninhibited workings of the market. During the screening process, for example, the EU objected to product charges since those ultimately act as subsidies for production costs. (Since product charges are supposed to be pumped back to the companies for extending their reusing and recycling capacities, product charges could be seen as a form of subsidy.) The EU also prohibits obligatory deposits on packaging materials since those may function as hidden import duties, as Belgium's recent example shows.

There is also the question of which set of laws and policy tools can actually guarantee that Hungary fulfils its promise to achieve a 50 per cent recycling rate by 2006 (in accordance with the temporary exemptions EU granted exemptions). Considering all these constraints – in quotas, policy tools and principles – and the existing diversity in practices, EU-conformity turns out to be a slippery concept. Thus, it is adaptable to political agendas of all kinds.

Why get upset about laws? They can be changed, after all. While laws and policies are indeed reversible, their effects may not be. The post-socialist transition and Hungary's subsequent Europeanisation is not only about markets and law. These radical transformations require and call for a new kind of culture and a new kind of subjectivity. The way in which a post-socialist society joins the West will produce a certain kind of consumer, employee or manager. The thrifty and waste-conscious material culture that developed under state socialism today may seem as backward and contrary to 'European' consumerism, rendering preventative waste policies less attractive and later achievable only at a much greater social cost. It is hardly a surprise that, despite the 5th Environmental Action Programme, waste generation is on the increase in the EU. Similarly in Hungary, short of explicit policy tools to rein in consumption, it seems that even the more modest waste-efficiency-increasing plans will fail.

Conclusion

The EU sends mixed messages to Hungary around waste policy issues. On the face of it, the EU is for preventative waste policies and is concerned about the environmental effects of economic growth. This is the position espoused in the Environmental Action Plans, the directives, and the programmatic studies. However, in terms of actual practices, particularly investments and economic and infrastructure requirements, the EU stands for unsustainable development and for putting economic interests before environmental interests. Given the contradictory sets of expectations placed on candidate countries, it is not surprising that environmental harmonisation was lagging and that business interests were dominating the legislative process in the individual accession countries.

The European Commission has loudly solicited NGO participation in the accession negotiations. But as these NGOs now lament, this has been merely a symbolic gesture. First, NGOs were excluded from committees deciding ISPA and SAPARD grants, where numerous decisions with long-term consequences were made. Second, the meetings between the Environmental DG and green NGOs of the region have been futile. The NGOs, armed with data and expert studies in the best spirit of professionalism, demanded introducing accession requirements to safeguard their countries' biodiversity from the adverse ecological effects of meeting other EU requirements, in the areas of infrastructure (road construction) and agriculture. Yet, the Commission repeatedly prevented its comments and opinions made at these meetings from being printed in the resulting summaries of the minutes. They 'did not want to be held accountable later for not fulfilling promises made there' (anonymous interview, Hungary, 2000). Ultimately, there were no practical steps taken to enforce NGO-suggested accession conditions. Symbolic gestures for civil society involvement produce symbolic results. Whether NGOs and social movements have more leverage once inside the EU structure remains to be seen. Their potential role in holding the EU to its principles will surely remain crucial. If this potential is realised, this would be a true step towards ecological modernisation, especially in its non-technocratic and democratic version.

NOTES

1. Reform economists argued already in the 1970s that harder budget constraints and more independence for enterprises in investment decisions – in sum, a free market – would force them to use natural resources more sparingly and with greater care. For detailed discussions and evaluations of the reform economists' approach to environmental problems, see DeBardeleben [*1985*] and Gille [*1997*].
2. According to a Hungarian official of the Hungarian Mission in Brussels, it was not until high-ranking politicians in candidate countries started openly advocating the 'possibility of

life outside the EU', that is in 1999, that the Commission sped up the negotiations (interview with Gábor Baranyai, Brussels, 10 July 2003).

3. What created a problem was not the weight but the type of wastes: owing to the increase in light-weight plastic packaging the volume of municipal waste did not increase, but capacities for the recovery of this type of waste in such amounts were insufficient.
4. Product charges on transportation fuel had been in effect since 1992 [*Lehoczki and Balogh, 1997*].
5. The adoption of the environmental chapter of the *acquis* started in 1999. I am using data from 2000, which did not considerably differ from data in 1995, based on which, if necessary, the environmental criteria for accession and funding targets were or could have been decided. The following data are compiled from the National Waste Management Plan. As the members of the Hulladék Munkaszövetség, the most important Hungarian NGO dealing with waste issues pointed out, the data are incomplete and do not add up [*Kukabúvár, 2002*]. Indeed, I found it hard to navigate the data provided. In light of these valid criticisms, I report only the basic figures.
6. According to Hungarian lawmakers this is made up of agricultural and forestry residues entirely recycled through biological cycles.
7. Owing to the inconsistency in Hungarian data collection, in which some authorities categorised data according to the hazardousness of waste and others according to their origins, one cannot tell exactly how much of hazardous waste is generated in production versus in consumption. The totals of these data do not add up to 100 per cent.
8. Roughly one-quarter of it is comprised of red mud resulting from aluminium production. Since red mud is not recovered, its large share in hazardous wastes does not explain the relatively high recycling rate for hazardous wastes (see below).
9. Reliable and comparable data on incineration are very hard to come by even according to the EU's own offices.
10. An example is the waste dump registry initiated by the Alliance of Technical Professionals in the 1980s [*Gille, 2001*].
11. Between 1988 and 1995 there were an estimated 18 million tons minimum of annual incinerator capacity proposed just in Russia, the Baltics, Hungary, Poland, the Czech Republic and Slovakia, with about 93 per cent of these capacities offered for export by Western countries. Put another way, about 187 facilities were proposed in the region, with Germany and Austria leading the way, accounting for 30 per cent of the offers. My calculation is based on Greenpeace data from Gluszynski and Kruszewska [*1996*].
12. This is not to imply that North American technologies are environmentally friendlier.
13. This was pointed out by several participants of the meeting of the Parliamentary committee evaluating the draft of the NWMP in June 2001 [*Parliamentary Committee Minutes, 2001*]. However, the final Plan does not establish a clear and concrete preference for recycling.

REFERENCES

Andersen, Mikael Skou (2002), 'Ecological Modernization or Subversion?', *American Behavioral Scientist*, Vol.45, No.9, pp.1394–416.

Andrews, Richard N.L. (1993), 'Environmental Policy in the Czech and Slovak Republic', in A. Vári and P. Tamás (eds.), *Environment and Democratic Transition: Policy and Politics in Central and Eastern Europe*, Boston: Kluwer Academic Publishers, pp.5–48.

Árvai, József (1990), *Hulladékgazdálkodás* (Waste Management), Budapest: Budapesti Mûszaki Egyetem, Mérnöktovábbképzö Intézet.

Bochniarz, Zbigniew, Sándor Kerekes and József Kindler (eds.) (1994), *Designing Institutions for Sustainable Development in Hungary: Agenda for the Future*, Budapest: Környezettudományi Központ.

Davies, Steve (2000), *The Private Sector and Waste Management in Central and Eastern Europe 2000*, London: PSIRU. Available at www.psiru.org.

DeBardeleben, Joan (1985), *The Environment and Marxism-Leninism: The Soviet and East German Experience*, Boulder, CO: Westview Press.

Dryzek, John S. (1997), *The Politics of the Earth: Environmental Discourses*, Oxford: Oxford University Press.

Environment Daily (1999a), 'EU Environmental Accession Talks Begin', *Environment Daily*, No.663, 9 Dec. 1999. Available at www.environmentdaily.com.

Environment Daily (1999b), 'Environment Takes Back Seat in EU Enlargement', *Environment Daily*, No.624, 14 Oct. 1999. Available at www.environmentdaily.com.

Environment Daily (2000), 'Hungary's Environmental Performance Assessed', *Environment Daily*, No.742, 12 Apr. 2000. Available at www.environmentdaily.com.

European Environment Agency (1999), *Environment in the European Union at the Turn of the Century.* Available at http://waste.eionet.eu.int/.

Eurowaste (2000), *The Final Report on the Project – Waste Management Policies in Central and Eastern European Countries: Current Policies and Trends.* Available at http://www.eurowaste.org.

Friends of the Earth (2002), 'Enlargement Plans Still Sideline Environment', Press Release, 13 Dec. 2002, Copenhagen, Denmark.

Gervais, Caroline (n.d.), 'An Overview of European Waste and Resource Management Policy (Executive Summary)', Royal Society for Nature Conservation. http://www.forumforthefuture.org.uk/uploadstore/Ex%20Sum%20-%20An%20Overview%20of%20European%Waste%20and%20Resource%20Management%20Policy.pdf

Gille, Zsuzsa (1997), 'Two Pairs of Women's Boots for a Hectare of Land: Nature and the Construction of the Environmental Problem in State Socialism', *Capitalism, Nature, Socialism*, Vol.8, No.4, pp.1–21.

Gille, Zsuzsa (2000a), 'Cognitive Cartography in a European Wasteland: Multinationals and Greens Vie for Village Allegiance', in M. Burawoy, *et al.*, *Global Ethnography: Forces, Connections and Imaginations in a Postmodern World*, Berkeley, CA: University of California Press, pp.240–67.

Gille, Zsuzsa (2000b), 'Legacy of Waste or Wasted Legacy? The End of Industrial Ecology in Hungary', *Environmental Politics*, Vol.9, No.1, pp.203–34.

Gille, Zsuzsa (2001), 'Social and Spatial Inequalities in Hungarian Environmental Politics: A Historical Perspective', in P. Evans (ed.), *Livable Cities: The Politics of Urban Livelihood and Sustainability*, Berkeley, CA: University of California Press, pp.132–61.

Gluszynski, Pawel and Iza Kruszewska (1996), *Western Pyromania Moves East: A Case Study in Hazardous Technology Transfer.* Available at http://www.rec.hu/poland/wpa/pyro-toc.htm.

Hajer, Maarten (1995), *The Politics of Environmental Discourse: Ecological Modernization and the Policy Process*, Oxford: Clarendon Press.

Hughes, Gordon (1990), *Are the Costs of Cleaning Up Eastern Europe Exaggerated? Economic Reform and the Environment*, London: Centre for Economic Policy Research.

Kaderják, Péter and James Powell (eds.) (1997), *Economics for Environmental Policy in Transition Economies: An Analysis of the Hungarian Experience*, Cheltenham: Edward Elgar.

Kindler, József (1994), 'Evaluation of Economic, Social, and Political Preconditions for a Successful Implementation of the Institutional Reform', in Z. Bochniarz, R. Bolan, S. Kerekes and J. Kindler (eds.), *Designing Institutions for Sustainable Development in Hungary: Agenda for the Future*, Budapest: Környezettudományi Központ, pp.119–51.

KSH (Központi Statisztikai Hivatal – Central Bureau of Statistics) (1988), *Központi fejlesztési programok. A melléktermék- és hulladékhasznosítási program 1987: évi eredményei* (Central Development Programmes: The By-Product and Waste Reuse Programme), Budapest: Központi Statisztikai Hivatal.

Kukabúvár (2002), 'Mennyi hulladék képzõdik Magyarországon?' (How much waste is generated in Hungary?), *Kukabúvár*, Vol.8, No.3. Available at http://www.kukabuvar.hu/kb29/kb29_36html.

Lehoczki, Zsuzsa and Zsuzsanna Balogh, (1997), 'Hungary', in J. Klarer and B. Moldan (eds.), *The Environmental Challenge for Central European Economies in Transition*, Chichester: John Wiley and Sons, pp.131–92.

Lewis, Flora (1990), 'The Red Grime Line', *New York Times*, 10 Apr., p.A(21)L.

Manser, Roger (1993), *The Squandered Dividend: The Free Market and the Environment in Eastern Europe*, London: Earthscan Publications Ltd.

Mol, Arthur P.J. (1995), *The Refinement of Production: Ecological Modernization Theory and the Chemical Industry*, Utrecht: Van Arkel.

NÉTI (1987), 'Az anyag nem vész el, csak drágább lesz: Feltáratlan kincsesbányáink' (The material does not get lost only gets more expensive: Our unexplored treasure mines), *Mai Magazin,* Mar. 1987, p.6.

No author (2001), 'NGO Response from Candidate Countries'. Available at http://wwwczp.cuni.cz/6EAP/EENGO.html.

Parliamentary Committee Minutes (2001), 'Minutes of the June 13th meeting on the National Waste Management Plan of the Article 56 in the 2000 Waste Act'. Available at http://emil.alarmix.org/sajto/ jegyzokonyv2001jun13II.html.

Regional Environmental Center (1997), *Demand for Environmental Technologies in the Czech Republic, Hungary, Poland and Slovakia: A 'Showcase Europe' Market Research Project*, Budapest.

Solomon, Laurence (1990), 'The Best Earth Day Present: Freedom', *Wall Street Journal*, CCXV, 20 Apr. (Eastern Edition), p.A14.

Spaargen, Gert and Arthur P.J. Mol (1992), 'Sociology, Environment, and Modernity: Ecological Modernization as a Theory of Social Change', *Society and Natural Resources*, Vol.5, pp.323–44.

Stirith, Jernej (2000), 'Enlargement and the New EU Environmental Action Programme', in *Metamorphoses*, No.17, Budapest: Regional Environmental Center for Central and Eastern Europe, no pp.

Svastics, Kinga (1999), 'A hulladékok káros hatása elleni védelem jogi, műszaki és gazdasági szabályozása' (The legal, technical and economic regulation of protection against the harmful effects of wastes), in J. Árvai (ed.), *Hulladékgazdálkodás* (Waste Management), Budapest: Budapesti Müszaki Egyetem, Mérnöktovábbképzö Intézet, pp.140–45.

Svastics, Kinga (1999), 'Waste Management, Hungary, Industry Sector Analysis'. Available at http://www.tradeport.org/ts/countries/hungary/isa/isar0005.html.

Takáts, Attila (1990), 'A hulladékok káros hatása elleni Védelem jogi, műszaki és gazdasági szabályozása.' (The legal, technical and economic regulation of protection against harmful effects of wastes.) In *Hulladékgazdálkodás*. (Waste Management), edited by J. Árvai, 140–145. Budapest: Budapesti Műszaki: Egyetem Mérnöktovábbkepző Intézet.

Trade Partners UK (2001), 'Environment Market in Hungary'. Available at www.tradepartners.gov.uk/ text/environment/hungary/profile/overview.shtml.

Tudományos Ismeretterjesztõ Társulat (1980), *Környezetvédelmi elöadói segédanyag* (Textbook for Environmental Protection), Budapest: Tudományos Ismeretterjesztö Társulat.

White, Robert M. (1994), 'Preface', in B.R. Allenby and D.J. Richards (eds.), *The Greening of Industrial Eco-Systems*, Washington, DC: National Academy Press, pp.v–vi.

EU Enlargement and Sustainable Rural Development in Central and Eastern Europe

ANDREAS BECKMANN AND HENRIK DISSING

Rural development in Central and Eastern European (CEE) countries contains numerous political, socio-economic and ecological challenges for eastern enlargement of the European Union (EU). Although often thought to be the domain of governments, many of these challenges are being addressed on a daily basis by non-governmental organisations (NGOs) throughout the region. These organisations, most of which have their roots in the environmental movement, are not only leading efforts to protect the environment, but are also, arguably, leading forces for social and economic development in the rural areas in which they work.

This contribution looks at principles of sustainable rural development and the extent to which these principles have been applied in the ongoing process of accession to the European Union. We argue that the pre-accession initiatives undertaken by the EU, and relevant to rural areas, have largely missed the mark of sustainability not only as a consequence of lack of genuine commitment, but also owing to wrong approaches. We look at specific examples of the activities of regional NGOs and the practical models they have developed for addressing the problems and particular needs of rural areas in CEE post-communist societies.

Sustainable Rural Development – A Key Challenge for Enlargement

The need to achieve both economic and environmental sustainability is now a major challenge for rural areas of Central Europe. Success or failure in meeting this challenge will have profound economic, social, environmental and, ultimately, political consequences not only for the Central European countries themselves, but also for the EU as a whole. Enlargement will open unprecedented disparities within the EU, severely testing the EU's commitment, anchored in Article 158 of the Amsterdam Treaty, to promote 'overall harmonious development' and to reduce economic disparities between urban and rural areas [*European Union, 1997*]. Rather, it is likely that EU expansion will open unprecedented disparities. In 2000, gross domestic product (GDP) per capita, with purchasing price parity, was less than 70 per cent of the EU average in all accession countries but Cyprus, and under 50 per cent for seven of the accession countries [*Eurostat, 2003*]. Accession to the EU can also be expected to widen rifts that have opened within the accession countries themselves, especially between dynamic urban areas like Prague, Budapest, Krakow and Tallinn where purchasing power and economic structure are becoming comparable to those in many Western European countries, and lagging regions, most of which are rural [*Weise et al., 2001: 46–9*].

The increasing integration of the accession countries into the European Common Market will force fundamental structural changes to their rural economies. In contrast to Western Europe, where agriculture employs some 4–5 per cent of the workforce and is being increasingly replaced by other sectors, such as tourism and local production, agriculture clearly dominates the employment structure of rural areas. Some 27 per cent of the Polish population is still employed in agriculture. In Romania, where industrial collapse has pushed many people back on to the land, agriculture accounts for 42 per cent of total employment [*Weise et al., 2001: 37*]. A relatively large portion of this employment is on small family plots that are farmed for subsistence. Besides agriculture, the other major source of employment in many rural areas is resource extraction, especially of timber. While Western European societies have undergone similar, often painful, transitions from predominantly agrarian rural economies, the changes are being forced on rural societies in CEE states over a matter of years rather than decades or centuries.

At the same time, rural development and the form that it takes will have a critical impact on the environmental dimension of enlargement, and will challenge the EU's commitment to achieving sustainable development and halting biodiversity loss. The new member states contain the bulk of the continent's natural wealth, including the last great areas of wilderness and cultural landscapes [*Stanners et al., 1995*]. Preserving this natural heritage, for example through implementation of EU nature conservation legislation such as the Birds and Habitats Directives, will depend on at least the tacit support of local stakeholders, including farmers and communities – stakeholders for whom, under current circumstances, socio-economic development can be expected to be top priority, and whose support can scarcely be expected if they cannot make ends meet. The extensive semi-natural areas, including flowering meadows, grasslands and orchards that have been developed over centuries of human cultivation and comprise a large part of the biodiversity in CEE states, are already being threatened by changes in land use, including land abandonment and introduction of intensive agricultural practices [*Guttenstein and Torkler, 2002*]. At the same time, catastrophic flooding that has affected especially CEE countries and had an impact on Germany, Austria and other neighbouring countries has underlined the consequences – including, ultimately, socio-economic costs – of environmental degradation in the newest member states, not only for the countries themselves but also for Europe as a whole [*WWF DCP, 2002*].

The EU and Sustainable Rural Development

The sustainable development of rural areas is thus a central concern for the current enlargement, and will have far-reaching implications for the future of the continent as a whole. Such challenges, though unprecedented now in scale and scope, are not new to the EU, for which rural development has been a central concern for much of its history. The EU commits approximately 80 per cent of its budget to agriculture and regional development, most of it targeted at rural areas. Unfortunately, such lavish subsidies have not managed to solve persistent rural problems, including low income and high unemployment in lagging regions [*CEC, 2002*]. At the same time, rural areas in the existing EU member states have suffered a steady decline in biodiversity. In England, tree sparrows, once common in rural areas, have disappeared as their populations have plummeted 95 per cent over the past 30 years [*RSPB, 2002*]. In Ireland, water quality in the country's river systems has declined significantly since the country joined the Union [*Lucey and Doris, 2001*], while in Denmark there has been a precipitous decline in populations of wild plants and animals [*Wilhjelm Committee, 2001*].

Such figures have contributed to a growing realisation that a fundamentally different approach to rural development is required. An increasing number of voices are calling for an approach where the countryside is no longer seen narrowly as a factory for producing food, but as providing a multitude of functions, including recreation, work and living places, aesthetic values and environmental services, including water management and purification as well as ecological stability. Along with this has come a shift in emphasis from agriculture to much broader rural development that supports amenities and diversification of rural incomes while placing an emphasis on protecting and maintaining nature and environmental services, including valuing them as a significant asset for development [*Dissing, 2002; ten Brink, et al., 2002; MWTV, 2002*].

At the same time, while past thinking has focused on exogenous approaches to rural development, emphasising resources coming from outside to stimulate and support development, new approaches put more emphasis on endogenous approaches, developing existing local (especially human) resources [*Bryden, 2001; Terluin and Post, 2001; Terluin, 2002; OECD, 1999*]. In many cases, approaches emphasising large-scale initiatives, including construction of factories, amusement parks or motorways – are insensitive to the local context and risk doing as much harm as good. In fact, large-scale projects designed and implemented without appropriate consultation and consideration of needs of the local population can overwhelm local communities by stifling initiative, disrupting the traditional economic and social relations, as well as eroding cultural and natural values. This is especially relevant and important in the context of the fledgling civil societies of Central and Eastern Europe.

This changing approach to rural areas has been reflected in the development of EU policymaking over the past decade [*Dwyer et al., 2002*]. In addition to increasing concern over and commitment to the environment and achieving sustainable development, the EU has taken steps towards adopting a fundamentally new approach to rural areas. Since the mid-1990s, the European Commission, particularly EU Agriculture Commissioner Franz Fischler, has been among the leaders calling for fundamental rethinking of the EU's approach to rural areas, in light of not only deteriorating environmental quality but also the prospects for EU enlargement as well as pressures for more trade liberalisation. The Cork Declaration, signed by many of the participants of a major conference convened by Fischler in Cork, Ireland in 1996, is remarkable for its new vision for rural development policy:

> Sustainable rural development must be put at the top of the agenda of the European Union, and become the fundamental principle which underpins all rural policy in the immediate future and after enlargement. This aims at reversing rural out-migration, combating poverty, stimulating employment and equality of opportunity, and responding to growing requests for more quality, health, safety, personal development and leisure, and improving rural well-being. The need to preserve and improve the quality of the rural environment must be integrated into all Community policies that related to rural development [*CEC, 1996: 2*].

The Declaration calls for a rural development policy to support diversification of economic and social activity that focuses on providing the framework for self-sustaining private and community-based initiatives, and decentralising rural development policy as much as possible. 'The emphasis must be on participation and a "bottom-up" approach, which harnesses the creativity and solidarity of rural communities' the Declaration said. 'Rural development must be local and community-driven within a coherent European framework' [*CEC, 1996*].

EU Accession and Sustainable Rural Development

So far, though the direction of reforms within the Common Agricultural Policy towards sustainable approaches is clear, Fischler's ambitious plans have largely run aground on fierce opposition from a handful of member states. Incremental changes have been made, but the much touted rural development component, the so-called Second Pillar of the Common Agricultural Policy, still accounts for only ten per cent of the CAP's total funding. EU support for rural development in the accession countries has been somewhat different. Here the European Commission has been able to place relatively greater emphasis on rural development over agricultural support, and some important elements for sustainable development have been introduced. Nevertheless, the end result has been mixed – owing to confused policy objectives as well as insufficient commitment in both Brussels and especially the accession countries to sustainable rural development.

The EU has been a key factor in shaping rural development policy in the accession countries. Developing regional development plans, and the structures needed for their development and implementation, is a requirement for accession as well as for tapping the significant regional development funding that has been an important consideration for the accession countries. The long-term planning, careful monitoring and evaluation, and environmental standards, including requirements in some cases for environmental impact assessments, are new for the accession countries. For some of the accession countries, like the Czech Republic, accession to the EU has provided a strong push towards decentralisation, bringing decision making closer to the needs and concerns of rural populations with the introduction of a third, regional tier of government [*Beckmann, 1999a*].

EU accession also has been the driving force behind the very substantial investments in environment and rural development that are being undertaken – a total estimated cost of 80–110 billion Euro in terms of environmental costs alone [*CEC, 2001a*]. Poland alone faces an estimated bill between 22 and 43 billion Euro for coming into line with EU environmental legislation and standards. Only a small amount of this burden will be shouldered by the EU through various pre-accession funding instruments, with the lion's share covered by the accession countries themselves. Pressure and encouragement from the EU has also been a key factor in pushing more participative approaches to decision making in the accession countries – especially opening up committees for developing regional development plans to representatives of organisations (NGOs), or giving NGOs and the general public an opportunity to comment on various accession-related programmes and legislation such as the National and Regional Development Plans [*Stockiewicz et al., 2002*].

In addition, key pieces of legislation could have a substantial impact in the future, depending on how they are ultimately implemented. The Water Framework Directive, which EU member states and accession countries are now beginning to implement, calls for the achievement of 'good ecological status' in all river basins by the year 2015. The potentially revolutionary piece of legislation mandates the development of integrated river basin management involving all relevant stakeholders, thus promoting environmentally friendly land use, integrated development and participatory decision making. The Aarhus Convention, which European countries are now in the process of introducing, stipulates fundamental rights of citizens to have access to information and decision making regarding the environment. The EU's Habitats and Birds Directives could provide a powerful impulse for sustainable rural development in many areas. The so-called Natura 2000 network of nature conservation areas that the twin directives will establish will not only protect critical nature values (currently some 18 per cent of the territory of existing member states), but also provide financial support for sustainable activities of local stakeholders [*Jen et al., 2003*]. If designed and implemented appropriately, these legal structures have the potential to provide the foundation for an ecological infrastructure, the core element of sustainable development [*Dissing, 2003*].

A closer look at EU funding programmes, however, shows that such positive features do not paint the whole picture. The overall impact of accession on sustainable rural development in the candidate countries so far has been ambivalent at best, and even negative at worst. The so-called pre-accession funds (including PHARE, ISPA and SAPARD[1]) have been designed to prepare the candidate countries for accession to the EU. Logical enough – but also the heart of the problem, since in many and most respects, the EU's present way of doing things is fundamentally unsustainable, in both economic as well as biological terms.

The ISPA pre-accession fund provides investment for major infrastructure projects in transportation and environment – similar to the Cohesion Funds, which have been responsible for much of the environmental degradation that has been wrought in Spain, Portugal, Greece and Ireland over the past years [*Maier et al., 1999*]. Much of the environmental spending within the ISPA pre-accession fund has focused on supporting sewage and wastewater treatment projects. The emphasis, for largely administrative reasons, has been on big projects of 5 million Euro and upwards – a minimum limit that makes projects economical only for the largest of cities [*WWF EPO, 2002a*]. In response to complaints that the size limit failed to address the real challenges of water treatment in rural areas, the European Commission has agreed to permit communities to form associations in order to submit common projects. The result has been that small communities have come together to build traditional, over-dimensional sewage treatment plants, pumping their sewage for kilometres across fields to reach the treatment facility (interview with Miroslav Kundrata, Director of Environmental Partnership Foundation, Czech Republic, 17 May 2001). Left unconsidered are low-tech, low-cost and more sustainable solutions like biological (for example reed-bed) sewage treatment facilities that could be ideal solutions for the plethora of small rural communities that are the nub of the problem with wastewater throughout the region [*WWF EPO, 2002a*]. Unfortunately, the pressure to invest in big projects – regardless of the economic efficiency, let alone environmental sustainability of doing so – will only increase rather than decrease, since the relative 'absorption' of pre-accession funding will largely determine future levels of funding after accession [*Unterwurzacher, 2002*].

More disappointing than ISPA has been the SAPARD pre-accession fund, which has fallen well short of the (probably unrealistic) expectations that first accompanied its establishment. High initial expectations surrounded development of the programme, which, similar to the Rural Development Regulation, was at first expected to be innovative, promoting the long-term integrated and sustainable development of rural areas in the spirit of the Cork Declaration of 1996 [*Dwyer et al., 2002*]. However, throughout the region, implementation of the SAPARD programme has suffered long delays as accession countries have struggled with programming as well as the development and accreditation of administrative and control structures for the programme, which is partly designed to teach the accession countries how to use and manage future EU funds. Investments in infrastructure, processing and marketing take up by far the lion's share of investments [*Waliczky, 2002; Avis et al., 2001*]. Over the course of development of the SAPARD programme, increasing emphasis has been placed on centralised control and decision making within the programme – as preparation for management of future, much greater EU support – rather than the more decentralised approaches that had initially been suggested. Rather than presenting an opportunity to set the accession countries on a more sustainable path of rural development, SAPARD, like ISPA, has focused on preparing the newest member states to join the EU – and adopt the same, problematic policies that apply there [*Dwyer et al., 2002*]. Ironically, this has been happening at a time when a string of crises, from mad cows to sick pigs, have underlined the failure of existing EU agriculture and rural development policies, and existing member states have begun considering fundamental reforms to their policies in these areas.

However misguided the EU pre-accession programmes have been with respect to fostering sustainable rural development, it is unfair to pin all of the blame on the EU. Much responsibility rests on the candidate countries themselves. Programming for the pre-accession funds has been undertaken by the accession country governments, within general guidelines and with the approval of the European Commission. In terms of actual project development, the European Commission has depended on submissions made by the candidate countries. EU funding guidelines emphasise sustainability, sensitivity to social and cultural concerns, subsidiarity, and the need for public participation in the design and implementation of rural development projects. Yet Central and Eastern European policymakers continue to favour top-down, large-scale projects of the kind developed under the previous communist regimes. Indeed, there is a common adage in many CEE countries that refers to projects that have been kept for decades 'in the drawer', and are now being dusted off, embellished with the right words and phrases regarding sustainable development and public participation, and submitted for EU funding. Many of the transportation and flood control projects being undertaken were projected years ago by communist planners, but lay dormant owing to lack of funding. The references to sustainability and participation on these projects are window-dressing, and do not represent any fundamental rethinking of present approaches to rural development or the environment (interview with Juraj Flamik, Coordinator for Czech Greenways programme, February 2003).

What is needed to make progress towards truly sustainable development is a deeper, fundamental shift in paradigm. But the accession process is doing relatively little to change this paradigm. Most activity and investment is instrumental, targeted at implementation of the *acquis communautaire*, the body of legislation and standards that must be adopted by each accession country in order to join the EU, while relatively little investment is made in preparing the necessary foundations, including social structures and development of human resources, that is needed for long-term sustainable development [*Rotbergs and Schwartz, 2002*]. The PHARE programme, which is targeted at preparing institutions for enlargement, has invested in improving training and supporting public administration – certainly a clear need. It has also provided some support for civil society development, though its effectiveness has been limited by red tape and inflexibility. However valuable, such support has not been sufficient. Development of human resources and social infrastructure has lagged, increasing the risk that investments in physical infrastructure do more harm than good. Both the ISPA and SAPARD programmes have missed a significant opportunity to adopt a 'bottom-up' approach and thus build social infrastructure while investing in physical development.

The development of the SAPARD programme in Hungary provides perhaps the starkest example of the dynamics of policymaking within the accession countries, as well as the influence that the EU has had. It also suggests the extent of the missed opportunities for encouraging the initiative and involvement of local stakeholders and mobilising local resources.

The Rural Development Unit of the Hungarian Ministry of Agriculture originally intended to base SAPARD on local participation and integrated programming. To catalyse this process, the Unit launched a programme for voluntary rural micro-regions. More than 200 associations applied for the programme, covering 2,500 settlements and almost the entire area of the country. Micro-regions were encouraged to create broad partnerships and develop long-term, integrated development programmes based on local resources and external aid, sustainability and local participation. The 'micro-regional SAPARD' generated much enthusiasm, mobilised human capital and local resources, and raised high expectations throughout rural areas. Networks and partnerships were set up, local actors learned to deal with 'EU type' vocabulary and procedures [*Dwyer et al., 2002: 46*].

This whole process was largely overtaken and by-passed by the eventual outcome of official SAPARD planning and implementation. Under pressure partly from the European Commission, an at first parallel process for developing the official SAPARD plan focused on developing a centrally organised, top-down structure for administering the programme. In this process, traditional agricultural interests were the most influential. As a result, in its final outcome, the SAPARD programme is focused on processing and marketing, while the groundswell of micro-regions that have been mobilised have been left deeply frustrated [*Dwyer et al., 2002; Nemes, 2002*].

NGOs and Sustainable Rural Development

In promoting large-scale, exogenous development projects, policymakers in Central and Eastern Europe often argue that there are no feasible alternatives. A host of exciting development initiatives across rural areas of Central and Eastern Europe show that this is not the case. They present not only a practical basis for further efforts but also promising models for addressing the deep-seated problems and achieving sustainable development in rural areas of the region. In the next section we examine the grassroots initiatives for sustainable rural development in the White Carpathians and the work of an intermediary funding agency, the Environmental Partnership for Central Europe foundations. These examples illustrate some of the sustainable rural development activities that are taking place, demonstrate the role that NGOs are playing in these processes, and offer insight into methods that can be used more broadly to promote such initiatives in the particular conditions of the post-communist societies of Central and Eastern Europe.

Grassroots Initiatives for Sustainable Rural Development in the White Carpathians

Stretched along the Czech–Slovak border at the western edge of the Carpathian mountain range, the White Carpathians ('Bile Karpaty') are a rolling, patchwork landscape of forests, orchards, fields and the brilliant flowering meadows that are the region's hallmark. Numerous environmental organisations have been active on both sides of the Czech–Slovak border in promoting sustainable development [*Vesely, 2002; Carmin et al., 2003*]. Their initiatives in the region range from organic farming, development and marketing of local crafts and products, 'soft' and agro-tourism, land trusts, activities focused on maintaining and promoting cultural and natural heritage of the area, and a variety of community development projects. As a result of the efforts of CSOP Veronica, CSOP Hostetin (both chapters of the Czech Union for Nature Conservation, 'CSOP'), and others working in cooperation with the local government, the small village of Hostetin is quickly becoming a model for rural sustainability in the region. The community of 207 people now boasts a biological, reed-bed sewage treatment facility, biomass heating plant, a small juicing factory that produces high-quality organic juice and supports the cultivation of local varieties of fruit, do-it-yourself solar collectors on a number of local homes, and, soon, a centre for rural development, housed in what should be one of the first low-energy buildings in the Czech Republic [*Vesely, 2002*].

A number of the environmental groups have come together with local farmers and communities to establish the Traditions of the White Carpathians association in order to promote and market local products and handicrafts, including dried fruit and the natural apple juice that is produced in Hostetin. Both products are now marketed throughout the Czech Republic, the juice through the Carrefour supermarket chain, though the long-term aim is to sell primarily to a regional market. While the products contribute to the local economy as well as promote the region and help foster a sense of regional identity, the original motivation for their development was environmental. Production and marketing of the juice and dried fruit was the logical next step in long efforts by environmental groups to preserve the rich genofund of native fruit varieties that have developed in the region over centuries of cultivation. The environmentalists are counting that the relatively high prices offered for local fruit will encourage area residents to care for the traditional orchards in the region.

Taken together, these initiatives and others are creating an alternative – and increasingly very practical – vision for the sustainable development of the White Carpathians region, which has been bifurcated by a national border since the division of Czechoslovakia in 1993. A remarkable feature of these initiatives is the leading role played by NGOs. Organisations like CSOP have played key roles as visionaries, catalysts, facilitators, supporters and active intermediaries in the sustainable development of these areas. In each case, it has been NGOs that have provided the long-term vision for sustainable development in the respective areas – visions that have presented an alternative to dominant paradigms, and to which the NGOs have gradually won over local communities and leaders. Vision alone has not been enough; in fact, the practical work of turning the vision into reality has been critical to securing the commitment of local communities and leaders. In a number of cases, as with the establishment of information technology and information centres in communities on the Slovak side of the White Carpathians, NGOs themselves have initiated many of the activities, and transferred them to local stakeholders once they are up and running.

The NGOs have also played an essential role in encouraging and facilitating the initiatives of local people themselves through small and matching grants and through co-financing arrangements as well as in developing partnership between organisations and among sectors. The St Nicholas Day celebration initiated by CSOP Kosenka draws together a growing circle of individuals and organisations, from local government to businesses and a wide variety of cultural and folk groups. In the southern part of the White Carpathians, CSOP Bile Karpaty works hand in hand with the administration of the White Carpathians Protected Landscape Area as well as with local communities, farmers and landowners in landscape stewardship arrangements to maintain the area's valuable flowering meadows [*Carmin et al., 2003*]. Although the NGOs themselves are usually run on shoestring budgets, their extensive foreign contacts and fundraising expertise have been crucial for realising many of the projects. In Hostetin, for example, CSOP Veronica was key in securing support from the Dutch and Czech governments (through a Joint Implementation programme) for the biomass heating plant, from the county and national governments for the reed-bed sewage treatment facility, or Luxembourg support for construction of the juicing plant [*Vesely, 2002*]. The NGOs are also now among those best prepared to tap EU and national funds for rural development, and in many cases are actively helping communities to develop projects and submit proposals for funding.

NGOs have been important not only in introducing the vision of sustainable development, but also practical knowledge and technologies needed to make these visions reality [*Beckmann, 2000*]. A common feature of the projects initiated by NGOs is their integrated nature; a variety of individual activities contribute small parts to increasingly integrated approaches to regional development and environmental protection.

Activities are infused with a strong sense of local and regional identity, rooted in both cultural and natural heritage of the region. There is also a clear sense that sustainable solutions require local support and that the ultimate bottom line is securing local livelihoods. The organisations are looking for ways to harness the free market to produce social, cultural and ecological benefits. Together, these components come together in an overall vision of integrated rural development.

The sustainable development projects in the White Carpathians are by no means the only such initiatives that are being undertaken in Central and Eastern Europe. Indeed, similar examples can be found across the region, from the Vainemeri area of Estonia and the Liptov region of Slovakia, to the Ramet Gorge in central Romania (for examples, see EPCE [*2003*]; EPCE Romania [*2002*]; Estonian Fund for Nature [*2003*]). Though there are growing examples of local authorities who are leading efforts in this direction (such as the area of Csotkeny in central Hungary) [*Townsend, 2002*], it is still NGOs that are providing the main initiative, leadership and coordination for sustainable rural development across the region.

Many of these NGOs have come from an environmental background. The legacy of communism, a regime in which damage to nature was reflected in the damage wrought on society as a whole, contributed to a strong conviction among many environmentalists in CEE countries of the intimate link between social change and environmental reform. After the fall of the Iron Curtain, the link was promoted by trends in global thinking, especially the Agenda 21 declaration that was signed by world leaders at the first World Summit on Sustainable Development in 1992, as well as by foreign foundations, governments and non-profit organisations that became active in CEE countries. Creating a healthy environment required creating a healthy (that is, democratic) society, built from the grassroots. This has conditioned the engagement in development of local communities, and has shaped the strong streak of civic engagement that is characteristic of many environmental NGOs in the region. The NGOs have been in a good position to be effective. They have been able to mobilise strong networks of foreign contacts, which have helped with new ideas and technology as well as access to resources. At the same time, dependent on neither the vagaries of the ballot box nor the profit motive, NGOs have shown the patience necessary particularly in the still fluid conditions of CEE societies to pursue their long-term vision.

Nurturing Grassroots Initiatives for Sustainable Rural Development

The initiatives that are taking place in the White Carpathians suggest that small, interrelated initiatives can add up to greater change at a regional level. But how can such initiatives be encouraged, magnified and multiplied? Helpful of course are legal and programmatic frameworks that are conducive to and encourage initiative and involvement of local stakeholders. These have, arguably, not been a main factor in the development of many of the grassroots initiatives throughout the region – certainly not in the case of the White Carpathians, where social entrepreneurs have had to contend with centralised decision making, distant from the concerns of local populations, and where for much of the period national governments have been indifferent – even actively opposed – to grassroots initiatives and environmental concerns. As discussed earlier, accession to the EU is changing these conditions, but insufficiently. In terms of financial support, a not insignificant amount of EU and government funding has gone towards projects in the White Carpathians. But this has gone overwhelmingly towards physical infrastructure, including the erection of the biomass heating plant or sewage treatment facility in the village of Hostetin. Support for development of the social infrastructure in the region, including operational costs of NGOs – the vital force that have actually made these projects happen – has been negligible.

The main support and inspiration for bottom-up sustainable development initiatives has come not from governments or the EU, but from private foundations, first and foremost from the United States (for example, C. S. Mott Foundation, Open Society Fund, Rockefeller Brothers Fund, German Marshall Fund of the United States), but also government aid agencies from a handful of countries, especially the United States (for example, USAID), The Netherlands (for example, MATRA KAP fund) and United Kingdom (especially Know How Fund). These sources have provided not only much of the funding for the operations of individual groups, but also the capacity building and networking support that has helped develop inspiration, know-how and ideas. The considerable support in nominal terms that has been distributed for NGO activities largely through the EU's PHARE programme has been less effective in actual practice, as it is relatively inflexible and comes wrapped in red tape. In contrast to much of the support from private foundations, which has gone (for example) towards core support of organisations, EU funding for NGOs, for example through the PHARE-funded Access programme, has been instrumental, focused on tackling specific tasks related to the adoption and implementation of the requirements for accession, and has strongly favoured large, established organisations. Such an approach makes sense in terms of filling the specific requirements of the *acquis*. But it misses the deeper need for social transformation in the region, which is best served by bottom-up initiatives [*Rotbergs and Schwartz, 2002*].

One of the most effective vehicles in CEE countries for empowering rural stakeholders and nurturing bottom-up sustainable development initiatives is the Environmental Partnership for Central Europe (EPCE) foundations. Originally established by a group of mostly American private foundations (principally the C. S. Mott Foundation, Rockefeller Brothers Fund, and German Marshall Fund of the United States), the consortium now includes independent foundations in the Czech Republic, Slovakia, Poland, Hungary, Romania (from 2000), and, starting in September 2003, also in Bulgaria. The hallmark of the foundations' work is their active approach to grantmaking, providing small, targeted investments (averaging about US$2,800) that can encourage and facilitate, yet not spoil or overwhelm grassroots initiatives. These are combined with other forms of active support, including technical assistance and networking [*Tolles and Beckmann, 2000*].

Special programmes of the foundations have focused on fostering community-based initiatives and partnership action with regard to particular issues, including landscape stewardship (for example, establishment of public–private land trusts); the creation of greenways for tourism, recreation and local development; or the formation of local partnerships for sustainable development comprising local governments, private businesses and NGOs [*Beckmann, 2000; Tolles, 2000; EPCE Poland, 2003*]. Similar to the NGOs involved in rural development at the local level, at a higher level these foundations are working as visionaries, catalysts, intermediaries, initiators and social entrepreneurs – providing not only vision, but also initiative, access to ideas, know-how and capital, essential ingredients for sustainable development in the region. Over more than a decade, the Environmental Partnership for Central Europe consortium and other foundations and organisations have made a substantial contribution to the development of a social infrastructure in many rural areas in CEE countries that can now provide the foundations for effective use of larger rural development support.

It is worth putting the relative costs of this investment in social infrastructure into perspective. As an example, over a period of nine years between their establishment in 1991 and 2000, the EPCE foundations in the Czech Republic, Poland, Slovakia and Hungary supported over 1,300 citizens groups and communities in more than 3,000 initiatives. These initiatives have planted more than 132,600 trees, employed 1,341 people (full and part-time, over the course of a year), mobilised and involved over 15,500 volunteers, and created over 2,620 km of cycling paths and nature trails. The total budget for the foundations over this period was approximately US$15.4 million [*Tolles and Beckmann, 2000*]. Even taking into account significant differences in purchasing power and inflation over the period, it is still an astonishingly small amount – especially when compared with other investments such as the cost of building a single kilometre of highway, which in the Czech Republic in the mid-1990s totalled about US$3.3 million [*Spurny, 2003*].

Conclusions

More than a decade since the fall of the Iron Curtain, the post-communist societies of Central and Eastern Europe are still undergoing transformation. Considerable changes have already taken place, including the establishment of democratic systems of government and a free market economy, as well as greater openness to the global market. The ongoing process of accession to the EU is further supporting and accelerating these changes, and adding a new emphasis in public policy on the environment, sustainability, and public access to information and decision making. Yet in many parts of the region, and especially in rural areas, there still has not been a full transformation in institutions, social structures, human resources and culture, mores and values – development of a social infrastructure – that is needed to under-gird these changes. While dynamic urban centres like Prague and Krakow have adapted relatively well to the new realities, most of the rural areas of the region are still poorly prepared to face the new challenges and take full advantage of the opportunities.

In these circumstances, NGOs and foundations have been playing important roles as visionaries, catalysts and intermediaries, providing services that in other societies and contexts are often provided by public authorities or local business people. The NGOs have been remarkable not only as motors for local development, but also the form of development this has taken – integrated across different sectors of activity, linking economic development with social development, protection and promotion of cultural and natural heritage. Their efforts have focused on animating and empowering local actors and mobilising local resources. They have fostered the development of social networks and structures where local stakeholders can develop ideas, exchange information and pool scarce resources. Through such structures, local communities become empowered, can compete for national and international support as well as exploit new opportunities offered by an increasingly global market. The relative success, especially compared with the modest means available, of initiatives like those in the White Carpathians underline the effectiveness of this approach not only for promoting socio-economic development, but also ensuring the long-term sustainability of rural areas – and thus addressing the significant challenges faced by these areas as well as preparing them for accession to the EU and an increasingly global market.

The EU and national governments would do well to learn from and build on such initiatives. The record of the EU and national governments on addressing the deep-seated problems of rural areas has been mixed, too often focused on simply continuing existing structures and approaches. In both the accession countries as well as existing EU member states, public policies have neither been able to address the persistent problems of rural areas nor provide hope for the sustainable future of these areas. Relatively few initiatives have really struck out in a new and promising direction – the chief exception perhaps being the EU's LEADER programme, which is focused on fostering locally based rural development initiatives [*CEC, 2003b*].

This could change. The agreement reached between the EU and the accession countries for phasing in agriculture and rural development support has relatively accented rural development support. And discussions of reform to the Common Agricultural Policy could lead to further emphasis of the 'Second Pillar' of rural development. Fortunately, if and when such changes come, they will not need to start from scratch, and can learn from, be inspired by, and practically build on the many existing initiatives already developed by NGOs in Central and Eastern Europe.

NOTE

1. The EU has developed three so-called pre-accession funds to help the CEE accession countries prepare for membership in the EU. The Poland/Hungary Aid for the Reconstruction of the Economy (PHARE), which now applies to all accession countries in CEE, is designed to support the adoption and application of EU laws and standards as well as reinforce administrative and judicial capacity. The Instrument for Structural Policies for Pre-Accession (ISPA) supports investments in large infrastructure, evenly split between transportation and environment. The Special Accession Programme for Agriculture and Rural Development (SAPARD) focuses on preparing agriculture and rural areas in the accession countries for accession [*McGiffen, 2001*].

REFERENCES

Avis, Charlie, A. Beckmann, M. Cierna, C. Gheorghe, R. Elod, and L. Potozky (2001), *Results of an Independent NGO Evaluation of SAPARD: National Plans and Processes in Czech Republic, Hungary, Romania and Slovakia* (Summary Report), Vienna: WWF Danube Carpathian Programme (Carpathian Ecoregion Initiative).

Beckmann, Andreas (1999), 'The Big Yawn: Decentralization in the Czech Republic', *Central Europe Review (www.cer.cz)*, Vol.1, No.13 (20 Sept. 1999).

Beckmann, Andreas (2000), *Caring for the Land: A Decade of Promoting Landscape Stewardship in Central Europe*, Stare Mesto: NP Agentura.

Bryden, John M. (2001), 'DORA: Dynamics of Rural Areas, the International Comparison', Arkleton Centre for Rural Development Research, University of Aberdeen. Available at http://www.abdn.ac.uk/arkleton/doradocs/icfinal.pdf (10 Nov. 2002).

Carmin, JoAnn, Barbara Hicks and Andreas Beckmann (2003), 'Leveraging Local Action: Grassroots Initiatives and Transnational Collaboration in the Formation of the White Carpathian Euroregion', *International Sociology*, Vol.18, No.4. pp.703–25.

CEC (Commission of the European Communities) (1996), 'The Cork Declaration – A Living Countryside', Declaration from European Conference on Rural Development at Cork, Ireland, 7–9 Nov. 1996.

CEC (Commission of the European Communities) (2001a), 'Communication from the Commission: The Challenge of Environmental Financing in the Candidate Countries', COM(2001)304 final.

CEC (Commission of the European Communities) (2002), 'Communication from the Commission: First Progress Report on Economic and Social Cohesion', COM(2002)46 final.

CEC (Commission of the European Communities) (2003b), website for the LEADER programme. Available at http://www.rural-europe.aeidl.be/rural-en/index.html (13 Feb. 2003).

Dissing, Henrik (2002), 'Elaboration of a Business Model for Sustainable Rural Development in Areas with Abundance of Nature', paper presented at European Commission conference, 'Building Bridges, the Importance of Implementation, Integration and Information in New Environmental Governance', Aalborg, Denmark, 19–20 Sept. 2002.

Dissing, Henrik (2003), 'The Water Framework Directive could be the key. Sustainable Water Management in the Baltic Sea Region in an Enlarged EU', *Ökologisches Wirtschaften*, 1/2003, Vol.19, 2003/1, pp.22–23.

Dwyer, Janet and D. Baldock, G. Beaufoy, H. Bennett, P. Lowe and N. Ward (2002), *Europe's Rural Futures – The Nature of Rural Development II*, London: Institute of European Environmental Policy.

EPCE (Environmental Partnership for Central Europe Consortium) (2003), unpublished manuscript for book on rural development case studies from the Czech Republic, Slovakia, Poland and Hungary (planned publication in Oct. 2004).

EPCE Poland (Environmental Partnership for Central Europe – Poland) (2003), Information downloaded from the website. Available at http://www.epce.org.pl/ang_1/glowna_ramka_ang.htm (4 Mar. 2003).

EPCE Romania (Environmental Partnership for Central Europe – Romania) (2002), *The Preservation of Ramet Gorge*, information leaflet, Mircurea Ciuc, Romania: EPCE Romania.

Estonian Fund for Nature (2003), Information on Väinameri project presented on website of the Estonian Fund for Nature. Available at www.elfond.ee (8 Aug. 2003).

Eurostat (2003), *Towards an Enlarged Union: Key Indicators on Member States and Candidate Countries*, Brochure Series: Theme 1 General Statistics. Available at: http://europa.eu.int/comm/enlargement/docs/pdf/eurostatapril2003.pdf (December 2003).

European Union (1997), Treaty of Amsterdam, *Official Journal of the European Union*, C340, 10/11/1997: 86[RTF bookmark end: BM0].

Guttenstein, Elizabeth and Peter Torkler (2002), *Enlargement and Agriculture: Enriching Europe, Impoverishing our Rural Environment?* (position paper), Berlin: WWF Germany.

Jen, Sandra and E. Townsend, eds. (2003), *Progress on Preparation for Natura 2000 in Future EU Member States*, Brussels: WWF European Policy Office.

Lucey, John and Yvonne Doris (2001), *Biodiversity in Ireland: A Review of Habitats and Species*, Dublin: Irish Environmental Protection Agency.

McGiffen, Steven P. (2001), *The European Union: A Critical Guide*, London: Pluto Press.

Maier, Petra, J. Kowalzig, M. Rocholl, S. Rostock, P. Schepelmann (1999), *Billions for Sustainability?: EU Regional Policy and Accession*, Bonn: BUND (Friends of the Earth Germany).

MWTV (Ministerium für Wirtschaft, Technologie und Verkehr des Landes Schleswig-Holstein) (2002) , *Wirtschftsbericht 2002* (Economic Report for the Land of Schleswig-Holstein for 2002)'. Keil: Landesregierung Schleswig-Holstein.

Nemes, Gusztav (2002), Panelist on sustainable rural development in Central and Eastern Europe, remarks made at the conference organised by WWF Austria, 'Enlargement Enriches the EU', Vienna City Hall, 26 Sept. 2002.

OECD (1999), *Best Practices in Local Development*, Local Economic and Employment Development Programme, Paris: OECD.

Rotbergs, Ugis and Katrina Schwartz (2002), 'Enlargement and Environment: The Ongoing Experience of Latvia', paper presented at the conference organised by WWF Austria, 'Enlargement Enriches the EU', Vienna City Hall, 26 Sept. 2002.

RSPB (Royal Society for the Protection of Birds) (2002), *The State of the UK's Birds 2001*, Sandy: RSPB.

Spurny, Jaroslav (2003), 'Tajemstvi sifry D47 (The secret of D47)', *RESPEKT*, Vol.15, No.5, pp.13–15.

Stanners, David and Philippe Bourdau (eds.) (1995), *Europe's Environment: The Dobris Assessment*, Copenhagen: European Environmental Agency.

Stockiewicz, Magda, I. Malbasic and S. von Pohl (2002), *Billions for Sustainability?: Lessons Learned from the Use of Pre-accession Funds*, Brussels: Friends of the Earth Europe/CEE Bankwatch.

ten Brink, Patrick, Claire Monkhouse and Saskia Richartz (2002), *Promoting the Socio-Economic Benefits of Natura 2000*, London: Institute of European Environmental Policy (IEEP).

Terluin, Ida J. (2002), 'Rural Regions in Europe', doctoral thesis, Groningen: Rijskuniversiteit Groningen.

Terluin, Ida J. and Jaap Post (2001), 'Key Messages on Employment Dynamics in Leading and Lagging Rural Regions of the EU', paper for the workshop organised by the International Institute for Applied Analysis (IIASA), 'European Rural Development: Problems, Chances, Research Needs', Laxenburg, Austria, 7–9 May 2001.

Tolles, Robert and Andreas Beckmann (2000), *A Decade of Nurturing the Grassroots: The Environmental Partnership for Central Europe, 1991–2000*, Brno: Czech Environmental Partnership Foundation.

Townsend, Ellen (2002), 'Cötkeny – Hungary: A Journey into the Future Integrating Nature, Society and Local Economies' (case study/brochure), Brussels: WWF European Policy Office.

Unterwurzacher, Erich (DG Regio, European Commission) (2002), 'ISPA Review and Changes to Cohesion and Structural Funds', paper presented to NGO Dialogue No.7, Brussels, 19 Nov. 2002.

Vesely, Michal (2002), 'Model Projects for Sustainable Development in the White Carpathians', Master's thesis, Brno: Masaryk University (unpublished, in Czech).

Waliczky, Zoltan (BirdLife International) (2002), 'An Environmental Assessment of SAPARD Rural Development Plans', paper presented at conference organised by Friends of the Earth Europe, 'Billions for Sustainability?: Lessons Learned From the use of Pre-accession Funds', Brussels, 25 Nov. 2002.

Weise, Christian, J. Bachtler, R. Downes, I. McMaster and K. Toepel (2001), *The Impact of EU Enlargement on Cohesion*, Berlin and Glasgow: German Institute for Economic Research (DIW) and European Policies Research Centre (EPRC) (commissioned by European Commission as background study for the Second Cohesion Report).

Wilhjelm Committee (2001), *Danish Nature – Status, Trends and Recommendations for Future Biodiversity Policies*, Copenhagen: Danish Ministry of Environment.

WWF DCP (WWF Danube Carpathian Programme) (2002), *Managing Floods in Europe: The Answers Already Exist* (briefing paper), Vienna: WWF Danube Carpathian Programme.

WWF EPO (WWF European Policy Office) (2002a), *WWF Briefing Paper on ISPA*, Brussels: WWF European Policy Office.

Nuclear Power and EU Enlargement: The Case of Temelín

REGINA AXELROD

The controversy over the Temelín nuclear power plant (TNPP) in the Czech Republic was transformed from a domestic issue to an international one by the year 2001. Besides providing an opportunity to examine domestic politics and administrative practices in the Czech Republic, the Temelín case raised questions about the future of nuclear power in Central and Eastern European (CEE) countries – and the rest of Europe. What began as a bureaucratic decision in the 1980s by the communist government of Czechoslovakia to build a nuclear power plant became by the late 1990s a major controversy affecting the enlargement of the EU and a nightmare for the foreign relations of the Czech Republic. After providing a general introduction to nuclear power in Central and Eastern Europe, this contribution chronicles the origins and development of the TNPP.

By examining changes over time, it becomes evident that Temelín is deeply connected to the resurgence of the anti-nuclear movement in Europe, the process of enlargement of the European Union (EU), the integration of environment and energy policy consistent with the EU's 6th Environmental Action Programme, and bilateral relations between the Czech Republic and its neighbours. The Temelín case occurring during the EU enlargement process, offered an opportunity for the EU to play an unusual role: that of moderator in a bilateral dispute between a member and non-member state.

Nuclear Power in Central and Eastern Europe

In the wake of the 1986 Chernobyl disaster and the dramatic political changes in the communist systems of the CEE and former Soviet regions, the safety of nuclear power facilities remained an important international issue in newly constituted post-communist states [*Andonova, 2002; Dawson, 1996*]. In 1992, the G-7 countries (Canada, France, Germany, Italy, Japan, the UK and the US) agreed that Russian-designed nuclear power plants should be closed owing to safety concerns, and that financial assistance would be given to replace nuclear power with renewable and alternative energy sources [*GAO, 1994*]. However, CEE governments and their nuclear industries wanted to keep plants open to prevent them from losing their investments. Consequently, the policy changed from closing plants to upgrading them, giving life to the nuclear industry in the form of contracts for equipment, instrumentation and control systems (I&C), and nuclear waste storage facilities, (which are interim not permanent) [*Woodard, 1995*].[1] While Western Europe (particularly France and Belgium) had excess electricity to sell and the nuclear industry was anxious to find new markets, particularly in CEE countries and Asia, the policy to upgrade Russian-designed plants established a vast new market benefiting suppliers of nuclear technology, particularly US and European nuclear engineering companies. In fact, the ability of Western European and North American governments to achieve closure of Soviet/Russian-designed nuclear power plants across CEE and former Soviet regions proved quite limited [*Dawson, 1996; Chandler, 2000; Gutner, 2002*]. Furthermore, international assistance to upgrade nuclear power plant safety across these regions appears to have extended the working lifetimes of many of these facilities, breathing new life into the nuclear industry across CEE and former Soviet states [*Darst, 2002*].

The Origins of Temelín

As one of the nuclear power plants initially slated to be closed and then revitalised through a change in policy, Temelín represented a test for nuclear power interests across Europe. During the communist era, Czechoslovakia experienced high energy intensity, low energy prices, and inefficient energy production and electricity transmission, all of which distorted the economy. Because Czech heavy industry and chemical production required a reliable supply of electricity, nuclear power seemed to be a viable alternative that was consistent with the Stalinist model of building large projects. Temelín is located in the southern part of the Czech Republic, near the city of Ceske Budejovice, approximately 80 kilometres from the Austrian border. The decision for construction was approved in 1978 and construction began in 1986.

A review of Temelín's design after the 1986 accident and fire at the Chernobyl nuclear power plant resulted in construction being halted. In 1992 without adequate information on electric supply and demand, an absence of public debate on nuclear power, and uncertain government leadership, the decision about construction was left to the new government of Prime Minister Vaclav Klaus, which favoured completion of the TNPP.[2] Studies by the International Atomic Energy Agency (IAEA) found flaws in the design of Temelín, and recommended replacement of the I&C systems. There were also questions regarding the use of Russian fuel as well as the fuel cycle itself, which contributed to higher levels of radioactive waste than Western designs. After a controversial and questionable bidding process, in 1993, Westinghouse was awarded a contract to graft Western technology on to the Russian-designed reactors.

Enter Austria

The Austrian position towards Temelín is influenced by its proximity to the plant and the fact that it is a non-nuclear state. In 1978, by plebiscite, Austrians agreed to close their one completed nuclear plant, Zwenterdorf. Consequently, in the early 1990s, when the contract with Westinghouse to upgrade Temelín was being considered, Austrian officials began lobbying against the TNPP in the US Congress. Similarly, the Austrian state later opposed the completion of the Slovakian Mohovice nuclear power plant in 1998.

By 2000, the Austrian position was complicated because of the nature of its coalition government. The far right Freedom Party (FPOe), headed by Jorg Haider, was vehemently opposed to Temelín, as was the Austrian Vice-Chancellor Susanne Reiss-Passer (FPOe), who compared Temelín to Chernobyl because of unpredictable risks associated with nuclear power plants (*Czech National Newswire (CTK)*, 26 April 2001). If Haider pulled out of the coalition shared with Chancellor Wolfgang Schussel's centre-right People's Party, the government could fall. All four political parties opposed Temelín, but it was the populist and xenophobic stance of the FPOe that threatened to prevent Czech accession to the EU. In addition, Austria is a federal state with provinces that have their own governments and legislatures. Both Upper Austria and Lower Austria have taken independent actions in efforts to influence the federal government and working with Austrian and international environmental NGOs opposed to TNPP. Upper Austrian Greens wanted direct contact with Czech officials and argued that Temelín should be closed pending a new environmental impact assessment (EIA).

Austria's strategy for opposing Temelín was to 'widen the scope of conflict' [*Schattschneider, 1960*] to involve other European states and international NGOs and to provide information to various publics. This strategy also involved launching a campaign against nuclear power in Eastern and Western Europe – making the issue greater than Temelín – a position of the Social Democrats (*CTK*, 19 September 2001). Chancellor Schussel agreed, stating that 'Europeanising Temelín will be the only way leading to EU standards for nuclear power stations' (*CTK*, 3 September 2001). In September 2000, the Austrian Parliament approved a resolution asking their government to block Czech entry into the EU because of

Temelín. Chancellor Schussel demanded that Temelín comply with safety standards valid in EU states. The problem here was that there exists no EU competency for nuclear power plant regulation, probably because a number of the nuclear states, including France and the United Kingdom (UK), are wary of opening a Pandora's box of regulatory debates. In fact, EU member states (and publics) remain quite divided on nuclear power issues. Seven of the 15 member states have nuclear power plants, and eight of the 12 candidate states are nuclear. On the other hand, countries such as Austria have totally banned nuclear power while Sweden and Germany are officially engaged in phasing out their nuclear power facilities. As a result, there is a lack of agreement among the 15 member states about both the future of nuclear power in the EU as well as standards of safety. Austria wanted criteria to be developed and applied to all EU nuclear power plants.

In October 2000, when nuclear fuel was activated in the first Temelín reactor, Austria moved to widen the controversy to Brussels. Austrian officials argued that states should have a role in protecting their citizens from an environmental disaster originating in another state. It was a position the EU could take seriously. Yet, Austria had no legitimate political role in the launching of Temelín or receiving assurances that it would be safe. Although there is another Czech nuclear power plant at Dukovany, it was easier to oppose Temelín than advocate the closing of an existing nuclear plant.

In the autumn of 2000, anti-Temelín forces set up blockades on the borders between the Czech Republic and Austria to increase public attention on the issue. The Czech reaction was that the blockades impinged on trade and free movement of persons – the protests themselves were not the problem. The FPOe lobbied hard to get the government to withhold approval of the Czech energy chapter, threatening to oust Chancellor Schussel. The Austrian Social Democratic Party (SPOe) disagreed, arguing that neighbours should not be held hostage over nuclear power safety issues and suggested that Austria find allies in the EU interested in seeking unified safety standards for the entire EU.

Austria soon changed its strategy from demanding the closure of Temelín, to blocking the closing of the Czech energy chapter in the accession negotiations – the FPOe position. This move could have jeopardised the entire accession process, since a veto of any of the 31 chapters by even a single EU member state would prevent accession to the EU. When Czech officials decided to go ahead with the completion of Temelín, they never thought the issue would rise to the level of potentially blocking Czech accession to the EU. The veto of one state could do so, which is what Austrian officials were threatening.

A Unique Role for the EU

NGOs engaged in anti-Temelín activity in Germany and in Austria, while political parties and regional governments became increasingly vocal. Owing to the impact of highway blockades between Austria and the Czech Republic and the emergence of the issue of nuclear safety as part of the accession negotiations, the EU became an important player. Mediating between two states that do not have equal status – one a member state and the other a candidate state – was a new role for the EU. Questions surfaced regarding whether it was, in fact, an appropriate role for the EU or whether the matter should have been left to bilateral resolution between the Czech Republic and Austria. In reality, however, bilateral negotiations were not proving successful even though the foreign ministers of the two countries seemed to share the same perspective.

At the request of the Czech foreign minister, Jan Kavan, the Commission offered its good offices to act as mediator at the end of 2000, when the Austrian blockades caused heightened diplomatic tensions. It was becoming increasingly difficult for the Austrian government to control the emotional demonstrations. Both Austria and the Czech Republic agreed to the mediation. Diplomatic contacts between the governments increased and it was agreed that the two heads of state would meet in December 2000. The result was the Melk Agreement, the result of many hours of tedious negotiation (Melk, Austria is where the agreement was signed). The Czech Republic agreed to an EIA with EU participation. Austria said it would cease threatening to block the closing of the energy and environmental chapters of the *acquis communautaire* and continue negotiations on both. It agreed to protect the borders from further blockades. As an early warning system for extraordinary events, a hotline was established from Temelín to the Austrian Federal Atom Centre at the Interior Ministry to supply updated studies on breakdowns and uncontrolled release of radioactivity. However, NGOs opposing Temelín were shocked when EU Enlargement Commissioner Gunter Verheugen suggested that Temelín would 'probably be the safest nuclear plant in Europe' (*Prague Post*, 29 November 2000).

Enlargement negotiations provided an opportunity to focus on nuclear power safety. The December 2000 Melk Process was undertaken specifically to examine nuclear safety issues and facilitate an exchange of information about Temelín [*Commission, 2001*]. EU Commission President Romano Prodi rejected Austrian threats to hold up Czech accession. 'Veto should only be used if vital interests of a country are at stake' (*CTK*, 6 January 2002). He acknowledged the critical role of the Commission in mediating the conflict and was concerned that the controversy could become quite serious. German Foreign Minister Joschka Fisher agreed, arguing against any 'artificial' delays in the enlargement process because of Temelín.

In reality, the EU became officially involved earlier in July 2000, when the Enlargement Group of the Committee of Permanent Representatives of the Council (COREPER) charged the Atomic Questions Group (AQG) – a permanent Council body – to prepare a position concerning 'a high level of nuclear safety in the candidate countries'.[3] Many prior European Councils had encouraged high levels of nuclear safety. In the absence of competence for energy and, more specifically, nuclear power in the *acquis*, legislation covering nuclear safety, except for levels of ionising radiation, transportation of nuclear fuel, and emergency preparedness derived from the Euratom Treaty and IAEA agreements, does not exist. While it was agreed that the EU would monitor Temelín until the accession of the Czech Republic, the position of the EU was that the responsibility for safe operations of a nuclear plant belonged to the country where the facility was located.

The result of these efforts was a 'non-paper' by AQG to COREPER in July 2001 that describes non-binding or 'soft' laws based on voluntary cooperation among EU nuclear states. Owing to the historical differences in their nuclear regulatory procedures and installations, these states strongly support only general rules of safety. The report also reviewed all nuclear candidate states. Based primarily on submitted documents – not on-site visits or comments by non-regulators – it admitted to being of limited scope. It did, however, note concerns at the Temelín nuclear power plant about embrittlement and integrity of the 'vessel beltline area welds' because of high nickel content, as well as weakness in fire prevention. Regarding the critical issue of nuclear waste disposal, the report said that in both Czech nuclear power plants – Dukovany and Temelín – long-term facilities were lacking. Spent nuclear fuel would be stored on-site at Temelín for ten years and then, most likely, transferred to Dukovany. The lack of an existing strategy to deal with nuclear waste is an issue of grave concern to Temelín opponents (and to the anti-nuclear power opponents in general). The report questioned the wisdom of increasing nuclear capability without resolving the issue of long-term nuclear waste disposal.

Recent years have also witnessed increased interest within the EU (especially the European Parliament) and among NGOs about the future of nuclear power because of uncertainties of climate change, policy commitments and renewed concerns about nuclear safety. Austria, however, was the driving force making nuclear power an issue within the context of EU enlargement. Enlargement negotiations provided an opportunity to focus on nuclear power safety. The December 2000 'Melk Process' was undertaken to specifically examine nuclear safety issues and facilitate an exchange of information about Temelín [*European Commission, 2001*]. While there were hearings and meetings in both countries, the Melk Process did not proceed smoothly. Czech Environmental Minister Milos Kuzvart doubted that the new EIA could be completed by May 2001, as agreed to in the Melk Agreement. Rudi Anschoher, Upper Austria Green Party leader, wanted the Temelín plant to close while the review took place.

The Commission on the Assessment of Environmental Impact of the TNPP released its report on 31 July 2001, based on its assessment of nuclear safety at Temelín as part of the Melk Process, and utilising the Directive on Environment Assessment of Public and Private Projects No. 85/337/EEC and No. 97/11/EC [*Commission, 2001*]. The members of the Commission included four Czechs, two representatives from the EU, and observers from Germany and Austria. The Melk Protocol established this expert mission to assist in identifying safety issues. Normally an EIA is done before a project is begun. Although the EIA was guided by existing EU legislation, this was a special case because it was retroactive. The actual document was prepared by the Czech Environmental Ministry. The Commission concluded that the environmental impacts were considered to be insignificant and acceptable.

Between February 2001 and July 2001, in a parallel process, there were ongoing discussions between the EU, Czech nuclear experts and Austria. Twenty-nine issues of Austrian concern were identified and addressed by the Czech Republic. Chancellor Schussel as well as Austrian environmental groups said that the Czechs did not provide sufficient documentation. As a result, one hearing in May 2001 was postponed. There were unpleasant words between the Czech Minister of Industry and Trade (MIT) Miroslav Gregr, who said Austrian demands were 'nonsense', and Upper Austria Governor Josef Puehinger, who called Gregr 'ignorant' (*CTK*, 24 April 2001, 27 April 2001).

Jan Kavan, the Czech Foreign Minister, told critics, 'We would shut down Temelín only if it were objectively proved that it does not comply with fundamental safety criteria' (*CTK*, 25 March 2001). At the same time, Temelín again suffered turbine problems that worried the Austrians, who again called for a 'zero variant' – consideration of an option closing Temelín. The German Environment Minister Juergen Tritten, also a long-time opponent of Temelín, asked for the closing of the plant. More demonstrations were threatened by the Upper Austrian Greens. Austrian Finance Minister Karl-Heinz Grosser said the Czech Republic should abandon Temelín while Austrian and German Greens called upon EU countries to boycott electricity from the plant (*CTK*, 16 June 2001).

Upper Austrian Commissioner for Nuclear Facilities Bordering Austria, Radko Pavlovec, said the Commission's report was deficient (*BBC*, 11 April 2001). The FPOe reaction was that the document was a provocation (*CTK*,

12 April 2001). Chancellor Schussel asked the Czechs for more information. Lower Austria said documentation about crisis scenarios was deficient and that Temelín constituted a real threat to countries neighbouring the Czech Republic (*CTK*, 8 June 2001). The Czech Foreign Ministry responded by asking if the Austrians were questioning the sovereign right of the Czechs to determine their own energy policy. The Czechs did agree to respond to the 'zero option' and provided additional information. This, however, did not satisfy the governors of Upper Austria, Lower Austria and Salzburg, who announced that they would file a lawsuit for potential damages. German Environment Minister Tritten pulled out of the meetings on Temelín to disassociate himself from any conclusions of the Commission report. Environmental NGOs argued that the EIA failed to consider a crash of an airplane or the method of liquidation of stored radioactive waste.

Austria submitted a report to COREPER criticising the shortcomings of nuclear plants in candidate countries, including the Czech Republic and Slovakia, making nuclear safety an issue for consideration in accession. However, Enlargement Commissioner Guenther Verheugen, who brokered the Melk Agreement, warned that Austria could not prevent the construction of a nuclear power plant in a neighbouring county (*CTK*, 22 June 2001). The Czech Foreign Minister Jan Kavan indicated that he understood the concerns of the Germans and the Austrians because of their closeness to Temelín, stating, 'We perceive the fears of our neighbours' citizens as understandable, but because we do not feel them justified, we will do everything to dispel them and assure the people that the plant is safe' (*Austrian News Organisation Report*, 29 August 2001).

Austrian Greens interpreted the remarks as sympathetic to their cause – that building a nuclear plant close to borders is unacceptable. Some Austrian Temelín opponents suggested giving the Czech Republic money to close the plant or purchase the plant. There also was a suggestion of an international conference to discuss the possibilities of closing the plant. Chancellor Schussel asked EU President Romano Prodi to make Temelín a European issue as a means for leading the way to EU standards for nuclear power plants (*CTK*, 11 September 2001). Upper Austria's Governor agreed, stating that this was not a bilateral problem with the Czech Republic. Austrian Greens maintained that Temelín was a European problem and should be resolved at a European level. A serious accident would affect not only Austria, but all of Europe. Commissioner Verheugen said there would not be an international conference unless the Czechs supported it, which they did not. In January 2001, almost a million Austrians signed a 'Veto Temelín' petition organised by Jorg Haider's FPOe. It demanded an Austrian veto to Czech accession if Temelín was not shut (*CTK*, 20 August 2001). FPOe also called for an Austrian referendum on Temelín. The pressure was relentless. The opponents argued that keeping the Czechs outside EU reduced the opportunity to make the plant safer, since Temelín would probably go online anyway. Schussel concluded by playing his trump card, stating that the energy chapter would not be closed until 'all safety and environment aspects of the Temelín nuclear power plant are assessed' (*CTK*, 11 November 2001).

The European Parliament, a strong supporter of environmental issues, passed a draft resolution in July 2001, recommending the phasing out of Temelín and hosting an international conference on the issue. It tried to convince the European Commission that Temelín was a failed investment. At the September 2001 plenary session of the Parliament, it was suggested that the EU finance the closure and dismantling costs of Temelín. The plenary session also advocated increased use of sustainable energy sources [*EC, 2001*]. This position was supported by all Austrian parties. The non-binding resolution was passed on 5 October 2001, recommending that as problems continue to come to light in the nuclear and non-nuclear section of the plant, the 'zero option' should be considered.[4] Resolution supporters hoped that the Commission would consider the Parliament's position seriously. This was the first time an EU institution tied Temelín to accession.

The German Approach

The German Environment Minister questioned the economic sustainability of Temelín and reiterated his position that Temelín would not meet German standards (*CTK*, 14 February 2001), or be viable in Germany (*Agence France Presse*, 2 November 2000). The enormous cost overruns even surpassed the break-even point established by Ceske energetic zavody (CEZ), the utility. The German anti-Temelín movement included environmental NGOs who joined with counterparts in Austria and the Czech Republic. The strategy was to force an in-depth EIA and raise public awareness through protests and the boycott of Czech-imported nuclear energy – a position supported by Environment Minister Juergen Tritten and Economics Minister Werner Muller. In July 2001, the German government formally asked the Czech government to revise its decision to operationalise Temelín (*Financial Times*, 8 July 2001). The Bavarian Economics Minister Otto Wiesheu complained that a boycott was a violation of free market competition as well as Czech sovereignty (*CTK*, 31 May 2001). Nonetheless, E.ON, a German power company, said it would cancel contracts with CEZ to import electricity. Meanwhile, Bavarian border towns launched a campaign to stop Temelín with petitions, in February 2002, Bavaria asked the Czech Republic to close Temelín (*CTK*, 14 May 2002).

A difficulty with the boycott strategy is the inability to distinguish between sources of electricity. Other German companies kept the CEZ contracts and purchased electricity indirectly through ENRON (*CTK*, 1 June 2001). Germany never threatened to block Czech accession over Temelín, although it is committed to close its own nuclear plants within 20 years. The Czechs were very aware of the anti-nuclear feeling in the Bundestag and tried to be responsive to inquiries. A study by the German Society for the Safety of Nuclear Facilities and Reactors said Temelín met international safety standards except for problems that could result from a break in the feeding water pipes (*CTK*, 19 December 2000).

Relentless Austria

Austrian Chancellor Schussel was in the awkward position of criticising the EU for lacking uniform nuclear energy standards, while demanding that Temelín comply with safety standards valid in EU countries. Since there are no EU standards, which national standards should apply? German, French and British standards are not the same. Czechs officials argued that the EU could not apply pressure to candidate states about nuclear power because it lacked the competency to do so with existing members. However, the EU position was that it could force an EIA on non-members even though it was not called for in EU legislation.

The conclusions of the Melk Process issued on 29 November 2001, defined a follow-up process. The agreement between the Czech Republic, Austria and the EU was 130 pages long. Each state recognised the sovereign right to its own energy policy, but there would be joint monitoring and cooperation to increase energy efficiency. In late November 2001, Chancellor Schussel changed his position regarding closing the Czech energy chapter. The Austrian Foreign Minister Benita Ferrero-Waldner implied that the energy chapter could be reopened, but she did not receive support from other foreign ministers. The Austrian Parliament passed a resolution giving it the right to reopen it in the future. This, however, would be highly unusual requiring the support of the Commission, which was supporting the Czech position. However, the Austrian Vice-Chancellor, Susanne Reiss-Passer (FPOe) still maintained that Austria take a stronger stand without fear of being isolated in the EU.

Why did Austria finally abandon a veto of Czech accession? First, Austria lacked support in the EU Council. Second, Chancellor Schussel risked jeopardising the strength of his coalition in a long, difficult and unpleasant fight. Having just recently been isolated by EU bodies and member states following the inclusion of Haider's right-wing FPOe in the government, Austrian officials were loath to risk being the 'outsider' again and being subject to reprisals in the European Council. EU Commission President Prodi rejected demands for safety guarantees at the EU level. There was no legal basis for stopping Temelín. Finally, the proposed conference on nuclear power at the EU level was rejected by the Commission, as it deferred to the Czech Republic. In April 2002, the government of Upper Austria brought suit against CEZ in an Austrian court. The court rejected the claim saying it did not have the right to rule because the Czech Republic was sovereign – possessing the right to make decisions concerning its own territory. The Upper Austrian government is appealing (*CTK*, 26 April 2002). These factors give rise to the need for an examination of 'sovereignty' and the relationship between EU and member states and candidate states. The Temelín case also casts doubt on the effectiveness of the veto, if a vetoing state risks isolation and accompanying retribution.

At the December 2002 Copenhagen Summit, at which the CEE states were invited to join the EU, Austrian officials wanted to embed a protocol to the accession treaty with the Czech Republic making the Melk Protocol subject to international law and subject to enforcement by the European Court of Justice. Lacking an EU nuclear energy policy and given the influence of the nuclear states, the attempt failed. Nuclear member states may have feared that such a move might put other nuclear power plants under European Court jurisdiction with possible lawsuits initiated by anti-nuclear groups. However, Austrian right- and left-wing parties argued that without enforcement mechanisms, the Melk Agreement was meaningless. Chancellor Wolfgang Schussel and Prime Minister Vladimir Spidla did agree on a declaration to be attached to the Czech Accession Treaty pledging the fulfilment of the Melk Agreement. It remains a bilateral agreement and not subject to international law. However, Austrians may turn to other strategies such as the International Court Justice, petitions or a national plebiscite.

Temelín Problems Continue

Most of the shutdowns and delays at Temelín were due to problems in the non-nuclear system. From the time CEZ put Temelín in test mode in October 2000, failures and shutdowns plagued the utility and the State Office of Nuclear Safety (SONS). Western European Nuclear Regulators Association (WENRA, the EU's nuclear safety advisory body) reported some safety concerns on the basis of the different safety concepts in Eastern and Western technology, which did, and would, continue to cause technical problems and delays (*CTK*, 20 November 2000). WENRA's President, however, left the decision of whether nuclear safety should be a factor in Czech accession, to the EU member states (*CTK*, 9 November 2000). In

November 2000, there was an automatic shutdown; in December, there was a failure of condensation pumps. After Temelín was connected to the grid, there was an oil leakage in a valve that controlled the amount of steam going into the turbine. The same situation happened in January 2001, but with a fire causing a two-week shutdown. A crack was found in one of the pipes and it was replaced (*CTK*, 17 January 2001). Because of excessive vibrations, 44 steel rings were welded to the vibrating ducts. The plant was closed again in March 2001, when another leaking oil control valve caused tens of litres of oil to escape from the primary circuit.

In March 2001, Skoda Energy adjusted some control valves trying to eliminate vibrations in the steam pipes leading to the turbines. The problem continued into May, when Temelín was to go back online. In June 2001, new control valves to eliminate vibrations on the intake piping were replaced by Skoda Energy. Temelín was shut for three months during the summer of 2001. During this time, Upper Austria issued a study by Hanover physicist Helmut Hirsch warning that the reactor vessel could become brittle due to high nickel content, that there were possible defects in the steam pressure piping, and that the containment of the primary cooling system was below Western norms (*BBC*, 2001). He concluded that Temelín did not meet EU safety standards and that the probability of an accident was 100 times higher than at other modern plants. Problems continued and Temelín was shut again in September of 2001 for 13 hours because of instability of turbine rotations. The response of a CEZ official was that Temelín could run at 100 per cent output and begin commercial operation, but it would be like driving in a fog (*CTK*, 20 September 2001)! Through 2001 and 2002, there were more closures. In mid-January 2002, technical malfunctions caused the plant to discontinue testing at 100 per cent capacity. A two-month shutdown occurred prior to June 2002. Problems continued into 2003 as Unit 1 experienced additional shutdowns. After Unit 2 was launched in May 2002, it too had technical problems such as requiring that its turbo-set rotors be replaced twice. Although both units have been connected to the grid, by early 2003 they were still not contributing a continuous and reliable energy supply.

NGOs Play a Hand

Generally, Czech NGOs were never successful in challenging the government position favouring Temelín except for a brief period in early 1998, when the transitional government's environment minister publicly opposed the plant, opened a media debate and authorised a cost/benefit study comparing energy futures with and without Temelín. Although NGOs participated in creative demonstrations in Prague and at Temelín, and attempted to raise public awareness and provided information, it was the intervention of foreign NGOs and green political parties which forced the public hearings and EIA within the context of the EU accession process. During the Melk Process NGOs had a formal role in presenting their views.

The groups most successful and active in opposing Temelín were Hnuti Duha (Rainbow Coalition), South Bohemian Mothers, and Greenpeace. Rainbow and Greenpeace were on a police subversive list issued by the Prague Police in June 1996, until they protested and were removed (also on the list was Children of the Earth). Many of Duha's activities included demonstrations and protests that gained media attention and had a national presence. South Bohemian Mothers, headquartered in Ceske Budejovice, near Temelín, was more focused on the plant itself and organising the local community. It boycotted the Melk Process hearings to draw attention to the insufficient time for expert analyses and lack of serious consideration of the zero-option for Temelín. In 2001 it was successful in its lawsuit against the local District Authority in Ceske Budejovice for approving Temelín without an EIA and successfully defended itself again charges of 'harming the reputation' of CEZ as well as the Minister of Trade and Industry, Miroslav Gregr.

The Czech Perspective on Privatisation of CEZ

While Temelín was portrayed as an opportunity to retire coal-fired plants, in fact, not much progress has been made. There is no plan for reducing coal mining or retiring old coal plants. Many plants cannot be retrofitted to reduce emissions. In 2001, government officials stated that when the CEZ utility is privatised, the new owner must guarantee the purchase of 28 million tons of coal from Czech miners over 15 years. It also stipulated a level of output from coal-fired plants to be maintained to meet anticipated growth in electric demand, along with nuclear power (*CTK*, 17 December 2001). This will hurt the development of environmentally benign energy sources and conservation. The Czech Republic now exports electricity without Temelín online. The surplus when Temelín comes online will be even greater, making the plant a revenue producer. Temelín is critical to the privatisation of CEZ, which has a monopoly of production and distribution. CEZ is being sold as a bloc, perhaps to prevent the less marketable items from being stranded and the various components from being separated.

CEZ needs Temelín operational and in good condition if the government is to receive a good price. It has been suggested by Radko Pavlovec, Upper Austria Government Commissioner for Nuclear Power Facilities in Other Countries, that there has been pressure to move forward on Temelín because of pressure to privatise by MIT. The short list of potential buyers was

Electrobel of Belgium, Enel (Italy), Iberdola (Spain), a consortium of NRG Energy (US) and International Power (Great Britain), and Electricité de France (EdF). The latter was the favourite. In December 2001, Prime Minister Milos Zeman cancelled the tender, citing underbidding of prospective buyers. There were questions about the bid of EdF which was late and never opened, because the government made public Enel's bid giving EdF an advantage in a second round of bidding (*Prague Post*, 28 December 2001). EdF demanded that the Czech government be responsible for any damages incurred if the Czech Republic ceased to use nuclear power in the next 20 years as well as the responsibility for disposing of nuclear waste. Privatisation has nevertheless been postponed. For EU accession, it is important that the government divest of monopolies.

A primary concern of anti-nuclear activists is that when Temelín is finally sold to a foreign company, the Czechs will lose oversight over the safety of the plant. In such a scenario, Temelín could be producing surplus electricity to supply Europe while the Czech Republic incurs environmental risks without the ability to closely monitor and control a nuclear power plant within its borders.

Energy Policy in the Czech Republic

The Czech Republic has been trying to move closer to EU policy in the energy sector. Over 75 per cent of electricity is generated from fossil fuels, 3 per cent from hydro, 20 per cent from nuclear, and an insignificant amount from renewable resources. Given the pressure to reduce air pollution from coal mining and coal burning, coal is not projected to have a long-term future unless environmental regulations are modified. In the 1990s, the government encouraged the public to switch from coal to electricity by subsidising the price of electricity. This increased demand was used as a justification for completing Temelín. Demand is forecast to continue to grow. MIT Minister Miroslav Gregr has proposed that more nuclear plants could be built (in North Moravia) to meet these projections. The Czech government has also stated that any new plants built after 2015 will have to use primary sources other than coal.[5] With nuclear power cast as a strategy to comply with the UN Framework Convention on reduction of greenhouse gases, it appears that a nuclear future is part of the country's long-term energy policy. In spring 2003, the Minister of Industry and Trade proposed a draft plan that would double the size of Temelín. It was met with criticism from opponents of nuclear power in the Czech Republic and Austria who instead support more financial support for renewable energy (*CTK*, 6 March 2003, 6 June 2003).

The Environment Ministry projects that renewable energy, which accounts for 2 per cent of the energy sector, will increase to 4–6 per cent by

2010. The development of this sector is one of the objectives of the 6th Environmental Action Programme of the Commission. The stated goal of the government is, 'creating a well-functioning, non-discriminating, transparent and motivating system of support and power savings, effective use of renewable energy sources, and co-generation of electricity and heat'.[6] While there are references to sustainable development and its significance in EU policy, the government admits there has been no improvement in the business, or public, approach to energy savings or renewable energy sources.[7] The MIT has a lower projection for renewable development of 1.5 per cent to 3–6 per cent by 2010, and 4–8 per cent by 2020. There are plans for energy savings programmes by the State Energy Agency. Because they estimate that more funds will be needed than are available, they are looking to the EU and World Bank for support. The new Energy Law and the Law on Energy Management, which came into force in January 2001, established rules for business operations using environmentally sound practices. There is also government support for energy audits, efficiency standards, labelling of appliances, and co-generation [*Ministry of the Environment, 2001*]. However, since energy prices are still below world market rate, there is little incentive to conserve energy [*Kramer, 1999*].

The mining of uranium has supported the nuclear power industry. Run by the state company Diamo, it employs about 1,000 workers. The EU would have liked the market opened to other sources of uranium, however, in the New Energy Act of 2001, Diamo has a two-year contract to continue to supply CEZ with uranium ore. The Czech Republic said it would lift the ban on imported uranium by 2002. The Environment Minister and the Foreign Minister were concerned that these restrictions would undermine Czech credibility in the EU (*Prague Post*, 15 November 2000).

Conclusion

The dynamics of energy and environmental policymaking in the case of Temelín provides a unique lens for examining the relationship between candidate states and the EU, as well as issues pertaining to the future of nuclear power in Europe. Many actors from within the Czech Republic, neighbouring states and the EU played supporting roles. Internally, the Czechs had to decide whether to continue a technically and economically questionable project. NGOs continued to gain strength as they found foreign allies, especially in Austria and Germany. The Environment Ministry and MIT were often at loggerheads, and decisions were ultimately made at the highest levels of government. Energy policy was made by MIT without consideration of or integration of environmental goals such as the development of sustainable means to comply with the EU 6th

Environmental Action Programme. The government vote in the Czech Republic in 1999 to continue Temelín was very close. Afterwards, ministers opposing Temelín had to support it publicly even as the cost escalated beyond all projections and delays in getting the plant operational mounted. These problems came as no surprise to many anti-nuclear NGOs as well as a number of scientists and environmentalists. Consequently, Temelín is regarded with pride by some Czechs and is perceived as a monumental blunder by others.

The Temelín case illustrates the limits of existing environmental policy not only in the Czech Republic, but among the member states of the EU where the long-term impact of nuclear energy has not been considered fully. Similarly, the World Bank has also met with mixed results in its attempts to close Soviet-designed nuclear power plants in Slovakia and Ukraine [*Gutner, 2002*]. EU approval of Temelín, while keeping the issue separate from Czech accession, overlooked difficult issues concerning nuclear safety and the desirability of an enhanced nuclear future. EU funds for nuclear power compete with commitments to support renewable energy. That the Commissioner for Energy and Deputy Chair of the Commission, Loyola de Palacio looks favourably on nuclear power and has indicated that the Commission will set safety standards, may open an EU-wide debate about the appropriate energy mix necessary for meeting sustainable environmental goals [see *European Commission, 2002*]. Based on this pending support, Bulgaria is considering building a new nuclear plant to compensate for the loss of its Kozloduy plant. At the same time, Finland is considering new nuclear power, Sweden is rethinking closing its plants, and Germany may be dragging its feet in closing its nuclear power plants. Yet, some attempts to set EU-wide minimum safety standards based on those from the International Atomic Energy Association is moving forward, partially as a result of the enlargement process [*European Commission, 2003*].

Temelín became an international issue when Austria and NGOs challenged its completion. Austrians, and later Germans with memories of Chernobyl, tried to stop construction of the plant and continued to oppose its operationalisation, supporting the sovereign right of a state (such as Austria) to protect its citizens from potential harm. Local government officials took independent action as well as pressuring their federal government. In Austria, almost all political parties eventually opposed Temelín, but advocated different strategies. The most adamant was the far-right Freedom Party (FPOe) of Jorg Haider. Both Germany and Austria have non-nuclear power policies, achieved by referenda in each state, so there was ample reservoir for anti-nuclear sentiment.

The intense bilateral negotiations over Temelín between the Czech Republic and Austria coincided with, or could be considered to be, the result of the Czech accession process. The Czech position was that if the plant was deemed unsafe by EU standards it could be closed. The Czechs argue that their plant has been scrutinised more than any Western European one. The problem was that there was no guidance from the EU because it could not agree on a nuclear policy. Standards for high nuclear safety are also lacking. The Austrians threatened to veto both the environment and energy chapters unless a new and comprehensive assessment was made of Temelín. The goal was to close Temelín or delay Czech accession. This was interpreted as extreme pressure or blackmail by most Czechs. Austrian opposition to Temelín was also perceived as outside interference threatening sovereignty. A few Czech leaders used the issue to arouse populist and nationalist resentment as well as old historical Austro-Hungarian antagonisms. As the plant became an object in Czech accession to the EU, the Austrians hoped this would be an opportunity for the EU to take a position on the future of nuclear power. Austria's aim was to raise questions, such as, is nuclear power consistent with sustainable development? What of long-term waste disposal and decommissioning?

Austria also raised the issue of cross-border environmental impact and sovereignty to public attention. Is a state free to decide how it will produce electricity? Is the answer yes for current member states and no for candidate states? If the majority of Austrians thought its government should try to convince its neighbours to abandon nuclear energy, does not Austria have the responsibility to do so? When the Czechs refused to stop construction pending an EIA, Austria increased its diplomatic pressure bilaterally and within the EU. For the Austrians, there was little room to compromise – the plant is either opened or closed.

Candidate states, in general, became more aggressive in the accession negotiations (for example, agricultural policy parity with Western Europe) because of the time schedule. In the future, a candidate state may want to use something similar to the Melk Process to expand the scope of conflict to garner support for its position from other candidate states or even member states. On the other hand, EU bodies (when unanimity has existed and when funds for closure were promised and provided) have forced candidate states such as Bulgaria and Lithuania to accelerate the closure of a small number of nuclear power plants deemed quite dangerous [*Gutner, 2002*]. The EU made termination of an unsafe nuclear power plant in Bulgaria a condition to begin EU accession negotiations [*REC, 2002*]. Without the spectre of EU membership it would have been much more difficult to close unsafe plants. Even so, Bulgarian officials and nuclear power interests continue to discuss the scheduling closing of a number of reactors in Bulgaria [*Andonova, 2002*]. These debates continue, at least in part, because Bulgaria has electricity export opportunities.

The EU could use the accession process to increase transparency in candidate states and support NGO pressure on their governments for information on environmental impacts of energy. Down the road, other candidate states will need guidance on energy policy. Decisions made during the accession process will affect the EU, especially after these states, some of which have Russian-designed nuclear power plants, achieve membership. However, expanding the scope of conflict could backfire if the EU goes down the path of harmonising nuclear standards, using the lowest common denominator.

NOTES

1. For a brief discussion of nuclear power in Central Europe, see Kramer [*1995*].
2. For a more detailed discussion of environmental policy and Temelín, see Axelrod [*1999*].
3. Contribution of the Commission services to question no.1 of the questionnaire submitted on 13 Sept. 2000 by the Presidency of the Atomic Questions Group in the framework of the mandate received from the COREPER on 26 July 2000 – Non-Paper (29 Sept. 2000), p.2.
4. See European Parliament resolution on the Czech Republic's application for membership of the European Union and the state of negotiations (COM(2000)703-C5-0603/2000-1997/2180(cos)) minutes of 9 May 2001.
5. See 'Energy Policy' pursuant to the decision of the Government of the Czech Republic of 23 June 1999, No.632 and §14 of the Act 244/1992 Col. On the appreciation of influences on the environment, p.26.
6. Ibid, p.5.
7. Ibid, p.4.

REFERENCES

Andonova, Liliana (2002), 'The Challenge and Opportunities for Reforming Bulgaria's Energy Sector', *Environment*, Vol.44, No.10, pp.8–19.

Axelrod, Regina (1999), 'Democracy and Nuclear Power in the Czech Republic', in N. Vig and R. Axelrod (eds.), *The Global Environment: Institutions, Law, and Policy*, Washington: Congressional Quarterly, pp.279–99.

Commission for the Temelín Nuclear Power Plant Environmental Impacts Assessment (2001), 'Standpoint of the Commission for the Temelín Nuclear Power Plant Environmental Impacts Assessment', Prague, 15 July, p.2.

Chandler, William (2000), *Energy and Environment in Transition Economies*, Boulder, CO: Westview Press.

Darst, Robert (2002), *Smokestack Diplomacy: Cooperation and Conflict in East-West Environmental Politics*, Cambridge, MA: MIT Press.

Dawson, Jane (1996), *Eco-Nationalism: Anti-Nuclear Activism and National Identity in Russia, Lithuania and Ukraine*, Durham, NC: Duke University Press.

EC (Environment Commission) (2001), *Environment for Europeans*, 8 October.

European Commission (2001), 'Commission Transmits Working Paper on Temelín to the Austrian and Czech Authorities', Press Release 1P/01/1161, Brussels, 31 July.

European Commission (2002), 'Communication from the Commission to the Council and the European Parliament: Nuclear Safety in the European Union', Brussels, 11 June 2002, COM(2002)605 final.

European Commission (2003), 'Nuclear Energy: The Commission Proposes a Community Approach to the Safety of Facilities and Waste', Press Release, IP/03/2003.

GAO (General Accounting Office) (1994), 'Nuclear Safety: International Assistance Efforts to Make Soviet Designed Nuclear Reactor Safer', Report to Congressional Requestees, GAO/RECD, 94-234, Sept.

Gutner, Tamar L. (2002), *Banking of the Environment: Multilateral Development Banks and Their Environmental Performance in Central and Eastern Europe*, Cambridge, MA: MIT Press.

Kramer, John (1999), 'Energy and Environment in Central and Eastern Europe', *Problems of Post-Communism*, No.6, pp 47–56.

Kramer, John M. (1995), 'Energy and the Environment in Eastern Europe', in Joan DeBardeleben and John Hannigan (eds.), *Environmental Security and Quality after Communism*, Boulder: Westview, pp.89–104.

Ministry of the Environment (2001), *State Environmental Policy 2001*, Prague, Czech Republic.

REC (Regional Environmental Center for Central and Eastern Europe) (2002), *The Bulletin*, Vol.11, No.1, p.5.

Schattschneider, E.E. (1960), *The Semi-Sovereign People: A Realist's View of Democracy in America*, New York: Holt, Rinehart, and Winston.

Woodard, Colin (1995), 'Western Vendors Move East', *Transition*, Vol.1, No.21, p.24.

PART III

CIVIL SOCIETY IN AN ENLARGED EU

Eurocratising Enlargement? EU Elites and NGO Participation in European Environmental Policy

LARS K. HALLSTROM

Environmental policymaking in the European Union (EU) is increasingly complex, and subject to an increased range of actors attempting to influence European decision making. With over 900 Euro-groups and thousands of nationally oriented interests in Brussels, a wide range of business, professional and public interest groups attempt to influence EU-level policies.[1] Relevant to both the study and practices of integration, research on the interaction between citizens and policymakers has shifted away from examining the contribution of private interests to integration, to investigating their effects on both European and national governance [*Kohler-Koch, 1997*]. This contribution addresses questions about how interaction occurs between private and public spheres in the EU, and examines the changing character of policymaking as the EU enlarges to include Central and Eastern European (CEE) states. It also attempts to place environmental organisations and interests within the context of the democratic legitimacy of the EU as a regulatory regime and source of environmental policy.

By examining environmental governance through the eyes of EU actors, this study provides a unique perspective on the environmental policy process. Based on a series of over 90 intensive interviews with EU officials during the early 1990s and 2000, it presents elite understandings and expectations of public input into environmental policymaking and integration at the European level.[2] Policy and political convergence raises both broad theoretical questions and more precise empirical issues regarding public opinion, participatory sentiment, and knowledge in public policy. In this piece, rather than address Robert Dahl's [*1961*] question of 'who governs', I examine who participates, who allows them to do so, and why.

This contribution focuses on how environmental policy functions in Brussels, the perceptions and experiences of EU policy elites, and their perspectives on the role for CEE interests and interest groups at the European level. While others have examined issues of formal access, opportunity structures and institutional constraints upon interest behaviour in the EU [*Kitschelt, 1986; Kriesi et al., 1992; Nentwich, 1996; Richardson, 1994*], I emphasise the informal side of EU policymaking. Specifically, I pursue the thesis that the evolution of informal policymaking at the EU level, combined with the technical and informational preferences of actors in institutions such as the European Commission, limits the role for non-business and non-governmental organisations (NGOs) in environmental policy. This has negative implications for the democratic legitimacy of the EU, and as the EU has increased influence over environmental policy in CEE countries, it also reduces the opportunities for environmental NGOs to be a source of public opinion and grassroots knowledge. While formal mechanisms may provide symbolic access, the influence and access of NGOs in CEE states is limited. Officials prefer to rely on input from actors with expert knowledge and engage in top-down modes of decision making, rather than including NGOs as a potential source of environmental policy. As the CEE states move towards EU membership, they become part of a 'Eurocratic' policymaking and administrative tradition that favours technocratic and centralised environmental policy.

Informal Policymaking in the European Union

The EU places few institutionalised constraints on participation in the policy process. Indeed, the opposite is true. While the formal procedures of EU institutions provide a degree of structure and a recognisable hierarchy, it is the political decision making and negotiating that takes place around and through these structures, usually informally, that is often the determinant of policy outcomes [*Middlemas, 1995*]. This combination of formal and informal factors in policymaking is far from unique to the EU, but the importance of informal political processes is increased in this emerging system. While formal procedures and institutions have become increasingly complicated,[3] these structures are continuously contested and redefined, according to a changing economic and political context [*Wallace and Young, 1997*]. Informal contacts with desk officers and officials in the Commission are disproportionately important inputs into public policy, but this importance is not consistent or predictable. This creates issues of legitimacy, access and, ultimately, influence for those actors and groups attempting to shape policy, since the connections between input and influence may not be visible or even available to interests.[4] While this style of governance does permit a highly flexible method of governing, particularly as it must include multiple member-state cultures and ministries, it can limit certain avenues for public opinion and knowledge.

Stoker [*1998: 187*] writes that, 'governance refers to the development of governing styles in which boundaries between and within public and private sectors have become blurred'. Based on the interaction of multiple levels of actors, the essence of informal policymaking in the EU lies in the development of largely unregulated 'haggling' relationships [see *Börzel and Risse, 2000*]. These relationships link diverse actors such as pressure groups, members of academe, groups and parties from both national and European legislative institutions, the media and the groups functioning within governance in the CEE states, but are neither stable nor predictable. This has led to not only a new class of EU-oriented bureaucrats with strong ties to each other, but what a former Deputy Director of the Bundesbank called 'a new type of state' built around 'the middle range' of administrative structures, those who have knowledge 'from the interior' (interview with former European Commissioner, date unavailable). While there are few institutional constraints on who can access or interact with this middle range, the symbolic openness of the EU is countered by an information-collecting and policymaking style that has a preference for inputs from EU-level industrial federations and their scientific or technical resources.

As Wallace and Young [*1997: 11*] point out, 'One of the defining characteristics of "Brussels" is the openness of opportunities for access, opportunities taken by battalions of national officials, by many representatives of organised industrial interests ... and a range of other interest groups ...' yet how this openness translates into influence on the formulation of public policy is only partially understood. In integrated policy areas such as the environment, the ways actors gain legitimacy and access to the policy process are a key element of European policymaking. As the political structures of the EU do not totally determine those who participate, there are two vital factors in who gains access to policymaking. These include not only the characteristics of the actors attempting to gain influence, such as access to information, credibility, professional expertise and financial resources, but also the perceptions, experiences and opinions of those whom interests attempt to influence, particularly those within the European Commission [*Middlemas, 1995*].[5]

Although the Commission has gained the reputation of being highly accessible to certain interest groups, the methods and use of hearings, advisory committees and expert consultants are far removed from any participatory version of policymaking. Interests at the pan-European level or on a more localised or regional scale must rely on an unstructured, *ad hoc*, and informally developed set of contacts to obtain input into the policy process. The Commission declared its commitment to ensuring opportunities of access and equal treatment to all interest organisations, regardless of size or financial strength in 1993, but these factors obviously figure in the access decision. While professional lobbies are viewed with some derision, there are considerable benefits that can be achieved by maintaining a permanent office in Brussels and attempting to gain or develop insider status. This obviously favours those interest groups with greater financial resources and size, allowing them to buy or rent office space, pay and train staff, and build a policy presence [*Greenwood, 1997*]. As a consequence, while the European Citizen Actions Service (ECAS) does provide resources for those organisations unable to afford placing a representative in Brussels, organisations with greater access to resources will often have, if not greater, then qualitatively better access to policymakers in the Commission.

Informal interaction is often seen as the best means of addressing European problems without attracting undue or unfavourable attention, and may even contribute to the Commission's European mission. Repeated contacts and informal meetings with interests and NGOs give such groups a better understanding of how the EU functions, the nature of the dominant policy discourse, and even the appropriate reactions to policy suggestions. Groups that develop and maintain consistent contacts with actors in the Commission can even converge in their views and opinions, becoming more *communautaire* than their national counterparts (interview with member of DG III, 18 November 1992). For environmental NGOs from CEE states that provide more public sentiment or local knowledge than expertise, and typically have limited financial resources, their role in EU policymaking is highly dependent on both their resources and the views of European policy elites.

With the trend towards centralising certain policy domains to the European level, the Commission is increasingly reliant upon external interests as sources of expertise and information. This quasi-dependent relationship has a key impact on the content of proposals and policies generated by the Commission. Groups that have evolved into 'insider' status have considerable potential to influence public policy [*Butt Phillip, 1985*]. At the same time, the increasing speed, scope and demands for legislative and regulatory proposals coming from the Commission can draw officials into semi-clientelistic relationships with certain interests. Usually, these groups are multinational firms or national organisations that have a permanent office in Brussels and a decision-making structure that allows for quick response. These factors place a number of interests at a disadvantage, particularly those smaller or localised groups that must rely on an organisational tie-in with a Euro-group based in Brussels. As a result, despite the enormity and complexity of decision making in Brussels, the most successful interests are those who display the usual professional profile. This profile includes advance intelligence, resources both informational and financial, contacts, and the capacity to put forward professional, rational and technical arguments that meet the information requirements of European institutions such as the Commission [*Mazey and Richardson, 1993*].

EU officials see many environmental groups as trying to counter-balance the number and resources of industrial or economic interests by working through the European network. This is a way of gaining additional resources and experience in Brussels, which is particularly important for groups from CEE countries. However, environmental groups are often seen by EU officials as poorly trained and therefore weak, while business interests are seen in a more positive light. When discussing public concern over environmental issues in the EU, one very senior member of Directorate General (DG) Environment said:

> One hopes that in a democracy that [environmental issues] make their way through to the decision makers, but we have to remember that the decision makers are also heavily influenced by the economic effect of people who actually provide the jobs and make the money, and they're not always open to environmental improvement, particularly if you're telling them you are moving toward a situation where their product is now seen as dangerous for the environment … I think we will continue to have the situation where we have many of the economic actors pleading in favour of continuation of what they're doing (interview in Brussels, 8 March 2000).

When this insight is combined with the Commission's preference for both majority and European-level opinion and expertise, the image is of multiple points of access, but with informal and informational (rather than structural) barriers to policy involvement. Not only does access have to be won, it must be maintained and defended over time. This can place groups with limited resources or personnel at a disadvantage, reducing the reasons for a European official to allow contact. This was exemplified by another member of Directorate General (DG) Environment, who described the calculus of policymaking access:

> The environmental groups are very weak. Now, the biggest European environmental organisation, the European Environmental Bureau, which puts [*sic*] 150 national associations, has a permanent staff of three persons. This means that for me, as an official, if I want to have, say, the top official on waste from genetically modified organisms, laboratories or plants or so on, and I call the Chemical Association, within one minute I have their specialist on the phone, and within one day, let's say three days, I have the top specialist in the world, coming

> either to me, or if I want I can pay him a visit and get the top information, most precise, and so on. If I address the Environmental Bureau, and they don't have a specialist of that nature … of course on that they are generalist, so they are not really a means of information for me, and that means that this side of information is practically under-developed (interview in Brussels, 15 March 2000).

Such informational needs, described by Greenwood [*1997*] as the 'management deficit', minimise the probability of modifying ingrained patterns of interest representation. This is particularly true in those Directorates General and policy areas where expertise and public opinion must often mix, as in DG Environment. While there is awareness that some balance should be achieved between economic, social and other interests, there is little desire to formalise contacts with 'the exterior'. As a member of the Cabinet of the President of the Commission pointed out, the DGs must have the maximum flexibility possible. While it was commonly felt that everyone should have a chance to 'get in touch' and not limit the potential of multiple sources of information, there is also a common understanding that certain DGs have very well-developed relationships with industrial and economic actors (interview with member of DG III (Chemicals) December 1992).

This places smaller, local or regional environmental groups at a distinct disadvantage. Not only does the Commission prefer to interact with Euro-level groups, but the relative absence of 'bargaining chips' such as economic influence or public status for such groups creates a further barrier.[6] The challenges are even greater for interests from CEE countries, both public and private. While the benefits of maintaining an office in Brussels are considerable, costs can be prohibitive for CEE interests. Coming from economies where mean incomes measure less than 20 per cent of those in Belgium, it is often simply too expensive to maintain offices in a city like Prague, let alone in Brussels (interview with members of the Czech Union of Nature Conservation, May 2000). As a result, few environmental groups from countries such as the Czech Republic or Poland maintain a formal affiliation with EU-based federations. Such groups are disadvantaged in the informal policy game in Brussels, as they rarely possess the technical knowledge, credentials, experience or resources required by the EU. As a result, although there are resources available to EU officials to improve the inclusion of environmental organisations, there is little interest in using them beyond a technocratic capacity.

The Regional Environmental Center for Eastern and Central Europe (REC) and the European Commission have long maintained extensive databases listing non-governmental organisations, environmental groups and environmentally oriented experts, and the REC functions as a publishing and communications hub for numerous CEE environmental organisations. However, the data collected for this study show little interest or concern with these resources on the part of EU elites. Of the 90 EU elites interviewed, no mention was made of the goals, benefits or costs of these information and participatory resources.

Given the fact that the Commission was both a founding body for the REC, and is responsible for the interest group database, for NGOs from CEE countries, such findings are particularly problematic. While the purposes of the two resources differ, the goal of improving communication and participation in public policy is the same. Although entry into the databases is voluntary, and the difficulties some CEE environmental NGOs have faced in maintaining offices or even telephone numbers can make contact difficult, such resources are a potentially important means for NGOs to gain access into EU-level policy networks. However, despite the numerous references made by EU elites to the importance of interests and the benefits of communication between themselves and the 'exterior' in order to gain information and policy expertise (interview with member of the Cabinet of the President of the Commission, 17 February 1993), the failure of these elites even to mention the REC and Commission database points to a number of possibilities regarding the practical or informal role for non-governmental organisations in EU environmental policymaking: (1) there may be a general lack of awareness among the officials interviewed as to the existence of the REC and database; (2) officials may be aware of these resources, but they may be perceived to provide little value, accuracy or efficiency; (3) already established patterns of interest intermediation and NGO involvement may render the database and REC irrelevant; and (4) EU officials may actually wish to avoid increased contacts with interests and NGOs, particularly as their numbers and policy scope continue to increase.

Despite the presence of pro-EU and participation-oriented groups in many member states, many of the officials interviewed look upon increased interest intermediation in the European Union as both a problem and a benefit. Problems lie in the increased demands on officials' time, and the difficulties associated with reconciling multiple and divergent perspectives on environmental issues and policies. For those working in Brussels, this is combined with a consistent recognition of the cultural and political issues facing public policy in CEE countries. Combined with the policy and environmental issues faced in CEE states, this has led to a consensus that environmental policy is and should be increasingly designed and monitored at the European level [*Hallstrom, 2003a*].

Despite this, environmental groups in Central and Eastern European states such as the Czech Republic and Poland do play roles in both the integration and development of national environmental policies [see *Bell, this volume; Hicks, this volume*]. There are numerous environmental organisations throughout CEE countries engaged in activities such as the clean-up of historic sites and national parks, the preservation of indigenous species, attempting to monitor the enforcement of environmental conservation policies and attempting to garner public support through demonstrations, petitions and protest-oriented activities [*Hallstrom, 2003b; Fric, 1999*]. However, on the whole the activities of environmental NGOs in these countries have changed substantially since the early 1990s, often moving away from more political and policy-oriented activities towards small-scale, localised and often preservation-based projects.

Environmental groups provided an organisational and intellectual training ground for many dissidents in CEE countries in the 1970s and 1980s, and as democratic reforms were adopted, resolving the major environmental failures of the Communist Party and central planning became a major goal. As a result, early in the transition period, environmental groups and environmentalism exercised some influence over policymaking. However, the combination of a shift in policymaking impetus to Brussels and neo-liberal political and economic reforms has led to an emphasis on privatisation and the consolidation of the party system, with little space for civil society and non-governmental organisations. Public involvement and participation in environmental politics and policy declined substantially during the 1990s, and many environmental groups claim to be increasingly separated from both national and European-level politics and policymaking [*Hallstrom, 2003b*].

The view of CEE environmental NGOs as generally weak and of limited use by EU officials is not entirely unfounded, but hinges largely on a very specific set of ideas and preferences about the role(s) for citizen-based groups in the integration and policymaking process. Environmental NGOs that are not consistent with these ideas and preferences, particularly those that do not bring technical expertise or knowledge to the policy process, are typically viewed as recipients, rather than providers, of policy-relevant information. As a result, some environmental groups, such as Children of the Earth in the Czech Republic, have started to draw on in-group expertise and education in order to gain better access to environmental policymaking, essentially trying to pursue an environmental agenda on two different fronts. They have lobbied the Czech government to resist the construction of highways and transportation infrastructure (an important element to EU membership), while also participating in environmental impact assessments (EIAs) as ecological experts. While these strategies do improve the role of environmental NGOs at the national and possibly even the European level, the shift towards expertise-based activity does little to maintain the grassroots and often participatory impetus on which they once relied [*Fagin and Jehlička, 1998*]. Combined with the shifting focus of national officials towards the *acquis* and EU membership, environmental policy and politics have shifted away from the CEE publics. CEE environmental NGOs have largely recognised this, and the difficulties of trying to gain influence and access to policy in Brussels. As a result, these groups have chosen either to adopt a professional and formalised pattern of interaction with policymakers, or to refocus their efforts at a small-scale and localised level. With the patterns of interaction between EU, national and sub-national levels of governance still emerging in CEE countries, the increasingly localised efforts of environmental groups often fall under the radar of EU-level officials, creating an inaccurate image of a completely, rather than only partially, disengaged public.

Environmental NGOs and the European Commission

In conjunction with the desire for a Europe-based approach to environmental policy, a former member of Directorate General III noted several roles that national and sectoral interests can play in Europe. Not only can they contribute to the decision-making process by providing positive or negative stimuli and disseminating information throughout the Union, they may also assist in the guarantee of subsidiarity. The increasing demands of multi-level regulation in certain policy areas may well benefit from organised interests capable of moving and communicating between policy levels, both monitoring and assisting under the subsidiarity principle (interview in December 1992). With influence over public policy and political decisions functioning at multiple levels, the presence of a plurality of interests is perhaps well suited to the creation of efficient European policymaking, and is a way to decrease dependence on 'specific national administrative connections, expert reports, or worse, the press' (interview 18 November 1992). However, it is also consistent with a perception of organised groups as the recipient, rather than the bearer, of policy-relevant information and Brussels as the primary source of public policy. As the process of environmental policy harmonisation continues, national-level ministries and NGOs become less important to the initial stages of the policy process. Instead, they are valued as actors improving the legitimacy of the EU institutions and policies, while at the same time improving the efficiency of policy integration and enforcement.

Interests and lobbies can also be seen in a positive light in terms of their information sharing capabilities. As Noel Muyelle pointed out, even citizens in Brussels, a city literally at the heart of European politics, are often unsure and misinformed as to the purposes, roles and direction of the European

Union. Despite the development of European institutions, citizens of Belgian cities such as Gand or Oostende 'do not understand what they are talking about [in regard to the EU]' [*Helbig, 2000: 2*]. As a result, organised interests can bridge this gap, providing the general population with information about EU policies, activities and opinion.[7] Furthermore, such groups can often answer questions, indirectly serving as a mouthpiece for the Commission or other EU bodies in areas where representation and direct access can be limited.

This role can indeed be valuable for the legitimacy and effectiveness of EU policies. However, it is seen predominantly in output, rather than input-based terms. An upper-level official in DG Environment put it neatly when he spoke of meeting with farmer groups 'to explain to them where we are trying to go, rather than in any negotiating terms'. He then went on to explain his actions as a form of lobbying on his part: 'That's why I spend a lot of time talking to farmer's groups, so that they understand that the kind of things that might be proposed over a period of time are not anti-farmer, but are maybe pro-farmer in a much longer term' (interview in Brussels, 8 March 2000). Such explanations are not seen as a way of gaining public input into environmental policy, but rather a way of persuading such groups of the validity and importance of adopting and incorporating the EU position. While interests and formal lobbies do often attempt to create input or influence into policy decisions, there is a strong official impetus towards top-down flows of information and expert knowledge, rather than more discursive or participatory interaction.

In terms of the CEE countries, this relationship is compounded by the awareness among EU policymakers of the short timeframe and limited policy experiences with which many groups, and particularly environmental NGOs, must contend. Although there has been a shift towards greater professionalism among NGOs in many countries in the region [*Jancar-Webster, 1998; Fagin and Jehlička, 1998; Fagin and Tickle, 2001; Carmin and Hicks, 2002*], a number of individuals in Brussels point to the socio-cultural effects of communism, and their continued presence ten years after the fall of the communist regime. These effects include the still-limited administrative capacity of these states to implement and enforce the *acquis*, but also the deliberative gap between decision makers, administration and the public in CEE countries (interviews with members of DG Enlargement, 28 March 2000). Formally, the Associate Agreement states that CEE countries are considered consolidated liberal democracies, but informally many EU elites are aware of the civil and democratic issues facing these states. These issues, such as the perceived weakness of environmental NGOs, public environmental concern, and a distrust of government, have led EU elites to assume the need for a strong EU presence, despite the

European democratic deficits. As a long-standing official in the Commission noted:

> [the political development of] these countries needs to do some catching up with the rest of the European states … I think it's our job, and perhaps we can try to do it through administrative contacts and NGOs, and that's to say we're here, and perhaps to develop capabilities at the same time (interview 5 April 2000).

This points to a number of characteristics of European environmental policy officials. While open to interest participation in EU environmental policy, many see little role for public knowledge, experience or input. Similarly, while both formal and informal meetings and contacts are important, EU elites prefer to interact with professional and technical organisations, rather than grassroots or civil groups. Finally, both official and informal contact with Central and Eastern European environmental interests and NGOs is limited. Not only do many groups lack the resources to maintain contact with the European Commission, but EU elites understand the conditional nature of the environmental *acquis*. As a result of this and the cultural effects of communism, EU elites often rely of pre-established patterns of interaction with NGOs, rather than deal with the weaker CEE groups (interview with member of DG Environment, 8 March 2000):

> What we have discovered is that the environmental NGOs with the exception of Hungary and to a lesser extent the Czech Republic are very weak, and there's no build-up of feeling insofar as concern for the natural environment or for agriculture and its relationship to the environment, so therefore the natural lobby within each country doesn't exist … we used our own background sources of NGOs to give us some kind of feeling as to their views …

One EU official who worked extensively with CEE countries felt that communist cultural policies effectively retarded or eliminated much of the administrative capacity of such countries. This is seen as particularly problematic in terms of enforcing new or EU-level directives. Not only is there often little real desire or perceived need by CEE actors for the effective installation of the *acquis*, but much of the knowledge necessary to the creation of large-scale environmental policy management is also lacking. This compounds the weakness of many CEE environmental ministries relative to economic concerns, and their inability to compete with dominant economic and political interests. As a result, environmental politics and policy stemming from domestic sources such as environmental ministries in CEE states can be viewed with some concern by EU elites.

Not only are these ministries often weak and ineffective, usually as a combination of inexperience, the attraction of more powerful political positions, and political patronage, but they can even have a negative impact on environmental quality. Such was the case in Poland, where the experienced ministers following 1989 (namely, Kaminski, Nowicki and Kozlowski) were eventually replaced by the former Deputy Chair of Rural Solidarity, Zygmunt Hortmanowicz. Hortmanowicz has since been criticised for an excessive and inappropriate use of gifts and patronage to serve Rural Solidarity rather than environmental needs. He also pursued appointments (such as Director General of Forests) that led to negative environmental outcomes. For example, this Director General of Forests issued a decree to decrease the adult elk population in Poland to zero [*Millard, 1998*]. Such experiences do not inspire confidence in policy actors at the EU level, and certainly do little to defray the belief that environmental policies should be made and controlled in Brussels. As one DG Enlargement official noted, 'My opinion is that we have a win-win solution for all sides, taking over the environmental *acquis*, because it means better environment, it means better technology, more cost efficiency, it means new investment coming also from the EU, so they win, we win.'

Eurocratising Enlargement?

There is a distinct form of environmental policy that has emerged in the EU. Officials see few viable domestic sources of counter-pressure in the form of environmental interests and lobbies. However, they are also aware of the difficulties facing groups attempting to function at the European level, and are concerned with the limited input and participatory opportunities available for environmental opinion. As a senior official in the Commission observed, the attempts made by the EU to address these elements of the democratic deficit have been feeble, owing to the under-recognition of the power elements of integration. Without creating a greater tie between European public opinion and the decision-making structures of the Union, he foresees only a continuation and expansion of the 'Eurocracy' rather than the socio-political unity necessary for 'greater political Union'. Unless steps are actively taken to continuously emphasise the importance of public opinion and participation in the CEE states, there are concerns within the EU that these countries will continue to face an under-development of civil society. Without this socio-cultural presence to counter-balance economic demands and development, there are fears that environmental demands will be swept aside in order to meet the economic and consumer needs of the public (interview in Brussels, 15 March 2000). Indeed, a number of environmental organisations face declining memberships in terms of younger (18 to 35 years old) members as they are drawn to careers and travel opportunities unavailable under communism (interviews in Prague, May 2000).

As this official went on to note, the relationships between state, (civil) society and environment are particularly problematic for bridging the democratic deficits present at both European and domestic levels in CEE countries. Not only did state socialism deliberately pursue normalisation and cultural control, but following 1989 it quickly became apparent that contrary to communist doctrine and propaganda, the state, or state-owned companies, was the largest polluter in countries such as the former East Germany and Poland. As a result, the political socialisation for several decades in these states was that legislation was 'one thing, and the economic and social day-to-day reality was something else'. This taught the public that laws were something often completely separate from the practice of living, but now a new approach and understanding of legality, regulation and the role of the state is required. This should be 'some sort of civic approach toward law making and what they might mean and what the obligations of everybody mean … we cannot protect the environment against or without the participation and resources of those who are concerned' (interview in Brussels, March 2000).

With weak environmental ministries and NGOs that have marginal national-level influence, officials involved in the enlargement negotiations are well aware of the difficult situation the position in CEE states can create. While optimistic as to the rural, agricultural and environmental policy development plans in certain countries, Poland in particular stood as a prime example of a negotiating stance one official described as (interview in Brussels, 8 March 2000):

> No, this is our plan, we don't see any need for further development, we don't see any need for an agri-environment, in other words we're not prepared to look forward, whatever number of years it is, when we arrive in the Community and to imagine what the scene … will be in the year 2010. All we want is an open door into your subsidy system and a grab of all your subsidy system without thinking what that subsidy system might be in 2010.

This perspective indicates an awareness in the EU of the complexity of problems facing environmental policy in the CEE region and also a concern with the motivations behind such a stance. Contact with environmental organisations, as well as policy officials, may help improve mutual understanding and lessen such tensions, but as this same official later admitted, he and his staff have little if any contact or exposure to environmental NGOs from these countries. Officials in the Commission see the public as having a very limited impact on environmental policy, one usually confined to implementation and enforcement. Aside from a number of complaints about procedures where 'regional NGOs as well as this and that and even those who doubt Community money and so on …' can participate in the monitoring and enforcement of policies, contact with CEE environmental NGOs is of variable quality (interview in Brussels, 16 March 2000).

While almost every individual interviewed for this project claims to see the strengthening of CEE NGOs and participation in environmental policy as a necessary and positive step, this support is much more symbolic than applied. A number of participants viewed meetings with policy stakeholders as frequent (ten to twenty per month), yet problematic because of their overtly political or agenda-driven nature. Several others had no real interaction with environmental interests at all, despite their position in the policymaking process, nor did they see this as a problem. Environmental interests in Brussels are still relatively rare, and only a very few CEE groups are members of European environmental federations. These groups and organisations therefore fall towards the bottom of an already peripheral list of policy interests. Environmental groups have made considerable inroads into the European political sphere, yet grassroots organisations generally have few of the resources desired by Commission officials. This is particularly true for many CEE environmental groups, and they are usually considered ill equipped to compete on the European policy field.

Conclusion

The findings of this study point towards a strengthening of the role for technical expertise and the 'Eurocratisation' of CEE environmental policymaking, but to the detriment of democratic policymaking and the involvement of environmental NGOs. Based on the informal patterns and opinions of EU policy elites, environmental NGOs generally do not have many opportunities to participate in the development or negotiation of policy. As a member of DG Enlargement stated, public input into European politics is important, at least symbolically. However, there is little concern with fostering this input from non-expert sources in practice, and when actors such as environmental NGOs are to be included in the policymaking process, it is preferred, if not often expected, that they will simply transmit the policy decisions made in Brussels back to the public.

Despite an increase in both the number of groups attempting to influence policy in Brussels and the willingness of EU officials to meet with such groups as integration accelerated through the 1990s, the flow of policy information remains largely top-down. While civil groups and organisations from across Europe have increased in both numbers and access to members of the DG Environment, it is unlikely that these groups have any direct input into the content of European environmental policy. This is particularly true for environmental NGOs from CEE countries. Since groups from this region rarely have the resources to travel to Brussels or establish a presence there, they must either rely on Euro-level federations, or compete with other groups attempting to meet with a desk officer travelling to their region of the country. As a result, while NGO participation on the whole is quite high in Brussels, it is built around a model of non-scientific NGOs as recipients of policy information. Their presence is beneficial, but not as a source of citizen experience or knowledge. Rather, they are a means of disseminating EU decisions, garnering public support for the EU, and increasing public knowledge about the European institutions and policies. In terms of CEE countries, this points to a decreased public or deliberative role for environmental organisations, and the transformation of CEE environmental policymaking into a largely EU and industrial-based enterprise.

This is tied directly to the continuing desire for expert and technical knowledge in the Commission. Rather than involve citizen perspectives and experiences to bolster the legitimacy and even quality of environmental policies, EU officials generally prefer to use non-governmental sources for their formal expertise as scientists, chemists and ecologists. This professional interaction is compounded by a preference for Euro-level federations and majority perspectives. Such federations can give a filtered and 'European' perspective, but they often lack direct support or membership from the public.

Heavily reliant on external expertise and with little room for grassroots or civic input from CEE environmental NGOs, the European Union that emerged during the 1990s contributed to democratic deficits on both sides of the former Iron Curtain. Rather than drawing on the NGOs from CEE countries to foster democratic inclusion and possibly even improve policy harmonisation, policy integration during the 1990s reinforced the prevalence of economic or industrial interests at the European level. This pattern of development provides information to the public, but does little to encourage citizen-based interaction and input with the decisions being made. As a result, environmental policy officials have become increasingly removed from the general public, and public opinion or participation is a limited factor in environmental policy.

For environmental groups in CEE countries, the combination of barriers to policymaking is formidable. Many in CEE countries recognise Brussels as the new location of policymaking and seek access at that level, yet are constrained by existing patterns of interest intermediation. Time, money and policy expertise place economic interests in an advantaged position, and a more participatory or democratic approach to environmental policy holds little practical attraction for business, national or EU officials. While EU officials claim to be open to input from NGOs, the preference for expert or technical input, rather than more participatory, grassroots and public-based inputs affects the democratic legitimacy of both European and national policymaking bodies. It also forces environmental groups to abandon more dissident elements in favour of professionalisation if they wish to 'succeed', or turn towards small-scale, highly localised projects [*Hallstrom, 2003b*].

While the shift in environmental policymaking to Brussels has had both positive and negative democratic effects on the policy orientations and opinions of CEE policy officials and NGOs, not only are these effects different in each post-communist state, but the processes of enlargement and integration may have also laid the groundwork for a more diverse and democratically rigorous form of policymaking in the European Union. Hyperpluralism and the fragmentation of civil society were problems during much of the 1990s [*Korbonski, 1996*], yet as states such as Hungary, the Czech Republic and Poland move towards membership in the EU in 2004, the combinations of environmental organisations that are emerging in CEE countries point to the possibility of truly multi-level governance in the European Union. This form of (environmental) governance can benefit from transnational organisations such as Greenpeace, EarthFirst! or Children of the Earth that meet EU-level preferences owing to substantial membership rolls and communication networks, while other, more localised, regional or state-level groups can foster grassroots involvement, increase environmental awareness, and promote local participation in policy development, implementation and enforcement.

While it is doubtful that environmental NGOs from CEE countries will become major players in EU environmental policy, the resilience of numerous environmental groups, and the success of environmental education projects such as the Czech Environmental Management Centre in Prague, points to cause for some optimism. Basic economic and political reforms dominated political life during the 1990s, yet as EU membership draws near, there is the possibility of a greater role for both environmentalism and environmental groups in EU and national-level policy. However, this role hinges on the attitudes, orientations and informal behaviours of not only national-level but increasingly EU officials.

As this contribution suggests, throughout much of the accession process, EU officials have tended to reinforce a technocratic and top-down perspective, where environmental NGOs should either provide technical expertise, or improve the legitimacy of EU policies by spreading information to the public. Viewing contact with environmental groups that do not meet these two goals as positive has been rare, and as a result only limited steps have been taken encouraging the participation of civil society actors from CEE countries. Until policy elites adopt a more inclusive model of policymaking, not only are the democratic benefits of participatory policymaking lost, but the legitimacy of both national and EU-level environmental institutions and policies remain diminished. Without these reforms, and as occurred in CEE countries during the 1990s [*Hallstrom 2003b*], the shift from state-based to European environmental policymaking can push environmentalism away from politics, ultimately limiting the development of national and European civil societies and slowing the emergence of a European *demos*.

NOTES

1. There are no exact figures as to the number of formal interests attempting to function at the European level. However, a Commission estimate in 1992 placed this figure at 3,000, and more recent estimates are substantially higher. Just under 80 per cent were business or professionally oriented, with 20 per cent categorised as public interest (environmental, consumer and civil/social groups). Many were established after 1984, and 75 per cent maintain offices in Brussels [*Greenwood, 1997: 179*]. 'Euro' groups grew from 300 in 1970 [*Butt Phillip, 1985*], 525 in 1990 [*Mazey and Richardson, 1993*], to over 900 in 2001 (see also Greenwood, Grote and Ronit [*1992*]).
2. The data for this study were collected from 93 intensive interviews with European Union and national officials. In addition to the interviews conducted by the author in 2000, the study utilises interviews conducted in the early 1990s and held at the Middlemas Archive at the University of Sussex, Brighton. Interviewees included members of the European Parliament, representatives of different levels and sections of Directorates General 3, 11, 15, 16 and Enlargement, former European Commissioners, members of the Delors Cabinet, representatives of the Secretary General of the Commission, a former Director General from the Council Secretariat, and Directors of the Bank of Italy and the DeutschesBundesbank. Interviews were tape-recorded and transcribed by the author. Transcripts were then categorised and divided on the basis of subject matter (enlargement, environmental policy, democracy, CEE NGOs, political culture). Owing to space constraints, these interviews are used here in a journalistic fashion.
3. This density can be symbolised by the web of committees, known as comitology, which has evolved to advise the Commission regarding policy formulation and implementation.
4. Even prior to the Treaty of Rome in 1956, a model of European administration and public policy was apparent. Wallace and Smith note [*1995: 143*], 'Enlightened administration on behalf of uninformed publics, in cooperation with affected interests and subject to the approval of national governments was therefore the compromise again struck in the Treaties of Rome', yet there has also been an awareness of the need for citizen input into European policies. Even in the late 1950s, it was argued that, 'It is high time, therefore, that the peoples be drawn into the venture, and that they grasp what is at stake …' [*Dehousse Report, 1969: Para. 22*]. Despite this, much of the literature addressing the development of the European Communities has stressed the elite-bargaining character of the treaties. This challenges the legitimacy of the EU institutions, particularly as the scope and scale of their decisions expands [*Wallace and Smith, 1995; Quermonne, 1990, in Hayward, 1995*].
5. It is erroneous to refer to a 'European' pattern of interest intermediation. Policymaking arrangements can and do differ between policy sectors. Influence is often dependent on the logic of policy, which does of course vary from policy sector to policy sector [*Cawson, 1992; Lowi, 1964*].
6. 'Bargaining chips' include information, economic influence, prestige, implementory capacity and interest 'cohesion' across multiple groups.

7. The Mouvement européen–Belgique (European–Belgian Movement). This group organises informational weekends, conferences and retreats for two purposes. The first is to further access for Belgians into the European political system. The second purpose of such meetings is to increase knowledge and information among the public about the EU and its policies. By doing so, they contribute to the legitimacy of the European Union as a political entity, and citizens gain a better understanding of the EU, as well as an opportunity to participate [*Helbig, 2000*].

REFERENCES

Börzel, Tanja A. and Thomas Risse (2000), 'Who is Afraid of a European Federation? How to Constitutionalise a Multi-Level Governance System', in Joerges Christian, Yves Meny and J.H.H. Weiler (eds.), *What Kind of Constitution for What Kind of Polity? Responses to Joschka Fischer*, Cambridge, MA: The Robert Schuman Centre for Advanced Studies at the European University Institute, Florence.

Butt Philip, Alan (1985), *Pressure Groups in the European Community*, University Association for Contemporary European Studies. London.

Carmin, JoAnn and Barbara Hicks (2002), 'International Triggering Events, Transnational Networks, and the Development of the Czech and Polish Environmental Movements', *Mobilization*, Vol.7, No.2, pp.304–24.

Cawson, Alan (1992), 'Interests, Groups and Public Policy-Making: The Case of the European Consumer Electronics Industry', in Justin Greenwood, J.R. Grote and Karsten Ronit (eds.), *Organised Interests and the European Community*, London: Sage, pp.49–61.

Dahl, Robert A. (1961), *Who Governs?*, New Haven: Yale University Press.

Dehousse Report (1969), *Pour l'election de PEG au suffrage universel direct*, Dehousse Report, PEG, Luxembourg.

Fagin, Adam and Petr Jehlička (1998), 'Sustainable Development in the Czech Republic: A Doomed Process?', in Susan Baker and Petr Jehlièka (eds.), *Dilemmas of Transition: The Environment, Democracy and Economic Reform in East Central Europe*, London: Frank Cass, pp.113–28.

Fagin, Adam and Andrew Tickle (2001), 'Globalisation and the Building of Civil Society in Central and Eastern Europe: Environmental Mobilisations as a Case Study', paper presented at the ECPR Conference, Sept 6–8, 2001 at the University of Kent, Canterbury.

Fric, Pavol (1999), *Activities and Needs of the Non-Profit Organizations in the Czech Republic*, Prague: ICN (Information Center for Foundations and Other Not-For-Profit Organisations).

Greenwood, Justin (1997), *Representing Interests in the European Union*, New York: St Martin's Press.

Greenwood, Justin, Jürgen R. Grote and Karsten Ronit (eds.) (1992), *Organised Interests and the European Community*, London: Sage Publications.

Hallstrom, Lars K. (2003a), 'Support for European Federalism? An Elite View', *Journal of European Integration*, Vol.25, No.1, pp.51–72.

Hayward, Jack (ed.) (1995), *The Crisis of Representation in Europe*, London: Frank Cass.

Helbig, Danielle (2000), 'Noel Muylle – Fonctionnaire et militant europeen', *Commission En Direct*, 31 Mar.–8 Apr., sec. People, p.6.

Jancar-Webster, Barbara (1998), 'Environmental Movement and Social Change in the Transition Countries', *Environmental Politics*, Vol.7, No.1, pp.69–92.

Joerges, Christian, Yves Meny and J.H.H. Weiler (eds.) (2000), *What Kind of Constitution for What Kind of Polity? Responses to Joschka Fischer*, Cambridge, MA: The Robert Schuman Centre for Advanced Studies at the European University Institute, Florence. The Jean Monnet Chair. Harvard Law School.

Kitschelt, Herbert P. (1986), 'Political Opportunity Structures and Political Protest: Anti-Nuclear Movements in Four Democracies', *British Journal of Political Sciences*, Vol.16, pp.57–85.

Kohler-Koch, Beate (1997), 'Organised Interests in the EC and the European Parliament', *European Integration Online*, Vol.1, No.009.

Korbonski, Andrejz (1996), 'How Much is Enough? Excessive Pluralism as the Cause of Poland's Socio-Economic Crisis', *International Political Science Review*, Vol.17, No.3, pp.297–306.

Kriesi, Hanspeter, Ruud Koopmans, Jan Willem Duyvendak and Marco G. Giugni (1992), 'New Social Movements and Political Opportunities in Western Europe', *European Journal of Political Research*, Vol.22, pp.219–44.

Lowi, Theodore (1964), 'American Business, Public Policy, Case Studies and Political Theory', *World Politics*, Vol.16, pp.677–715.

Mazey, Sonia and Jeremy Richardson (eds.) (1993), *Lobbying in the European Community*, Oxford: Oxford University Press.

Middlemas, Keith (ed.) (1995), *Orchestrating Europe: The Informal Politics of the European Union 1973–1995*, London: Fontana Press.

Millard, Francis (1998), 'Environmental Policy in Poland', *Environmental Politics*, Vol.7, No.1, pp.145–61.

Nentwich, Michael (1996), 'Opportunity Structures for Citizens' Participation: The Case of the EU', *European Integration Online*, Vol.0, No.1.

Quermonne, Jean-Louis (1990), 'Existe-t-il un modele politique europeen?', *Revue Francais De Science Politique*, Vol.40, No.2, pp.192–211.

Richardson, Jeremy (1994), 'EU Water Policy: Uncertain Agendas, Shifting Networks and Complex Coalitions', *Environmental Politics*, Vol.3, No.4, pp.139–67.

Stoker, Gerry (1998), 'Governance as Theory: Five Propositions', *Social Science Journal*, Vol.155, March, pp.187–95.

Wallace, Helen and Alasdair Young (eds.) (1997), *Participation and Policy-Making in the European Union*, Oxford: Clarendon Press.

Wallace, William and Julie Smith (1995), 'Democracy or Technocracy? European Integration and the Problem of Popular Consent', in J. Hayward (ed.), *West European Politics, Special Issue on The Crisis of Representation in Europe*, London: Frank Cass, Vol.18, No.3, pp.137–57.

Further up the Learning Curve: NGOs from Transition to Brussels

RUTH GREENSPAN BELL

Accession to the European Community is a certainty for some countries of Central Europe and on the horizon for others. What impacts will membership have on the environment and on the numerous environmental non-governmental organisations (NGOs) and environmental advocacy groups of the region? The EU has said that merely conforming laws will not be sufficient, and that it will demand implementation of the adopted directives by the accession states, although phase-ins are being offered as a solution to some of the more difficult to achieve requirements. Will this mean that improved environmental conditions and more forceful and effective environmental protection efforts will be the outcome of joining the EU? The EU's environmental directives impose tremendous costs on, and demand institutional skills of, small, inexperienced and under-funded environmental ministries that have had great difficulty realising their existing laws and regulations. The enormity of the tasks imposed by EU membership on frail institutions is a looming problem. In March 2003, Romania, for example, estimated the cost of ecological reconstruction in the range of $20–21 billion [*Radio Free Europe, 2003*].

For environmental NGOs, EU accession might diminish their power in some respects and enhance it in others. On the one hand, the environmental agenda that the accession countries must address by virtue of EU membership was set in Brussels, rather than in each country or in response to domestic demand, and demands related to the environment from Brussels will continue into the foreseeable future. It will be more difficult for domestic NGOs to lobby successfully for national-level environmental initiatives that are inconsistent with or do not enhance the objectives set out in EU directives, simply because there will be few resources and little residual energy for adding additional tasks to an already very full agenda dictated by the EU. The NGOs have the option of trying to shape the EU agenda, sometimes in partnership with other European NGOs. But to do so, Central and Eastern European (CEE) NGOs that have only recently gone up a significant learning curve will need to develop additional new skills, expertise and perhaps new sources of funding. This is because, among other things, they will be required to learn to interact effectively with Brussels [see also, *Hicks, this volume*].

On the other hand, EU membership creates opportunities for NGOs and other organisations to police governments on their progress in implementing the directives. NGOs can take advantage of these opportunities to act as monitors and advocates; if they learn well, they can use EU mechanisms to force their governments, in effect, to implement the many EU-consistent domestic environmental laws and to embarrass them when they do not. The first few years of EU membership will show whether Central European NGOs prove to be more like the northern European NGOs that have learned the advantages of raising EU environmental law issues in domestic litigation, or the southern European NGOs whose track record is very different.

Environmental NGOs in Central and Eastern Europe

Environmental organisations, operating either under the auspices of the government or in a way that didn't violate the bounds of acceptability, could be found in most countries in the former Soviet bloc. It is well known now that many of these groups were active in the limited public debate that was possible before 1989 and in the period leading up to the transition [*Welsh and Tickle, 1998; Jehlička, 2001*]. Famous examples include the Polish Ecology Club (Polski Klub Ekologiczny), the activism that developed to oppose the Hungarian–Czech plan to build the Gabickovo-Nagymaros Dam, and the rise in Bulgaria of ecological organisations such as the Green Party and Ekoglasnost, whose platforms stressed decentralised government and a strong role for the individual in determining quality of life and preservation of the environment. Conservation groups did exist under the old regimes. But the ecological movement had a different character; environmental activism was often safe-harbour and a surrogate for otherwise dangerous political activism [*Welsh and Tickle, 1998; Jancar-Webster, 1993, 1998*]. Some environmental activists were looking for political change, not merely environmental reform. There is a body of opinion that the pre-1989 environment ministries and laws were created, at least in part, to quiet criticisms from environmental groups as well as to respond to international pressure [*Carmin and Hicks, 2002*].

With the fall of the communist governments and the opportunity to form associations independent of the state, a more varied NGO community has developed and the very definition of environmental organisations has changed. There is no longer a need for environmental organisations to provide cover for protest groups. Post-1989, many of the organisations whose purpose is to represent the environmental 'public interest' have refined and diversified their tactics and approach and many now more closely mirror similar organisations in the West [*Carmin and Hicks, 2002; Jancar-Webster, 1998; Hicks, 2001*]. In the new conditions, NGOs have grown in their understanding of what roles they play in advancing the purposes of environmental protection. Previously, it was anathema to work constructively with the government. Instead, demonstrations and street protests were their means of making their views known. Now, many NGOs lobby, provide information to legislators and the public, litigate, and comment on legislative proposals – in other words, the NGO community, or at least part of it, interacts with government, rather than shuns or fights it.

US foundations actively supported environmental NGOs from the start of the transition. One of the best examples was a very productive alliance between the German Marshall Fund, Rockefeller Brothers Fund, Charles Stewart Mott Foundation and Atlantic Philanthropies that created a small grant programme in CEE countries after the fall of communism for environmental awareness, reform and the development of civil society in the Czech Republic, Hungary, Poland and Slovakia. This initiative – the Environmental Partnership for Central Europe – has evolved into independent foundations in the four countries with a fifth foundation added in Romania. The foundations make grants and act as a resource and catalyst for the broader environmental community in their countries and throughout the region to address country-specific problems and problems common to the region [see *Beckmann and Dissing, this volume*].

Examples abound of environmental NGOs interacting productively with their governments to institute change. For instance, NGOs from throughout CEE countries were active in the international negotiation and domestic implementation planning for the UN/ECE Convention on Access to Information, Public Participation in Decision-Making and Access to Justice in Environmental Matters, popularly known as the 'Aarhus Convention'.[1] NGOs have also been active in more local debates, as demonstrated by a Polish NGO that lobbied at a city and national level on transportation issues; the result was the establishment of a Warsaw Transportation Roundtable to make recommendations to the City government for improved transportation and urban planning policies and implementation plans. Also in Poland, an NGO successfully campaigned to change the structure of the country's electricity/telephone poles, which annually killed some 30–50,000 migrating birds. NGOs have been very active in the establishment of 'Right to Know' laws. A small public interest law group in Slovakia has successfully lobbied the Parliament and also brought environmental cases to the Constitutional Court. And a broad array of partners and local stakeholders worked to preserve southern Poland's Jura Highland, a natural and cultural treasure by developing a tourist route to alleviate recreation pressure on sensitive landscape, and a grant programme for schoolyard conservation projects. A final example is the Tisza Klub of Hungary, formed in response to the Yugoslav war devastation and subsequent cyanide contamination following a mine spill in upstream Romania, to protect the Tisza River, a tributary of the Danube.

The apparent belief of foundations and other funding institutions that the transition has been completed throughout the CEE region has reduced the amount of available aid for environmental efforts. Early in the transition, NGOs looked principally for their funding to Western foundations. The relationship has been productive in many cases, but also habit forming to the degree that some local NGOs have been characterised as 'grant junkies'. As external funding has been reduced or even terminated, many NGOs have encountered difficulties finding the financing to continue their efforts [*Jehlička, 2001*]. In 2002, for example, the Open Society Fund announced changes in its funding priorities, the effect of which will be to further reduce support for indigenous environmental NGOs. The NGOs' search for alternative sources of funding runs into the reality that CEE countries don't have either the tradition of this kind of giving, or the advantageous tax laws found in the United States, that encourage private contributions to the not-for-profit world by making them tax deductible. The prospects of significant amounts of private support for the NGO movement in CEE countries must be further discounted because the NGOs are located in poor societies.

On the other hand, some countries have found innovative ways to make support available to NGOs. In Hungary, for example, taxpayers may donate one per cent of their personal income tax bill to the charity of their choice, provided that the charity complies with certain legal requirements. The 1996 law allows donors to designate a non-governmental organisation, a national institution, a designated public foundation, a governmental programme (such as programmes for higher education) or a cultural institution. The Hungarian Tax Authority facilitates donations by providing a small form with the main tax form. Some NGOs have been adroit in promoting this possibility. Hungary's most prominent public interest environmental litigating entity, the Environmental Management and Law Association (EMLA) posts instructions on its website, encouraging supporters to 'write EMLA's complete name in Hungarian and EMLA's tax number as written below and enclose it in an envelope with your tax papers to be posted by you or your accountant. Write your name, address and personal tax number on the front of the envelope, seal it and sign the back of the envelope across the sealed flap.'

NGOs have had to move up a learning curve, a sometimes quite steep one, not only in their efforts to find sustenance for their activities, but also to develop the capacity to manage the funds they do receive. Their need to manage outside funding and be accountable for their spending has forced them to develop accounting proficiency and other new skills. Some NGOs survive because they are associated with, or affiliates of, international organisations such as Friends of the Earth, Greenpeace and WWF, but this affiliation can smother as well as nurture local NGOs. Certainly, the access to resources and savvy of transnational NGOs can help local NGOs become more effective advocates, and become more knowledgeable about interacting with the EU system to put environmental issues on the agenda or raise public awareness. But this relationship could also turn out to have been a Faustian bargain. Not only can 'home-grown' NGOs easily be overwhelmed and marginalised by the presence and force of the multinational organisations, but the issues of importance to the international organisation may be very different from those the locals want to pursue.

A few local NGOs – particularly those with environmental law skills – have found alternative sources of support for financial viability. They do contract work for the European Union and/or environment ministries in support of EU accession, or bid for contracts from the EU to provide, for example, environmental law drafting for other accession states or to support the rebuilding effort related to the Balkan war. The downside of these admirable efforts to stay afloat is that the NGOs' independence and advocacy function can be reduced or even compromised by its meeting the demands of the EU or other funders. NGOs will have to disentangle or at least reconcile roles that might otherwise be considered diametrically opposed – that of adviser to governments versus acting as the loyal opposition to push governments towards change.

Environmental Regulation and Compliance, Before and After 1989

Long before 1989, most countries of Central Europe had both environmental laws and institutions in place, although they were stronger on paper than in achieving pollution reduction. It was common to have an environment ministry, a framework environmental act, and a set of ambitious standards for discharge to air, water and soil. Most countries used fees and fines on emissions as basic tools of environmental protection. In practice, the effectiveness of any of these institutions and laws was open to question; the states' main objectives continued to be growth, production and national security. Pollution charges were paid out of the soft budgets of state enterprises and, therefore, had little chance of influencing enterprise behaviour.

A burst of environmental activity in the early 1990s, following the fall of communism, resulted in a number of new environmental laws, attention to the environmental aspects of privatisation, and experimentation with new initiatives [*Bell, 1992a, 1992b*]. An example of the latter is the effort to apply market-based environmental instruments such as emissions trading [*Bell, 2003; Bell and Russell, 2002, 2003*]. However, as the transition deepened, environmental activities essentially died down in most countries [*Jehlička, 2001*]. There were two reasons for this. Opposition leaders, who had previously worn environmental guises, were now able to move into the political mainstream. More urgently, other pressing issues took centre stage. CEE countries were faced with grave challenges as they sought to transform on all fronts – socially, economically and politically. There was little remaining energy to tackle difficult environmental problems. As a result, most countries had at least framework environmental laws on the books, but often without implementing laws or regulations, or much institutional commitment, until the process of EU accession began.

In recent years, the environmental agenda has been dominated by EU accession efforts. The desire to join the EU and to approximate and harmonise domestic laws has spawned a cottage industry that writes laws and adjusts policies to fit EU directives, not only for the environment, but for all matters of government. In some ways this has been helpful for the environmental movement, which might otherwise have been completely sidelined by competing demands. That environment is among the subjects of the existing EU directives has kept it on the agenda in countries where it might otherwise have been eclipsed. However, there are substantial questions about whether the countries' environmental approximation efforts are, or will be, mainly a formality or more deep-rooted, despite signals from Brussels that implementation will be required. Is the edifice that is being constructed a Potemkin Village, analogous to the pre-1989 structures?

What Must be Done to Join the EU

Reports from the EU repeatedly pinpoint the 'environment' as one of the most problematic areas for the harmonisation of legislation. This is not surprising. The task is to adapt about 200 environmental directives of varying complexity. Some are quite specific. Others leave considerable room for implementation discretion at the national level. Little room is left for negotiation on process and none on the substance of the directives [*European Commission, 1997*]. The process by which this happens is well known. The environment ministry in each country must compare existing domestic legislation with the EU *acquis communautaire* ('the screening process') and then bring domestic laws into line with those of the EU. To ensure that the process does not become entirely a paper exercise, the screening process includes assessment of the institutional, monitoring, reporting and communication needs of the relevant government agencies [*Lynch, 2000; Jacoby, 1993*].

The complexity of the process obscures a very real question: whether the countries have the wherewithal to implement and enforce so many new laws instituted so quickly. Hungary, for example, closed its environmental 'chapter' in June 2001, after instituting 26 new regulations in about seven months, mostly in the area of air pollution prevention and water management. Recognising to some extent that change cannot take place overnight, the EU granted extensions for the implementation of a number of specific environmental laws and regulations. In order to gain these concessions, Hungary had to provide a detailed proposal complete with schedules, target tables and financial particulars for each of the four different areas where concessions were granted [*Bell, forthcoming*].

The burst of energy that allowed Hungary to close its chapter responded to the European Commission's 'Agenda 2000' Progress Report that identified areas of progress and areas for additional work. The Report provides a glimpse of the range of environmental subjects that accession placed on the national agenda of each of the CEE countries, and a hint of the complexity and expense of turning promises into action. It noted improvement on such disparate issues as laws for nature protection, industrial pollution, petrol and diesel quality (including banning the sale of leaded petrol), waste management, environmental product charges for oil products, collection and processing systems for waste oils and used batteries, lawnmower and household appliance noise standards, large combustion plants, and control of major accidents. However, the November 2000 Report noted the need for additional work to align legislation for environmental impact assessment, access to information, waste management, quality of drinking and bathing water, wastewater treatment, prevention and reduction of industrial pollution, protection and reduction of environmental pollution by asbestos, and certain other noise emissions [*Europa, 1999, 2000*].

One extension granted to Hungary by the EU involves incineration of hazardous waste disposal. None of Hungary's biomedical waste processors and only two of 22 other hazardous waste incinerators are currently in compliance with the EU standards as of the time of this writing. The EU, which had only just implemented the relevant directive, recognised that reforms will take time, but their solution was to give Hungary until 2005 to upgrade these plants or to build new ones. A 7–14 year extension was granted for biological treatment of sewage water for larger settlements. Other concessions include upgrading of certain types of air filters and instituting a recycling system for packaging materials by 2005. It is estimated that these particular improvements alone will absorb about $8 billion or about 1.5 per cent of Hungary's GDP each year until 2015 [*Szabo, 2001*]. Ninety-three per cent of these funds must be generated by the country; assistance from the EU can cover the balance. As with all accession countries, the EU will closely scrutinise Hungary's development in these areas and has the right not to carry through the accession if it deems progress unsatisfactory.

The most important point to note here is simply how much energy it took for the accession countries to meet the demands of the EU on a purely formal level, namely that of law writing, and what resources must be dedicated to build the required wastewater treatment plans and incinerators. The effort for environmental approximation was only one of numerous similar exercises in the subject areas of labour, agriculture and a myriad of other subjects. The demands these laws place on governments will overshadow, if not eliminate entirely, domestic discretionary environmental expenditures and direct the environmental agenda of the countries in transition for the foreseeable future. And the complexity of the tasks involved presents a separate set of strains. Is a 7–14 year period for installation of biological treatment of sewage from larger cities a generous extension or an entirely unrealistic requirement, in view of the cost and complexity of the task? To meet this demand, the countries must not only construct expensive, sophisticated centralised wastewater treatment plants but also sewage connections to bring the waste to treatment.

To set these demands in context, it took the US about 28 years to almost double the number of people served by privately operated treatment works (POTWs) with secondary or greater levels of wastewater treatment, from 85.9 million in 1968 to 164.8 million in 1996. In the same time period, the number served by POTWs with less than secondary level of treatment (that is, discharge of raw waste or waste treated to a primary level) dropped from 54.6 million in 1968 to 17.2 million in 1996; the 17.2 figure includes about 5.1 million people served by systems that were legally exempted from the requirements of the Clean Water Act and allowed to discharge less than secondary treated effluent into deep, well-mixed ocean waters. The US is far bigger than any of the accession countries, but it is also a great deal richer. The total costs of municipal wastewater infrastructure and operations in this timeframe have been well over $200 billion with a comparable amount for operations and maintenance. Over a 25-year period, 1970 to 1995, EPA provided $61.1 billion in Federal Construction Grants Program funds towards this effort, and since 1988 has provided over $16.1 billion in support for state revolving loan funds for a wide range of water quality improvement projects [*Hudiburgh, Jr., 2003*]. Most of this construction took place without the distractions – economic and political transition and social change – that are a constant in Central Europe, and in a context of strong public attention and the two-edged sword of both EPA support and rigorous enforcement.

Moreover, in fairness, existing member states are having tremendous difficulty with these expensive, complex requirements. The so-called 'transposition deficit' of countries already members of the EU is followed in a document called the 'Internal Market Scoreboard'; it details how each country is doing in implementing the EU directives correctly and on time at a national level. The Scoreboard notes that for all member states, the transposition deficit for the environmental directives is much higher than for overall directives [*European Commission, 2001*].

How 'Real' is the EU Process for Progress towards Environmental Protection?

To this author, the process, and the speed, with which the leading accession countries are incorporating EU standards raises worrying concerns about whether the adoption of EU requirements will prove to be illusory or real. Environmental regulation in any country is not a task that can be simply assigned to the government to carry out. Typically, many people distributed throughout society must significantly change their institutional and personal practices to achieve the necessary result. Industrial facilities must put into place costly control technology, but beyond factories and treatment plants, even ordinary citizens must often take actions that might cause them inconvenience or expense. Reducing the flow of nutrients and toxics to the Danube or the Vistula, for example, will require the participation of industry, municipalities, large and small farmers, individuals who garden small personal plots, and a host of others. The mere fact that Parliaments have enacted new laws does not necessarily indicate whether there is a public willingness to undertake these efforts. Will citizens have sufficient trust and confidence in the rules and requirements they are asked to implement that they will assume the burdens that the requirements impose [*Bell, 1992a*]?

Normally in a democracy, questions of popular will are sorted out through the legislative process. If there is sufficient popular support for any particular environmental restriction, the legislature will act. If a culture of compliance exists, most factories and companies will follow the rules. Enforcement is the process of ensuring that even the most disgruntled will follow the rules.

Approximation, in contrast to the democratic process of law making, puts into place requirements that have been defined and adopted by the European Parliament. The current member states participated in the process of law adoption, but the accession states have not had this opportunity. The aspirants to membership are, in effect, adopting a bundle of already formulated rules, in return for which they gain the benefits of EU membership. They do this through their democratically elected legislatures, and national referenda have confirmed the voters' desire to become part of the EU. But the downside of this is that they have not been part of the participatory process that some sociologists argue is necessary for developing public trust and reinforcing the legitimacy of the decision process and that may, in turn, be connected with achieving implementation and compliance [*Tyler, 1990*].

Any country that wants to join the EU must confront this dilemma, but the conundrum has particular resonance, and irony, for the countries that have recently emerged from state socialism and are working hard to incorporate democratic processes into the legislative process. The process of EU accession – accept these rules or be denied membership – has arguable parallels to socialist practices of law creation, which also lacked public process and were essentially dictated. This is especially true for EU directives that are relatively prescriptive, perhaps less so for those that leave a certain range of discretion to national legislators in the interpretation stage, in which case the local Parliament can shape its laws to fit its domestic needs, so long as the result is consistent with the framework provided by the directive.

Prospectively, when the new member states have the opportunity to participate in future EU debates over new legislation, the concerns about legitimacy may diminish. In theory at least, the concerns and interests of Slovenia or Hungary will be considered along with those of Germany and France before laws are enacted. In fact, however, in view of the large amount of new directives being considered and the relatively small resources of the new member states, particularly the smallest ones, the accession countries may *de facto* have to concentrate their energies on only a handful of laws of greatest concern to them. There is reason to believe this already happens in the EU, with the richer countries better able to participate in the law creation process. In this case, a certain amount of law will be enacted with greater participation by the more powerful countries, even if each country (large, small, rich or poor) has only one vote in the final adoption process.

Nevertheless, popular sentiment in most of the CEE countries supports accession despite some seemingly unfavourable conditions that come as part of the commitment. The strong psychological and economic desire for full integration into the European Community may be sufficient to override democratic concerns about particular rules and requirements, and create the conditions under which domestic populations will undertake the necessary efforts to realise the specific requirements of the EU laws. In other words, if a sufficient number of Poles, Czechs, Slovenians or Hungarians want to join the European Community, and if compliance with the environmental directives is a necessary component of the 'package' to enter Europe and obtain the benefits of that relationship, that alone may be sufficient to substitute for the consensual process that accompanies rule creation in a democracy. Farmers along the Vistula River will control their animal waste and citizens of the Czech Republic will recycle their package paper and aluminium.

However the CEE accession countries respond to the resource and institutional challenges inherent in the EU's environmental directives, NGOs may need to rethink their priorities and the very way they go about effecting environmental change in their respective countries. If they seek to put environmental issues on their national agendas that are either inconsistent with EU directives, or deal with subject matter that the EU is not currently requiring of member states, they may learn that their countries have little patience to consider such additional environmental burdens, or lack the domestic resources to carry out such expensive demands. If governments are struggling to find the wherewithal to effectuate the EU's demands, they may not be very enthusiastic about taking on other new and arduous projects. What if NGOs find local conditions that militate against using EU solutions, or if they simply don't agree with the EU priorities? The results could be extremely frustrating.

Clever NGOs might find ways to repackage their initiatives so that they appear to fit more closely to the EU directives, or to use the directives and EU membership in a more aggressive fashion to force their own countries to put real resources into environmental programmes. A road map to the several ways they can do this is found in the four new levels of scrutiny that can be brought to bear on the specifics of compliance with EU directives that joining the EU provides. Before 1989, there were few remedies available when a Soviet bloc government failed to honour its own laws. Similarly, there was no oversight mechanism for frustrated citizens to invoke. This will change with EU membership.

Each of the European Community policing mechanisms has its own unique features and none of them is perfect. Some must be initiated by the Commission itself, while individuals can invoke others. Together, they can help ensure that environmental obligations are incorporated into practice.[2] In its role as the guardian of EU law, the first mechanism is the Commission's power to bring 'infringement proceedings' in the European Court of Justice against member states. If a member state found to have violated Community law has not remedied this situation within a reasonable time (that is, by bringing its national situation into substantive conformity with EC legislation),[3] the Commission can start a case seeking imposition of a daily penalty.[4] Individuals do not play a role in these proceedings, but individuals can bring infringements directly to the attention of the Commission by letter or may do so indirectly by their right of complaint to the Parliament.[5] There are weaknesses in this procedure, but it nevertheless exists. The European Commission has also made it clear that it will monitor member states on their implementation of directives and it has already issued some warnings.[6]

Another avenue of review is found in the European Court of Justice's doctrine of 'direct effect' [see also, *Kružíková, this volume*]. Individuals can invoke provisions of Community law against public authorities (but not individuals) before national courts. The provisions invoked must be adequately clear and precise.[7] Through this legal device, the main legal instrument for EC environmental law – the directive, which is an order to a member state to act – can create rights for individuals.[8] Because directives cannot be invoked against individuals, the Court has created the doctrine of 'sympathetic interpretation' to close the 'legal protection gap'. In cases involving individuals, national courts and authorities must interpret national legislation as much as possible in conformity with the EC legislation.

Finally, an individual can seek damages before a national court against an EU member state for violations of Community law caused by that state. This doctrine comes out of the groundbreaking 1991 *Francovich* case.[9] For the first time in the history of the Community, the European court ruled that it is a principle of Community law, inherent in the system of the EC Treaty, 'that the Member States are obliged to make good, loss and damage caused to individuals by breaches of Community law for which they can be held responsible'.[10] The *Francovich* case may hold the most promise as a complaint vehicle for domestic Central European constituencies unsatisfied with their countries' implementation of environmental directives.

To understand why this could advance the cause of genuine (rather than formal) environmental protection, it is useful to examine a somewhat parallel experience in the United States. When US federal agencies were sluggish in implementing environmental laws for which the Congress had set specific time deadlines in the laws, public interest environmental groups brought 'mandatory duty law suits', asking the courts to compel government agencies to act and to adhere to the directives set in federal law.[11] These cases were instrumental in moving the environmental agenda forward in the United States and forcing the EPA to act in a timely manner. The *Francovich* line of cases in Europe suggests an analogous strategy might be available for creative environmental NGOs of Central Europe.[12]

Without waiting for the European Commission to take action, NGOs can bring lawsuits to hold specific member states, including the new accession countries, responsible for their failure to implement the environmental directives, if these groups can meet the conditions set out in the *Francovich* case. The result of a successful case will be damages, *not* an injunction to the relevant government agency to act, as it is in the United States. Nevertheless, in the best case, NGOs dissatisfied with the performance of their countries can use this kind of tool to force governments towards active implementation. At a minimum, the very fact of a lawsuit can create unfavourable publicity and embarrassment for a state that has failed to meet its commitments, as it has for accusations that Belgium has not complied with water quality directives.

All these are new tools that will only exist by virtue of EU membership. Whether the CEE NGOs use them to reinforce the promises contained in the adoption of EU environmental directives depends on the domestic legal system and legal culture and their own creativity and will. Just because a legal tool exists does not guarantee that it will be used, and just because an issue exists does not mean it will resonate with the local population. The NGOs of the accession countries could follow the model of The Netherlands – where a large portion of national environmental cases contain an EC law component – or might be more like their peers in Spain, where there is a relative dearth of EU causes of action, indicating that citizens and the legal profession do not appear to have the same familiarity with Community law. If they take the former course, this kind of tactic could be a prod towards active implementation at a national level of the EU directives.

NGOs and the Private Sector as Watchdogs

Independent of the newly available EU mechanisms, the civil sector, with its diverse concerns and interests, could find other ways to play an independent role in policing its respective country's environmental performance and helping it achieve its goals. In the Soviet past, environmental compliance was almost completely a function of the centralised decision process, controlled by the government. State-owned industries and the officials of the environment ministry had no room for independent action, and the ability of individuals and associations to challenge official action or – more likely – inaction was severely limited. Since 1989, multiple centres of power and authority have developed, and industry's goals have changed.

An NGO watchdog function can take any number of forms. In the countries that have signed and ratified the Aarhus Convention, NGOs and others now have the legal right to ask for information about environmental performance from the government. This Convention, which was also signed by the EU, introduces procedures very similar to the United States' Freedom of Information Act (FOIA). It commits countries to enact and implement laws and institutions that will allow any person or organisation to ask for and receive environmental information in the possession of the government, with certain exemptions for sensitive and legitimately confidential information. FOIA-like laws were legislated in a number of transition countries throughout the 1990s, but the implementation track record left something to be desired. By the time of the ratification of Aarhus, many of these countries still needed to build government infrastructure, systems of records, ways to track and respond to citizen requests, and methods to ensure that requests were responded to in a timely fashion (or at all).

The NGOs that played an active role in the process of writing the Aarhus Convention have re-energised the demand for environmental information access. Although it is easy to say that freedom of information is an innovation in CEE countries, the fact is that it is not a European tradition. In Western Europe, only a handful of countries – including Finland, The Netherlands, Norway and Sweden – offer these opportunities to their citizens at a comparable level to the United States; FOIA was a late arrival in the UK, only beginning to come into effect in April 2002, although it was enacted earlier. Indeed, many commentators have noted the comparatively low levels of transparency, accountability and public involvement in decision making including environmental policy EU institutions.

EU participation in the Aarhus Convention led to revisions in its directive on environmental information (Directive 2003/4/EC (28 January 2003), repealing Council Directive 90/313/EEC (7 June 1990)). The Aarhus Convention and anticipation of Directive 2003/4/EC (28 January 2003) offered an opportunity to NGOs in Central and Eastern Europe and some have been active in pursuing it. Assisted and sometimes led by the Regional Environmental Center for Central and Eastern Europe (REC, located in Szentendre, Hungary), NGOs from throughout the region (including Ukraine to the east) organised together during the negotiation of the terms of the Convention and later to push for ratification and to pursue the rights contained in it. The author was present at strategy sessions in Szentendre, in which NGOs formulated domestic plans to push their governments towards both ratification and ultimately implementation of the Aarhus Convention.

Implementation has been a concern because the REC and many NGOs appreciate that a number of environmental laws on the books in Central and Eastern European countries have not gone much beyond the law enactment stage. They want the Aarhus Convention to be an exception. For example, a pilot implementation project in Hungary and Slovenia in 2000–2002, in which the author was deeply involved, demonstrated both NGO activism and commitment in the two countries, as well as some of the distinct differences in terms of preparedness between two of the more advanced countries of the region (both slated for early EU entry).[13] The project worked with government officials who were on the front lines of environmental information access and therefore tasked to respond to public requests for government-held environmental information, and with domestic NGOs who might be among the main requesters and users. The focus was on practical measures, such as the need for government record management and retrieval systems in order to be responsive to requests, as well as developing better understanding of the legal requirements.

As became clear in the Hungarian/Slovenian project, even these two countries presented very different challenges, at both the government and at the NGO level. Hungary had an established legal framework for public access to information earlier than Slovenia, and therefore a head start. With the basic laws in place, and 'public relations' departments in ministries assigned to interact with the public and accept information requests, but not always proficient in responding, the tasks in Hungary were focused on improving the government's understanding of its obligations and helping government personnel devise more effective ways of responding to requests. And Hungary, with a much larger population than Slovenia, already had more NGOs engaged in environmental public participation. There was a critical mass to work with. As a result, Hungary had the person-power and the experience to move forward more rapidly.

With project support, Hungarians created a detailed Hungarian-language guidance manual for public officials and an empowering citizen guide with sample letters, instructions on how to submit requests, and advice on how to protest incomplete responses. The effort of working together to devise solutions also helped build better cooperation between the Hungarian Ministry of Environment and the Ministry of Transport and Water Management (now integrated into one ministry) as well as between the ministries and NGOs. At the end of the effort, a Hungarian government manager of water-related data explained that she had begun to understand the broader ramifications of her job in developing a constituency for Danube pollution reduction.

In Slovenia, by contrast, the basics were not yet in place. Slovenia needed to refine and extend its laws. Because the government of Slovenia was very focused on EU accession, it was not difficult to get its attention to these issues, so long as they served the purpose of the accession process. But there were limits. Slovenia is a small country of approximately two million people. Its limited government resources were stretched by the huge demands of accession. In the environmental part of the accession, alone, the Aarhus component was but one of almost 200 environmental directives that needed to be adopted. The government's interest (and perhaps the available resources) could not manage much more than attention to the formalities – legal reform and not institutional implementation. Moreover, Slovenia did not boast the number and diversity of environmental NGOs of Hungary. The burden of pushing forward on these issues fell on a smaller number of very able, but over-extended environmental advocates. Thus, the focus in Slovenia was, of necessity, on improving the legal basis for public access to environmental information. Hopefully, manuals and citizen guides such as those developed in Hungary will follow after the initial issues are addressed.

This demonstrates how far NGOs have come in the transition, but also that there are still differences among them and in the countries in which they work, such that strategies and outcomes are not going to be predictably the same. The next phase of the described project, anticipated to begin in 2004, will bring government personnel and NGOs that participated in Phase I from Hungary and Slovenia with their counterparts in Serbia, Bulgaria, Croatia, Bosnia-Herzegovina and Romania, to share experience. The five new countries each have their own environmental NGOs, but their experience each has been very different in the 15 years of transition, ranging from brutal war to severe economic strains. Even if their experiences had been identical, each country historically carries the burden of distinct differences in their national and legal culture. These differences will help determine how NGOs interact and what they can accomplish within their own countries.

This study has focused on the roles and activities of environmental NGOs. However, it is important to acknowledge that taking action to promote environmental interests is not limited to these groups, but extends to a range of non-state actors including industry and industry associations. Under state socialism there was no need for industry to organise in order to represent its own interests before the legislature, government or the public – it *was* a branch of the government, and neither the legislature nor the public were important factors. Now, as in Western democracies, industry is distinct from government. Some countries have an environmental service sector including firms that provide environmental audits or clean up contaminated sites.

Private sector organisations interested in promoting environmental management or otherwise pursuing their economic and social interests have formed trade associations and organisations. These provide a forum for firms that are seeking to improve their environmental management, share information, or lobby for joint environmental interests and concerns. In the view of this author, industry and industry associations should also be counted among non-governmental voices speaking out about the environment. It seems likely that they, like industry in the United States, will begin to use the Freedom of Information Acts to obtain information about what plans their government has for future regulation, or to learn what it can about competitors.

NGOs, trade associations, private industry and ordinary citizens can all use the information they obtain on request from government, supplemented as necessary by independent information, to make complaints to legislators, to the press, to electronic communities through the Internet, or – in the case of NGOs and citizens – directly to polluting industries. The information can be used to prepare lawsuits, depending on the rules and traditions of the legal systems in each country. Increased transparency could make it harder for governments to avoid bringing enforcement actions. It will be more difficult than it was during the period of state socialism to obscure or hide poor environmental performance. Inherent conflicts of interest, such as those during state socialism, which encouraged government officials to decide in favour of economic performance and against environmental regulation, are more likely to be exposed. This might be an influence to reduce questionable transactions.

The changed perspective of industry could also be a factor in achieving some level of environmental progress. As it seeks to compete in a European or even a global market, industry in Central Europe will have to consider a variety of strategies. In today's world, some of these involve gaining customers by being perceived as better environmental managers. For example, ISO 14000 certification has already had some impact in Central Europe. ISO's Environmental Standards (EMS) are a series of voluntary standards and guideline reference documents that include environmental management systems, eco-labelling, environmental auditing, life cycle assessment, environmental performance evaluation, and environmental aspects in product standards. ISO certification is often demanded by purchasers or others in the supply chain and, as a result, many firms that wish to sell in the global marketplace are seeking ISO certification. ISO compliance and other forms of environmental compliance may also turn out to be important promotional tools in markets where companies advertise their environmental friendliness. An example of this was Mercedes Benz' advertising blitz in the early 1990s that each part of its cars is recyclable. Mercedes – like other German industry – was forced by German law to recycle and managed to turn the rather imposing regulatory requirements into a marketing tool that made them look environmentally friendly. Similarly, industry in Central Europe may find occasion to turn the burdens of environmental compliance to their advantage [*Bell, 1993*].

Another incentive in the private sector is the economic imperative of a 'level playing field'. Some multinationals require all their plants, wherever located, to meet company standards; the standards are based on the highest requirements the company must meet globally. These companies are not anxious to price themselves out of markets. If some companies are investing to meet environmental standards, they eventually will demand that others – particularly competitors – do so as well.

Other industries exist only in response to strong environmental regulation and, once established, often become a potent economic force to lobby for further regulation. Firms that manufacture pollution control equipment or manage waste, particularly hazardous waste, are in this category. The lifeblood of their industry is the requirement for control technology or safe, legal handling of wastes. When waste disposal came to be regulated in the United States under CERCLA and RCRA, industries were created and strengthened. The same could happen as the countries of Central Europe implement certain of the EU directives on waste disposal. When, in the early 1980s, an EPA administrator threatened cutbacks in implementation of US laws, important economic interests felt threatened and made their views known to Congress. It is difficult for US Congressmen to ignore pleas when constituents face economic harm. Perhaps even parliamentarians of the accession countries who are not fond of environmental regulation will be moved when industry comes in to tell them that reduced regulation will drive industry out of business. Thus, a variety of possibilities exist in which EU membership, or increased engagement in Western economies, may create possibilities for stronger environmental protection in the accession countries.

Prognosis for the Future?

EU membership and the expectations that come with being part of the democratic West will inevitably change the CEE countries, including their environmental programmes. The world in which they now exist is one where transparency will be emphasised over the 'twilight world of nods and winks' [*Wedel, 1989*] that was encouraged by the old communist system. In this dynamic, the NGO community – indeed, the entire non-state sector – is called on to become participants rather than mere protestors in the evolving civil society. As Jancar-Webster [*1998: 84*] notes, 'if cynicism is the proper attitude for the person living under Communism, positive activism is more appropriate for the person living under democratic institutions.' The question for observers and – more important – for the people in each country who breathe the air and drink the water – is whether or not the pressure to conform laws to EU requirements will ultimately provide a structure on which genuine environmental protection can be constructed.

It is most likely that EU accession will channel all government efforts towards achieving the dictates of EU law for the foreseeable future. This will probably reduce opportunities for local initiatives from NGOs and others, unless those efforts advance the purposes of the EU directives. The demands of the EU are too large, demanding and expensive to leave much space for local initiatives. Countries that have had difficulties issuing permits for their industrial facilities are now required, for example, to meet the European demand for integrated pollution prevention and control (IPPC). IPPC requires them to issue permits with best available techniques (BAT) requirements taking into account the entire environmental performance of the plant, including emissions to air, water and land, generation of waste, use of raw materials, energy efficiency, noise, prevention of accidents and risk management. This is an enormously difficult task for even the mature environmental protection regimes, much less those of Central Europe. The British experience on which the IPPC Directive is somewhat prefigured has demonstrated implementation challenges, including the difficulty in adopting clean technologies and technological innovations to meet BAT requirements [*Davies, 2001*]. In addition, the governments will be preoccupied with fulfilling the entire menu of EU directives, not only those for the environment. All in all, this will be an exhausting and costly exercise.

The most important task before each of these countries will be to develop a culture of compliance in which each of the critical players – cities, communities, individuals, factories and plants, farmers and gardeners – undertakes and enforces its responsibilities to carry out the myriad tasks to implement the many difficult and expensive EU environmental requirements. If creative NGOs in each of the accession countries use the new tools available to them as a consequence of EU accession, they can act to hold their governments accountable for the commitments they have made as part of the accession process. NGOs will have more tools at their disposal, if they choose to use them. But they will also have difficult decisions to make: should they agree to the agenda set in Brussels and work to enforce it, or try to introduce new issues for consideration?

NGOs have travelled a considerable distance since the early days of the transition. They are no longer simply protest groups, but many interact constructively with government, industry and communities to find and implement solutions to environmental problems. They engage in public outreach, and find ways to involve the public in environmental protection activities. But there is still considerable learning ahead, as they adjust to the challenges that enlargement puts to activism. Moreover, NGOs like other institutions in their societies necessarily are part of and reflect the conditions and status of their own country and its distinct culture. For example, their natural suspicion of industry may be intensified by having experienced the period of state socialism when industry and government were one and the same [*Bell and Fülöp, 2003*]. NGOs that have been successful in effecting actual change in their countries may have a different attitude from those who still feel embattled and alienated. In countries like Hungary and Poland, environmental NGOs played a prominent and constructive role earlier than 1989 and it was possible, albeit risky, to develop a non-governmental point of view; thus they have much history and experience to rely on. In others, the conditions were much harsher and the starting point for NGOs post-dated 1989 and the period of transition.

Fifteen years of experience has shown the futility of making predictions for a broad, undifferentiated 'Central Europe'. Countries that used to be spoken of as a group have taken many divergent paths and have had varying speeds in their transition. EU membership will give each of them, and their citizens and NGOs, certain opportunities and certain obligations. How they will use these is yet to be seen.

NOTES

1. The Aarhus Convention, signed in June 1998, enters into force in particular countries on ratification. Accession states are also required to put these conditions into place because an EU directive incorporates the Aarhus obligations, requiring them for member and accession states.
2. Jürgen Lefevere of FIELD provided valuable insights to help the author understand these provisions.
3. The authority for this is contained in Article 226 of the EC Treaty. Many of these proceedings concern whether member states have put national legislation in place to transpose EC directives.
4. The Commission starts dozens of such cases annually. It was reported on 16 March 2002 that the European Commission would take legal action against several member states for having incorrectly transposed Community legislation on air quality and emissions. The authority is contained in Article 228 but is very similar to a case brought under Article 226. The Court did this with Greece for the first time in 2001.
5. The Parliamentary complaint is on the basis of Article 194 of the EC Treaty. The Commission retains full discretion on deciding whether and when to start a case against a member state.
6. For example, the European Commission decided to send 'Reasoned Opinions' to France, Belgium, Germany, Italy, the United Kingdom, Greece, Spain and Portugal for failing to adopt and communicate the texts that are necessary to implement the 1999 Directive on consumer access to fuel economy and CO_2 emissions data, see *Rapid* (30 July 2001).
7. If there are questions of interpretation of EC law, the national court can (or in certain circumstances must) direct questions to the European Court of Justice in the form of a preliminary ruling. The ECJ responds to these questions and the national court then has to apply the answers in its ruling.
8. The rule is that where there is a discrepancy or contradiction between national law (including constitutional law) and EC law, EC law takes precedence and must be applied by national courts and authorities. However, directives cannot be invoked between individuals, only regulations and decisions, both of which are legal instruments that are directly applicable in the national legal order.
9. [1991] ECR I-5357, [1993] 2 CMLR 66.
10. *Francovich* itself involved Italy's failure to implement certain labour directives. The case established conditions which had to be met in order to obtain relief, namely (1) the result prescribed by the directive should entail the granting of rights to individuals; (2) the contents of those rights must be identified on the basis of the provisions of that directive; and (3) the

existence must be established of a causal link between the breach of the state's obligation and the loss and damage suffered by the injured parties.

11. For example, the US NGO, Natural Resources Defense Fund, brought a seminal lawsuit in the mid-1970s, which resulted in a US Federal District Court directing EPA on a specified schedule to promulgate effluent guidelines under the Clean Water Act. See, *Natural Resources Defense Council v. Train*, 6 ELR 20588 (D.D.C., June 9, 1976). The case was modified numerous times as the EPA struggled to issue effluent guideline rulemakings on the court-ordered schedule. An extremely high percentage of the rules promulgated by EPA are issued pursuant to court orders resulting from similar such suits.
12. See also Cases C-46/93 & C-48/93 *Brasserie du Pêcheur and Factortame III* [1996] ECR I-1029; [1996] 1 CMLR 889, where the Court essentially ruled, 'Where a breach of Community law by a Member State is attributable to the national legislature acting in a field in which it has a wide discretion to make legislative choices, individuals suffering loss or injury thereby, are entitled to reparation where the rule of Community law breached is intended to confer rights upon them, the breach is sufficiently serious and there is a direct causal link between the breach and the damage sustained by the individuals.'
13. A thorough explanation and examination of the project is found in Bell, Stewart and Nagy [*2002*] and in Bell and Fülöp [*2003*].

REFERENCES

Bell, Ruth Greenspan (1992a), 'Environmental Law Drafting in Central and Eastern Europe', *Environmental Law Reporter, News & Analysis*, Vol.XXII, No.12, pp.10701–4.

Bell, Ruth Greenspan (1992b), 'Industrial Privatization and the Environment in Poland', *Environmental Law Reporter, News & Analysis*, Vol.XXII, No.2, pp.10092–8.

Bell, Ruth Greenspan (1993), Exporting Environmental Protection, *Environmental Law Reporter, News & Analysis*, Vol. XXIII, No.9, pp.10599–605.

Bell, Ruth Greenspan (2003), 'Choosing Environmental Policy Instruments in the Real World', OECD. Available at http://www.oecd.org/pdf/M00042000/M00042064.pdf.

Bell, Ruth Greenspan (forthcoming), 'Hungary: Developing Institutions to Support Environmental Protection', in M. Auer (ed.), *To Restore Cursed Earth, Appraising Environmental Reforms in Central and Eastern Europe and the Former Soviet Union*, Lanham, MD: Rowman & Littlefield Press.

Bell, Ruth Greenspan and Sándor Fülöp (2003), 'Like Minds? Two Perspectives on International Environmental Joint Efforts', *Environmental Law Reporter*, Vol.XXXIII, No.5, pp.10344–51.

Bell, Ruth Greenspan and Clifford Russell (2002), 'Environmental Policy for Developing Countries', *Issues in Science and Technology*, No.3, pp.63–70.

Bell, Ruth Greenspan and Clifford Russell (2003), 'Ill-Considered Experiments: The Environmental Consensus and the Developing World', *Harvard International Review*, Winter, pp.20–25.

Bell, Ruth Greenspan, Jane B. Stewart and Magda Toth Nagy (2002), 'Fostering a Culture of Environmental Compliance Through Greater Public Involvement', *Environment Magazine*, Vol.44, No.8, pp.34–44.

Carmin, JoAnn and Barbara Hicks (2002), 'International Triggering Events, Transnational Networks, and the Development of the Czech and Polish Environmental Movements', *Mobilization: An International Journal*, Vol.7, No.2, pp.304–24.

Davies, Terry (2001), *Reforming Permitting*, Washington, DC: Resources for the Future.

Europa (1999), 'Regular Report from the Commission on Hungary's Progress Toward Accession'. Available at http://www.europa.eu.int/comm/enlargement/report_10_99/pdf/en/hungary_en.pdf.

Europa (2000), 'Regular Report from the Commission on Hungary's Progress Toward Accession'. Available at http://www.europa.eu.int/comm/enlargement/report_11_00/pdf/en/hu_en.pdf.

European Commission (1997), 'Commission Opinion on Hungary's Application for Membership of the European Union,' Document Drawn Up on the Basis of COM(97) 2001 Final: Bulletin of the European Union: Supplement. Brussels: European Union.

European Commission (2001), 'Internal Market Scoreboard'. Available at http://europa.eu.int/comm/internal_market/en/update/score/score8en.pdf.

Hicks, Barbara (2001), 'Environmental Movements and Current Environmental Issues in Central Europe: Hungary and the Czech Republic', *National Council for Eurasian and East European Research Working Papers*, Washington, DC: National Council for Eurasian and East European Research.

Hudiburgh, Jr., Gary W. (2003), US Environmental Protection Agency, personal conversation, July.

Jacoby, W. (1993), 'Priest and Penitent', *Review of European Community and International Environmental Law*, Vol.2, No.1, p.16.

Jancar-Webster, Barbara (1993), 'Eastern Europe and the Former Soviet Union', in S. Kamieniecki (ed.), *Environmental Politics in the International Arena: Movements, Parties, Organisations, and Policy*, Albany: State University of New York Press, pp.199–221.

Jancar-Webster, Barbara (1998), 'Environmental Movement and Social Change in the Transition Economies', in S. Baker and P. Jehlička (eds.), *Dilemmas of Transition*, Portland, OR: Frank Cass, pp.69–90.

Jehlička, Petr (2001), 'The New Subversives – Czech Environmentalism after 1989', in Helena Flam (ed.), *Pink, Purple, Green: Women's, Religious, Environmental and Gay/Lesbian Movements in Central Europe Today*, Boulder: East European Monographs, pp.81–103.

Lynch, D. (2000), 'Closing the Deception Gap,' *Journal of Environmental Development,* Vol, 12, pp.426–37.

Radio Free Europe/Radio Liberty Newsline (2003), 'Romania Needs $30 Billion for Ecological Reconstruction', Vol.7, No.47, Part II, 12 March.

Szabo, G. (2001), *Tisztulo Levego* (Clearing Air), HVG 6/9/2001.

Tyler, Tom R. (1990), *Why People Obey the Law*, New Haven: Yale University Press, 1990.

Wedel, Janine (1992), *The Unplanned Society: Poland During and After Communism*, New York: Columbia University Press.

Wedel, Janine (1998), *Collision and Collusion*, New York: St Martin's Press.

Welsh, Ian and Andrew Tickle (1998), 'The 1989 Revolutions and Environmental Politics in Central and Eastern Europe', in Andrew Tickle and Ian Welsh (eds.), *Environment and Society in Eastern Europe*, Essex: Longman, pp.1–29.

Setting Agendas and Shaping Activism: EU Influence on Central and Eastern European Environmental Movements

BARBARA HICKS

When asked what are the most pressing issues facing their movements, environmental activists in Poland, Hungary and the Czech Republic list the European Union (EU) among the top contenders, and they do this increasingly often and with increasing vehemence.[1] The EU, however, is much more than an 'issue' facing environmental movements in the post-communist countries of Central and Eastern Europe (CEE); it is also both a target for mobilisation and a source of external pressure shaping the evolution of environmental activism in the region. The central task of this study is to develop a framework for understanding the many ways in which EU accession is affecting the evolution of environmental movements in Central and Eastern Europe. Based on observations from field research in Poland, Hungary and the Czech Republic, I trace several of these paths of influence. Although the EU does not fully determine the course taken by environmental movements in the region, it strongly influences the development of these movements on two levels: (1) it sets significant portions of the issue agendas addressed by environmentalists; and (2) it helps to shape the means and conditions of activism itself.

As a way of reflecting on the significance of such influence, the last section of this study takes up the question of balancing domestic social interests and external influence in movement development. Striking a balance among the priorities of external actors, movement elites at the national level, and grassroots concerns is crucial to increasing the movements' effectiveness in institutional politics while at the same time strengthening their function as channels of citizen participation in the new democracies. The diffusion of norms, issues, modes of action and direct support from the EU and organisations based in EU countries has clearly increased the organisational efficacy, knowledge and skills of many movement actors. Whether these movements are strengthened as channels of citizen influence depends on the survival of grassroots groups, communication among organisations within a national movement sector, and the will and skill of national movement elites in bringing grassroots concerns to the fore. It also depends on the EU opening avenues for input into its own policies.

Environmental Movements in the Transition from State Socialism

Environmental movements began to emerge in the Soviet sphere of influence in the 1980s.[2] Taking advantage of greater formal attention of their regimes to the environment after the 1972 United Nations Conference on the Human Environment in Stockholm, activists started to press their governments on environmental issues, raising the engagement levels of some official organisations and forming independent groups of their own [*Carmin and Hicks, 2002*]. These activists and independent groups became the core of environmental movements in the early 1980s. The issues they addressed most strongly were those of industrial pollution and its effects on human health, but ecologists were also concerned with conservation and with nuclear energy (nuclear concerns were present even before, but especially after, the nuclear reactor fire at Chernobyl).

Poland, Hungary and Czechoslovakia all saw the development of environmental movements in the last years of communist rule, but the size, organisation and degree of independence of these movements varied according to the political and environmental conditions they faced. Poland's open atmosphere of the Solidarity period (1980–81), developed political opposition, and relatively relaxed political atmosphere after 1985 allowed for a rather large and vociferous set of movement actors with a couple of strong and many small independent organisations and increasingly critical and active official organisations [*Gliński, 1996; Hicks, 1996*]. Czechoslovakia's repressive regime severely limited the amount of independent activism in the country, and most environmental activism took place in official and semi-official conservation organisations, while a small circle of opposition-linked activists pressed more controversial issues (interview with Miroslav Vaněk, Prague, 23 January 2001). Hungary's relatively permissive regime sought to isolate the political opposition, but it did not prevent the development of a large Danube movement resisting the construction of the Gabčikovo-Nagymaros dams, nor did it stem other environmental activism at the local level and among university students (interviews with Éva Kutí, Budapest, 12 September 2000; Viktória Szirmai, Budapest, 26 June 2001; Zsuzsa Foltányi, Budapest, 25 May 2001).

Throughout the Soviet bloc, especially during the Gorbachev period, environmental activism became a comparatively safe venue for critique of the regime. Even some of the countries with more repressive regimes, such as East Germany, Bulgaria, Romania and parts of the Soviet Union, saw an increase in environmental activism. Given the fact that the regimes were directly responsible for all economic and conservation activity in their countries, ecological critiques were critiques of governments, their decision-making processes, and their policies. At the same time, focusing on specific issues about which the governments purported to be concerned was a safer strategy than engaging in direct political opposition.[3] Thus, the movements grew rapidly. Although they were dominated by those concerned primarily with ecology, they also included substantial groups of activists for whom critique of the regime was the major appeal to their engagement in environmentalism.

During the collapse of communist rule, environmental groups and issues mobilised a great deal of public support for change [*Jancar-Webster, 1993, 1998; Dawson, 1996*]. Seen as channels for genuine citizen participation, these groups became identified with a resurgent civil society and with true democratisation [*Fisher et al., 1992*]. Their grassroots nature, non-hierarchical and pluralistic organisation, and status as challengers to officialdom identified them as social initiatives in the state–society dichotomy of the day. They enjoyed popular credibility both for their stances and for their means of action, and this credibility translated into rapid growth in the number of organisations in the early transition period. The relative influence of the grassroots in the movements remains an important issue today, not only because of the role of social movements in bringing about democratic transitions, but also because they can serve as channels of citizen participation in the democracies set up after the transitions. With only new and under-developed political parties and a dearth of interest groups in these countries, social movements are still a favoured form of citizen action and participation to the degree that citizens engage in any political participation beyond voting. Indeed, Barbara Jancar-Webster points out that the shift towards professionalised NGOs that participate in government decision processes often distances environmentalists from potential popular bases who still mistrust government and favour a more critical, mobilisation politics [*1998: 83–4*].

After the initial wave of transition, the governments and societies of the region focused increasingly on institutional change and electoral politics and on economic transformation and the austerity that came with it. The environment fell down the list of policy priorities, even though public opinion polls still showed strong ecological values and support for improvement in the area of ecology.[4] The growth in environmental movements slowed as the number of groups stabilised in the latter half of the 1990s (interviews with Simona Šulcová, Prague, 24 January 2001; Małgorzata Koziarek, Warsaw, 12 November 1997). At the same time that environmental groups' visibility on the national political agenda dissipated, economic difficulties stemming from the market transitions challenged their ability to stay afloat. With weak voices and weak finances, many groups welcomed the aid and advice of external organisations. The danger, however, was that they became increasingly independent of their societal bases of support and increasingly dependent on organisations that also had their own agendas. While integrating into the transnational environmental sector and strengthening their participation in national policymaking, these groups, therefore, are struggling to remain responsive to grassroots concerns and to remain channels of citizen participation in politics.

EU Influence

The questions surrounding European Union – and, more generally, international – influence in the development of social movements in the transitioning CEE countries are complex.[5] They involve not only external and domestic influence over specific policy positions taken by movement actors, but also resources for activism and shaping of organisational development. Tensions between national interests and EU interests can be exploited by environmental movements, but they can also place activists in the uncomfortable position of resisting the EU or their government. Movements must also adapt to changing political contexts, often by adopting different means of activism. The advice of Western organisations is very helpful in making these adjustments, but it also does tend to reproduce the systems and movements that exist in the West. Some activists in the transitioning countries criticise this adaptation in the face of the simultaneous adoption of Western consumer capitalism, arguing that an institutional and policy-oriented approach modelled after Western movements has resulted in lost chances for fundamental critique and restructuring to avoid the environmental problems entrenched in the developed countries of the West.

To date, EU influence on the environmental movement sectors has produced mixed effects. On the positive side of the ledger, EU advice, resources and political pressure help build the capacity of environmental organisations, open government decision-making processes to social input, and provide a powerful ally in movements' efforts to improve environmental protection and keep ecology on the crowded transition agendas of their countries. Key organisations do indeed have a stronger ability to act now, and the EU is one of the forces contributing to this development. However, the predominant position the EU holds in defining high priority issues and setting guidelines that shape the political processes in these countries endangers the visibility of other environmental issues that are not EU priorities and favours strategies that engage the established political processes. General ecology-oriented groups with critical or more holistic perspectives have trouble drawing government attention to their agendas or even finding a place at the table with government and movement-based issue specialists hammering out the minutiae of regulations. Finally, the public perception of the inevitability of adopting EU policy may make it difficult to mobilise popular and even activists' support for campaigns focused on other issues and outcomes.

Turning to the question of *how* the EU has these effects on movement development, we need to examine several channels of influence. It is useful to distinguish between sway over the agendas of movements and pressures that shape the nature and organisation of activism. On both of these levels, the EU and European environmental organisations used to working in the context (and often with the support) of the EU exert both direct and indirect influence over the development of environmental movements in countries aspiring to EU membership. Table 1 sketches the most significant paths of influence. These paths are not discrete; they overlap and interact. Discerning their starting points and how they interact, however, allows us to understand EU influence in these movements.

Setting Agendas

The EU sets large portions of CEE environmental agendas on several levels. The most direct influence wielded by the EU is through its own *environmental agenda.* CEE countries have agreed to participate in the EU environmental agency, information network and programmes.[6] Beyond these formal agreements, the EU has pursued a policy of promoting the adoption of its standards for emissions and other environmental indicators, as well as developing monitoring systems and treatment facilities. Participation in these programmes and adoption of these policies and standards are at the heart of what EU membership means in the area of environmental protection. The EU also pays attention to the procedures by

TABLE 1

CHANNELS OF EU INFLUENCE ON CEE ENVIRONMENTAL ACTIVISM

	Setting Agendas	Shaping Means of Activism
Direct	EU environmental agenda Harmonisation of environmental laws EU policies and projects in other areas	Funding for group development and activities Organisational requirements for funding Laws on access to information and participation
Indirect	EU agenda for governments: Political standards Economic policies	Influence over agendas Shaping democratic processes

which environmentally relevant decisions are taken. One of the EU's primary efforts to strengthen environmental protection in Central and Eastern Europe has been to press these countries to develop environmental impact assessment processes with a public input phase. The EU has devoted considerable attention to this issue in the accession process as well as in its funding for environmental organisations. Access to information and environmental impact assessments are parts of the community framework provisions in the environmental chapter of the *acquis communautaire*. Because both the EU and environmental organisations in Central and Eastern Europe highlight the need for impact assessments with public information provisions, groups and citizens can now use this avenue to express their opinions and press for change. In addition to its policy initiatives and political pressure, the EU has several programmes to help fund the implementation of specific projects to reduce and monitor pollution.

Another mechanism for expanding the EU's environmental policy directly into this region is the *harmonisation of laws* that precedes accession. Poland, the Czech Republic and Hungary all began negotiation on the environment section of the *acquis* in December 1999, completed provisional agreements by October 2001 (Hungary and the Czech Republic in June 2001), and closed negotiations in December 2002. The provisional agreements contained exceptions that allowed governments a specified amount of extra time to achieve EU-wide standards in precisely delineated areas. By the time they closed negotiations on the environmental chapter, the Czech government had negotiated three transitional arrangements, the Hungarian government four, and the Polish government ten.[7] There is no question, however, that the leading accession countries have moved substantially towards the goals of the *acquis* and will attain them. Doing so reproduces the legal structure – hence the priorities, limits and approaches – of environmental protection in the West.

Movements' reactions to these policies have been mixed. Most activists support the general direction of EU efforts, which they see as strengthening environmental protection in their countries. Some are pleased to have the 'stick' of EU accession to keep their governments from ignoring the environment. However, there are strong differences about individual policies. For example, one Hungarian activist pointed out that certain EU standards for water contamination were actually lower than Hungarian standards, allowing the government to weaken regulations (interview with Hungarian activist, Budapest, November 2000). Other activists in the region have questioned the priority of certain goals over others or criticised end-of-pipe solutions and missed opportunities for truly green reconstruction. They argue that concentration on adopting the existing legislation and policies of the EU rather than on the fundamental reconstruction of their economies along sustainable lines wasted the opportunity to achieve far stronger protection of the environment.

While the policy-oriented groups debate the merits of particular measures in the harmonisation process, many NGOs focus on service provision, or projects in the field, and have become involved in the actual implementation of new programmes and improvements to meet EU standards. These groups have benefited from EU programmes supporting specific projects ranging across civil society development (the Access Programme), cross-border cooperation, and the shoring up of government implementation of newly harmonised laws and standards (Regional Environmental Accession Programme or REAP), all funded under the umbrella of PHARE (Poland and Hungary: Assistance for the Restructuring of the Economy).

Another direct channel of EU influence over the issues on which activists focus derives from the *EU's own actions – programmes, investments and policies – beyond the strictly environmental sphere*. For example, the EU transport programme's role in designing and organising funding for an entire highway network to link Western Europe to Eastern Europe and beyond has been a focus of different types of activism in the countries under discussion here. Groups in all three CEE countries have taken exception to specific highways or routes (interview with Dariusz Szwed, Warsaw, 17 June 1998), the priority of highway development over other means of transportation (interview with Vojtěch Kotecký, Prague, 14 March 2001), and the process by which the decisions were taken (interview with Magdelena Stoczkiewicz, Krakow, 18 February 1998). Indeed, several transportation-centred groups have focused their activism on the EU programme and their own governments' decisions to accept these plans.

Other areas of EU policy have also mobilised CEE environmental groups. European Investment Bank energy projects, for instance, were a main inspiration for the formation of the CEE Bankwatch network – one of the most successful examples of transnational networking and coordination originating with environmental organisations in the area. This network monitors and seeks to influence the decisions of major multilateral banks and organisations with respect to investments in CEE countries that are likely to have environmental consequences, particularly in the energy and transport sectors (interview with Magdelena Stoczkiewicz, Krakow, 18 February 1998). Agricultural policies have also raised attention [see *Beckmann and Dissing, this volume*]. Activities in these and other policy areas generate specific environmental issues as well as concerns about the lack of local input. In general, most of environmentalists' objections to the EU, hence many of the issues they take up, are responses to Union policies *outside* of the environmental sector rather than direct criticisms of the environmental policies promoted by the EU.

The EU also shapes the issue agendas of environmental movements in these countries indirectly, through its *political and economic agendas for their governments*. In order to accede to the EU, a process that has accelerated rapidly, the CEE countries first signed Europe Agreements, becoming Associate Members, and then Accession Agreements with the European Union. They are harmonising their laws with the EU's *acquis*, and have received aid and guidelines for restructuring their economies and their political and legal systems. The PHARE programme was the first major initiative to provide funds for restructuring and development, but several others now fund what the EU deems to be priority development projects to bring these countries up to European levels. ISPA (Instrument for Structural Policies for Pre-Accession) funds pre-accession projects in environment and transportation; LIFE (Financial Instrument for the Environment) funds environmental and nature protection programmes, including the Nature 2000 initiative; and SAPARD (Special Accession Programme for Agriculture and Rural Development) focuses on agricultural restructuring and other aspects of rural development, such as investments in infrastructure and water programmes.[8]

Although they do not have a direct effect on ecology, the EU has political programmes to promote democratic institutions and transparent governance. Transparency at the domestic level is also promoted by the level of information required by the EU for its own purposes at the international level [*Grigorescu, 2002*]. The EU's political programmes reinforce and add some issues to environmentalists' agendas. They also shape environmental movements in that they change general government decision-making processes and information policies, thus altering some of the channels and means for activism. A primary area of political reform has been transparency measures, such as freedom of information acts; these measures have a direct effect on activism. Given the importance of these policies to the way environmental movements pursue their goals, the information, transparency and governmental reform policies themselves have become issues on environmentalists' agendas. Indeed, environmental activists have often taken the leading role in pressing for such measures that strengthen the overall development of civil society (interviews with Dariusz Szwed, Warsaw, 17 June 1998; Zuzana Drhová, Prague, 16 March 2001). For example, the Czech Green Circle pressed the Ministry for the Environment to post drafts of pending legislation and regulations on its website (interview with Zuzana Drhová, Prague, 16 March 2001), while Polish environmental activists convinced their Ministry to appoint a liaison for contacts with the movement shortly after the transition (interview with

Teresa Orłosz, Warsaw, 27 July 1993). This liaison was an early inroad for citizen groups into the Polish ministries and later became part of a larger office of promotion and information. Environmental groups in all three countries, shored up by the EU's emphasis on transparency, were active in pressing for acts on freedom of information and environmental impact assessment [*REC, 1998*].

Economic policies promoted by the European Union often have direct effects on the environment, and thus indirectly provide issues for activists' agendas. In particular, Brussels has required these countries to open their land, consumer markets and industry to European investment, change agricultural standards and practices, and restructure industry. A whole host of issues has emerged, or been intensified, with these changes: urban sprawl, increased automobile traffic, built-over farmland, consumer packaging waste, new factories in some areas and old ones either abandoned or polluting in other areas. Along with 'EU accession', activists also identify these specific issues as critical, and they have increasingly focused attention on them.[9]

In sum, the European Union exerts powerful direct and indirect influences over the agendas of environmental activists in the post-communist countries of Central and Eastern Europe. As a signatory to many international conventions, the EU also strengthens the priority of those conventions (for example the Aarhus Convention and the Kyoto Protocols) in member and acceding countries. Although environmentalists focus on locally and nationally generated issues as well, the centrality of the EU agenda during the accession period shifts the percentage of time, effort and resources that groups can allocate to activism heavily towards European issues (interview with Vojtěch Kotecký, Prague, 14 March 2001).

Shaping the Means of Activism

EU input into, even origination of, the topics of activism is a means of influence over *what* CEE environmental movements do. The EU has become a major force in defining what environmental issues become politically relevant. However, the EU also shapes *how* these movements do what they do and, therefore, how they organise – in effect, what these movements are becoming. This influence is, of course, not total, or even always determining, but it is strong.

Before turning to the direct channels of EU influence over the activism of CEE environmental organisations, we should recognise the main indirect means by which EU influence reconfigures this activism. The *influence over agendas*, discussed in the previous section, plays a strong role in shaping the nature of activism and movements, not just the issues they address. As movements devote more attention to issues from the EU agendas, they structure their actions and organisations to accommodate that focus. They also shape their activism to address the institutional procedures and power of the EU and of their own governments as the latter adopt and implement EU-influenced policies.

By *shaping the democratic processes* via support for certain institutions and practices in individual countries, the EU is also affecting the immediate political context to which movements must orient their means of activism and ultimately their organisational structures. Thus, the combined effects of EU influence on environmental agendas and government reforms serve to push CEE environmental movements further in the direction of their Western counterparts. In particular, the movements have increased their emphasis on institutional procedures – lobbying, consulting on draft legislation, researching and writing reports and opinions, attending public meetings – in order to address their own governments' implementation of EU-initiated policies [*Fagin and Jehlička, 2001*]. Groups still protest (for example, tree-sitting in the Góra Św. Anny reserve in 1998 to resist the siting of one of Poland's new highways to link to the European network), but the sense of inevitability about the government accepting EU policies tends to lead to strategies of influencing the details, or changing policies at the margins, or sometimes simply demanding that their governments implement the policies. In order to be able to engage politicians and bureaucrats at this level, in turn, we see a premium on expertise and specialisation. This premium is then reflected in the overall structures of the movements in the region; the professionalised and specialised groups now make up a larger share of movement organisations and have become more influential in the movements over time.[10]

A more direct channel of influence over the way these movements develop is through *funding for activism and certain aspects of organisational development*. Some of the PHARE funds went to develop environmental movements in the area, either directly to organisations to carry out projects, or to intermediary funding organisations, such as the Regional Environmental Center (REC), that then distributed grants to individual groups. In the early years of the programme, some funding was devoted to developing the capacity of the 'third sector', an endeavour that included supporting organisations' projects, production and dissemination of information, and institutionalisation, as well as supporting networking efforts among the region's movements. Clearly, supporting certain groups and certain types of networking affects the movements that emerge. Once given funds for a project, though, an organisation does not always back EU positions fully. For example, REC received EU funding to do a study of environmentalists' opinions about accession, a study that ended in a rather sobering report from the EU's perspective [*REC, 2000c*]. However, most funding has gone to organisational development or projects that achieve material change in the environment along the lines envisioned by EU policy.

Another avenue through which EU funding directly shapes movements is underwriting projects of West European organisations that guide CEE movements. For example, EU funds were a significant part of the start-up costs for lobbying bureaux in Poland and the Czech Republic, promoted and guided by the Dutch organisation Milieukontakt Oost-Europa. The bureaux are sponsored by about a dozen organisations in each country's movement to facilitate the flow of information to the movement and coordinate activism in legislation and regulation on the issues where member organisations agree to such coordination (interviews with Dariusz Szwed, Warsaw, 17 June 1998; Zuzana Drhová, Prague, 16 March 2001). These types of projects increase transnational networking by the groups involved, and they also tend to increase the similarity between Western and Eastern movements as leaders of the former reproduce models of their own activism and facilitate the adaptation of leading environmental groups in the candidate countries to political systems that are themselves increasingly similar to Western democracies.

The *requirements for funding* also can alter internal procedures of the recipient organisation. To receive grants, organisations must be officially registered, and, in order to register, they must have at least a minimally hierarchical structure and by-laws. Even this amount of formal structure contradicts the philosophy and practice of some ecology groups. Another structuring influence of the funding process is tied to the actual use of funds. The person signing a grant application – regardless of whether this signature is on behalf of an organisation – becomes responsible for the project and, hence, the distribution of funds (interview with Małgorzata Koziarek, Warsaw, 12 November 1997). This responsibility may change the balance of power in the group (interview with Leszek Michno, Krakow, 18 February 1998). Even without change in their internal power balances, groups become more formalised and their divisions of labour more pronounced through these interactions. They also must devote more time to the projects that are funded, leaving fewer resources for any other initiatives or priorities they may have.

Finally, the European Union's measures concerning *laws on public participation and access* to policy processes help to shape the avenues of activity open to environmental organisations, both in their own countries and in the EU itself. The environmental impact assessment processes promoted by the EU have provisions for public information and participation. Reinforcing the emphasis on access to information and right to participate is the EU's signing of, and adherence to, the 1998 Aarhus Convention on Access to Information, Public Participation in Decision-Making and Access to Justice in Environmental Matters. Although the Aarhus Convention was an initiative of the UN Economic Commission for Europe, once the EU became a party to the agreement, it accepted the obligation of implementing it. This is also the case for the many other conventions that the EU has signed, including several on the environment.[11] While the EU encourages participatory policy processes in individual countries, its own policies often come as a given to applicant countries. The public discussion period and bargaining between national officials over EU policy is beyond, often prior to, the involvement of activists from aspiring countries. Indeed, the issue of opening EU policy processes in all areas to public participation was raised strongly at the Laeken Summit of the European Council in December 2001 and became the focus of the Convention on the Future of Europe and current attempts to draft and pass a constitution for the EU.

On balance, the EU has contributed to the strengthening of major groups and organisations in the environmental movements in the CEE region. Through its direct policies and through its influence on the governments of the region, the EU has encouraged participation in increasingly open policymaking processes. Concentration on participation in legislative and regulatory processes, in turn, promotes institutionalisation of non-governmental actors, professionalisation, specialisation, and coordination of action. While the overall impact of EU influences has been favourable to these sorts of organisations and activities, others face marginalisation. Some EU policies have sparked protest, but, in general, campaign-oriented activities take a back seat to policy formulation and consultation. In terms of overall balance, policy-oriented and issue-specific networks are on the rise, while the role of general action and alternative value groups in these environmental movements is diminishing (interviews with Simona Šulcová, Prague, 24 January 2001; Dariusz Szwed, Warsaw, 17 June 1998).

Balancing Social Interest and External Influence

Until now, this study has focused on one direction of influence. However, I do not want to overstate the case or claim that the European Union determines the nature of the area's environmental movements. The diversity and decentralisation of these movements mitigates some of the impact of EU shaping. Grassroots initiatives do still emerge, even though their influence on the general direction of movement activism or national policy is difficult to see. They tend to remain local, seek out networks concerned with similar issues (for example, local anti-incinerator groups looking for information and support from larger organisations specialising in waste management), or get picked up as cases by more professionalised issue-oriented groups at the national level seeking to forward their cause. The major groups with national profiles, even the more 'radical', campaign-oriented groups, are all engaged to some extent with EU issues. However, they also pursue other issues and domestic policies. Funding they receive from external organisations for specific projects provides some overhead with which they can continue other projects (interviews with Leszek Michno, Krakow, 18 February 1998; Laszlo Perneczky, Szentendre, 7 November 2000; Vojtěch Kotecký, 14 March 2001). Moreover, many of the EU issues are, at the same time, pressing domestic issues, so activists often do not face a strict 'either-or' dilemma in allocating their time and effort. Finally, movement elites at the national level play an important role in presenting domestic needs and positions to external funders and interpreting how external programmes should be understood and implemented in the specific contexts of their countries (for example, through the Study reported in REC 1997).

Still, two sets of questions remain if European Union policies regarding environmental activism in Central Europe and Eastern Europe are going to promote a balance in these movements between effective organisations addressing issues of transnational importance and those responding to citizen initiatives that press local issues and values. One set of questions concerns encouraging diverse and multi-level development of movements as well as openness to agenda items from below when supporting environmental activism. The other concerns opening the EU's own processes to participation and influence from below [see *Hallstrom, this volume*].

The EU can play a keener role in strengthening movements and their connections to society. As external support for general civil society development in this area dries up and non-governmental organisations are left to become self-sustaining, it is important to the social grounding of movements to make sure that at least some groups taking different approaches or addressing non-EU issues survive. As suggested by all of the country officers for the Regional Environmental Center and the Environmental Partnership for Central Europe's successor organisations who were interviewed, one way to do this is to ensure some continuing support for general capacity-building grants not tied to projects. These grants can sustain a variety of small initiatives and alternative types of groups long enough for them to achieve goals or build their own base of support. Key to achieving this balance in support is a granting decision process that casts its application nets wide and has decision makers truly knowledgeable of the movements and committed to their diversity – grant givers who can recognise the types of initiatives that are important to at least some citizens but not likely to be taken up by prominent national groups. Such a strategy may help the movement as a whole maintain social support and legitimacy as an actor in the political process, which is important for developing a civil society oriented towards participating in environmental decision making and monitoring the implementation of government policies.

EU policy processes have long been a source of discontent among non-governmental organisations in old member states. The EU is at this very moment involved in a debate about how to democratise itself. In the works are a constitution and strengthened provisions for public participation. If the general democratisation process in the EU bears fruit, member countries' citizens and environmental organisations will likely have more opportunities to influence EU policies and programmes. Movement organisations in candidate countries, however, will likely always have only very limited input into the policies the EU pursues in their countries, simply by virtue of being outside the Union when policy is made and of their own governments' drive to meet EU requirements. The uneven balance of power between the EU and an individual country slants the bargaining process towards outcomes favoured by Brussels; acceding governments have little room for manoeuvre, and often NGOs will not be the strongest candidates to benefit from the limited leeway a government might succeed in negotiating. In particular, the environment is not an area where candidate governments are likely to negotiate exceptions to EU policy when key industrial and agricultural sectors are on the line. Moreover, if they negotiate exceptions, they are likely to be delayed in implementing EU policy and not provision for other approaches to the problems at hand.

Nevertheless, CEE environmental organisations have pressed for input into EU programmes and policies in several ways. EU receptivity has varied. Since 1999, the European Commission's Directorate General Environment has been holding a series of Dialogue Meetings with environmental NGOs funded through PHARE and facilitated by the Regional Environmental Center. The meetings allow EU representatives to explain their policy, encourage NGOs to play the role of watchdog for the implementation of EU policies in their countries, and provide a forum for consulting with these NGOs about EU environmental policy. Although these meetings comprise a forum for dialogue, the summaries of their discussions suggest how much more of the dialogue is about pursuing EU policy and putting pressure on governments to do so than it is about what EU policy should be. This was particularly true of the earlier meetings; the later meetings, though still uneven, have shifted a bit more towards dialogue and, most recently, to consideration of broader international programmes.[12]

CEE environmental movements pursue other avenues of influence in European Union policy, but the asymmetry in power and resources argues for the need to adopt much stronger mechanisms for public input into both the formation and implementation of EU policy. When their countries finally gain membership, more institutional channels will be available to the movements as they follow their Western colleagues in building their political presence in the EU [*Marks and McAdam, 1999*]. This strategy of working at the EU level is already evident among activists in the candidate countries: CEE Bankwatch, for example, has had a representative working in Brussels since September 1999.

CEE environmental organisations are joining the chorus and initiatives to democratise the EU. To the extent that the EU strengthens the role played by the public and by NGOs in policymaking, the balance of influence between these movements and the EU over policy and programmes will improve. So also may the ability of national movement elites to address the concerns of their bases and thus ensure that these movements function as effective channels of citizen participation in the new democracies of Central and Eastern Europe.

NOTES

I would like to thank the International Research Exchanges Board (IREX) and the National Council for Eurasian and East European Research (NCEEER) for research support in 1997–98 and 2000–2001, respectively. The information and interpretations offered here, however, are solely my own and in no way reflect the views of these institutions.

1. I asked this question during interviews with Polish environmentalists in 1993 and 1997–98 and with leading environmental activists in all three countries in 2000–2001. Most of the statements in this contribution are based on the interviews, reports and other materials I collected during these research trips. Only those interviews and sources directly cited are included in the references to this contribution.
2. For more on the emergence of these movements and their activities under communist rule and in the early phases of regime transition see, among others: Singleton [*1985*]; French [*1990*]; DeBardeleben [*1991*]; Slocock [*1992*]; Fisher *et al.* [*1992*]; Fisher [*1993*]; Růžička [*1993*]; Jancar-Webster [*1993, 1998*]; Gliński [*1989, 1996*]; Hicks [*1996*]; Dawson [*1996*]; Szirmai [*1997*]; Tickle and Welsh [*1998*]; Baker and Jehlička [*1998*]. Major works on environmental protection and conditions under the Soviet-type regime include Volgyes [*1974*]; Fullenbach [*1981*]; Kramer [*1983*]; DeBardeleben [*1985*]; Ziegler [*1986*]; Jancar [*1987*]; Albrecht [*1987*]; Weiner [*1988*]; Pryde [*1991*]; Bochniarz [*1992*]; Feshbach and Friendly [*1992*]; Vaněk [*1996*].
3. 'Safer', however, does not mean 'safe', and many environmental activists took risks and suffered repression for their activities. In particular, activism only increased from non-existent to sparse under the most brutal regimes, e.g. Ceaușescu's Romania, where *any* unofficial activity was treated as opposition.
4. This trend was widely reported to me by activists and scholars in all three countries. The only systematic, region-wide opinion surveys, the Central and Eastern Eurobarometer (conducted annually in 1990–97, published in 1991–98) replaced by the Candidate Countries Eurobarometer (starting in October 2001), are carried out with support from the European

Commission and focus largely on the role and perception of the EU. They report, for example, that in 1997 36 per cent of the population of all candidate countries would like to know more about the EU's positions in the environment. This issue was ranked fourth after 'working and living conditions in the EU', human rights, 'EU–(our country) relations'. Among national decision makers, the environment also ranked fourth after 'EU–(our country) relations', PHARE and 'economic/monetary affairs' [*CEEB, 1998: annex figures 30 and 31*]. The Candidate Countries Barometer showed similar interest in the autumn of 2002. When read a list of actions the EU could undertake and asked whether each should be a priority, 64 per cent of respondents said the environment should be a priority. Polish respondents mirrored the average, while 71 per cent of Czechs and 75 per cent of Hungarians labelled the environment a priority [*CCEB, 2002: 132 and annex table 8.1*]. National surveys vary across time and country, but a similar picture of high environmental values emerges when piecing this evidence together.

5. Although this study is limited to EU influence over social movements, in particular environmental movements, the flow of norms and shaping of contexts, as well as actual legal and administrative harmonisation apply to the general political development of candidate and member countries. For a more general framework of the impact of international organisations on these states, see Schimmelfennig [*2002*].
6. On 18 June 2001, the European Council accepted agreements concluded with the Czech Republic, Poland, Hungary, Slovakia, Slovenia, Romania, Bulgaria, Latvia, Lithuania, Estonia, Cyprus, Malta and Turkey for their participation in the European Environment Agency and the European environment information and observation network [eionet]. They have all since joined eionet and have either joined or are just completing formalities to join the European Environment Agency.
7. For a record of negotiations by chapter and country, including lists of transitional arrangements, see http://europa.eu.int/comm/ enlargement/negotiations/chapters/chap22/index.htm.
8. For more on these programmes, see http://europa.eu.int/comm/enlargement/financial_assistance.htm.
9. These issues were cited most frequently in my interviews with movement leaders (see note 1).
10. My in-depth interviews with people involved in channelling external funding and expertise to movement groups, with movement activists in particular groups, and with scholars studying these movements produced unanimity on this point, regardless of the interview subjects' positions in the movement and their personal preferences for types of organisations and activism. One can also see more specialised groups in the recent REC Directory [*REC, 2001c*]. A further sign of professionalisation is the strong position of legal experts (for example, Environmental Management and Law Association – EMLA in Hungary and Polish Environmental Law Association or Towarzystwo Naukowe Prawa Ochrony Środowiska – TNPOŚ in Poland) and coordinated lobbying efforts in these movements. Environmental NGOs are also a more frequent source of legal and technical information for journalists (interview with Zsuzsa Foltányi, Budapest, 25 May 2001).
11. For a list of most major environmental agreements, see REC [*2000b: 6–8*].
12. To date, there have been seven such meetings, the last one taking up the programmes and initiatives reviewed at the 2002 UN World Summit on Sustainable Development. The Regional Environmental Center has published proceedings of the second, third, fourth, fifth and seventh meetings [*REC, 2000, 2001a, 2001b, 2002, 2003*]. They may be accessed at http://www.rec.org/REC/Publications/Full_List.html.

REFERENCES

Albrecht, Catherine (1987), 'Environmental Policies and Politics in Contemporary Czechoslovakia', *Studies in Comparative Communism*, Vol.20, No.3–4, pp.291–302.

Baker, Susan and Petr Jehlička (eds.) (1998), *Dilemmas of Transition: The Environment, Democracy, and Economic Reform in East Central Europe*, London: Frank Cass.

Bochniarz, Zbigniew (1992), 'The Ecological Disaster in Eastern Europe: Background, Current Aspects, and Suggestions for the Future', *Polish Review*, Vol.37, No.1, pp.5–25.

Carmin, JoAnn and Barbara Hicks (2002), 'International Triggering Events, Transnational Networks, and the Development of the Czech and Polish Environmental Movements', *Mobilization*, Vol.7, No.2, pp.304–24.

CCEB (Candidate Countries Eurobarometer) (2002), conducted by The Gallup Organisation, Hungary. Available at http://europa.eu.int/comm/public_opinion/archives/cceb/2002/cceb_2002_en.pdf (last accessed 30 Mar. 2003).

CEEB (Central and Eastern Eurobarometer) 8 (1998), Public Opinion Surveys and Research Unit of EU Directorate General X for Information, Communication, Culture, Audiovisual Media. Available at http://europa.eu.int/comm/public_opinion/archives/ceeb/ceeb8/ceeb18_38.pdf (last accessed 30 Mar. 2003).

Dawson, Jane I. (1996), *Eco-Nationalism: Anti-Nuclear Activism and National Identity in Russia, Lithuania, and Ukraine*, Durham and London: Duke University Press.

DeBardeleben, Joan (1985), *The Environment and Marxism-Leninism: The Soviet and East German Experiences*, Boulder and London: Westview Press.

DeBardeleben, Joan (ed.) (1991), *To Breathe Free: Eastern Europe's Environmental Crisis*, Baltimore: Johns Hopkins University Press.

Fagin, Adam and Petr Jehlička (2001), 'The Impact of EU Assistance on Czech Environmental Movement Capacity since 1990', paper presented at the Workshop on Environmental Challenges of EU Enlargement, Robert Schuman Centre for Advanced Studies, European University Institute, Florence, Italy, 25–26 May.

Fesbach, Murray and Alfred Friendly, Jr. (1992), *Ecocide in the USSR: Health and Nature under Siege*, New York: Basic Books.

Fisher, Duncan (1993), 'The Emergence of the Environmental Movement in Eastern Europe and its Role in the 1989 Revolutions', in Barbara Jancar-Webster (ed.), *Environmental Action in Eastern Europe: Responses to Crisis*, Armonk, NY and London: M.E. Sharpe, Inc., pp.89–113.

Fisher, Duncan, Clare Davis, Alex Juras and V. Pavlovic (eds.) (1992), *Civil Society and the Environment in Central and Eastern Europe*, London: Ecological Studies Institute.

French, Hillary F. (1990), 'Green Revolutions: Environmental Reconstruction in the Soviet Union and Eastern Europe', Worldwatch Paper 99, Washington, DC: Worldwatch Institute.

Fullenbach, Josef (1981), *European Environmental Policy: East and West*, Boston: Butterworths.

Gliński, Piotr (1989), 'Ruch ekologiczny w Polsce – stan obecny', *Prace Komitetu Obywatelskiego przy Przewodniczcym NSZZ Solidarność*, No.2, pp.31–77.

Gliński, Piotr (1996), *Polscy Zieloni: Ruch spoleczny w okresie przemian*, Warsaw: Wydawnictwo IFIS PAN.

Grigorescu, Alexandru (2002), 'Transferring Transparency: The Impact of European Institutions on East-Central Europe', in Ronald H. Linden (ed.), *Norms and Nannies: The Impact of International Organizations on the Central and East European States*, Lanham, MD: Rowman & Littlefield Publishers, Inc., pp.59–87.

Hicks, Barbara (1996), *Environmental Politics in Poland: A Social Movement Between Regime and Opposition*, New York: Columbia University Press.

Jancar, Barbara (1987), *Environmental Management in the Soviet Union and Yugoslavia: Structure and Regulation in Federal Communist States*, Durham: Duke University Press.

Jancar-Webster, Barbara (ed.) (1993), *Environmental Action in Eastern Europe: Responses to Crisis*, Armonk, NY and London: M.E. Sharpe, Inc.

Jancar-Webster, Barbara (1998), 'Environmental Movement and Social Change in the Transition Countries', *Environmental Politics*, Vol.7, No.1, pp.69–90.

Kramer, John M. (1983), 'The Environmental Crisis in Eastern Europe: The Price of Progress', *Slavic Review*, Vol.42, No.2, pp.204–20.

Marks, Gary and Doug McAdam (1999), 'On the Relationship of Political Opportunities to the Form of Collective Action: The Case of the European Union', in Donatella della Porta, Hanspeter Kriesi and Dieter Rucht (eds.), *Social Movements in a Globalizing World*, New York: St Martin's Press, pp.97–111.

Pryde, Philip R. (1991), *Environmental Management in the Soviet Union*, Cambridge: Cambridge University Press.

REC (Regional Environmental Center for Central and Eastern Europe) (1997), *Problems, Progress and Possibilities: A Needs Assessment of Environmental NGOs in Central and Eastern Europe*, Szentendre: REC.

REC (Regional Environmental Center for Central and Eastern Europe) (1998), *Doors to Democracy: Public Participation and the Environment in Central and Eastern Europe*, Szentendre: REC.

REC (Regional Environmental Center for Central and Eastern Europe) (2000a), *EC – NGO Dialogue Group: Summary of the Second Meeting*, Szentendre: REC.

REC (Regional Environmental Center for Central and Eastern Europe) (2000b), *Europe 'Agreening': 2000 Report on the Status and Implementation of Multilateral Environmental Agreements in the European Region*, Szentendre: REC.

REC (Regional Environmental Center for Central and Eastern Europe) (2000c), *Greener with Accession? Comparative Report on Public Perceptions of the EU Accession Process and the Environment in Hungary, FYR Macedonia and Romania*, Szentendre: REC.

REC (Regional Environmental Center for Central and Eastern Europe) (2001a), *DG ENV – NGO Dialogue Group: Summary of the Third Meeting*, Szentendre: REC (online publication date 2000).

REC (Regional Environmental Center for Central and Eastern Europe) (2001b), *DG ENV – NGO Dialogue Group: Summary of the Fourth Meeting*, Szentendre: REC

REC (Regional Environmental Center for Central and Eastern Europe) (2001c), *NGO Directory for Central and Eastern Europe, 2001*, Szentendre: REC. Available as a database at http://www.rec.org.

REC (Regional Environmental Center for Central and Eastern Europe) (2002), *DG ENV – NGO Dialogue Group: Summary of the Fifth Meeting*, Szentendre: REC.

REC (Regional Environmental Center for Central and Eastern Europe) (2003), *DG ENV – NGO Dialogue Group: Summary of the Seventh Meeting*, Szentendre: REC.

Růžička, Tomas (1993), 'Crisis of Environmental Movement in the Czech and Slovak Republics', *Conservation and Sustainable Development*, Vol.1, pp.41–53.

Schimmelfennig, Frank (2002), 'Introduction: The Impact of International Organizations on the Central and East European States – Conceptual and Theoretical Issues', in Ronald H. Linden (ed.), *Norms and Nannies: The Impact of International Organizations on the Central and East European States*, Lanham, MD: Rowman & Littlefield Publishers, Inc., pp.1–29.

Singleton, Fred (1985), 'Ecological Crisis in Eastern Europe: Do the Greens Threaten the Reds?', *Across Frontiers*, Vol.2, No.1, pp.5–10.

Slocock, Brian (1992), *The East European Environment Crisis: Its Extent, Impact and Solutions*, The Economist Intelligence Unit Special Report No. 2109, London: The Economist Intelligence Unit.

Szirmai, Viktória (1997), 'Protection of the Environment and the Position of Green Movements in Hungary', in Káty Láng-Pickvance, Nick Manning and Chris Pickvance (eds.), *Environmental and Housing Movements: Grassroots Experience in Hungary, Russia and Estonia*, Aldershot and Brookfield, VT: Avebury, pp.23–88.

Tickle, Andrew and Ian Welsh (eds.) (1998), *Environment and Society in Eastern Europe*, Harlow and New York: Addison Wesley Longman.

Vaněk, Miroslav (1996), *Nedalo se tady dýchat: Ekologie v Českých zemich v letech 1968 až 1989*, Prague: Ústav pro soudobé dějiny AV ČR.

Volgyes, Ivan (ed.) (1974), *Environmental Deterioration in the Soviet Union and Eastern Europe*, New York: Praeger Publishers.

Weiner, Douglas R. (1988), *Models of Nature: Ecology, Conservation and Cultural Revolution in Soviet Russia*, Bloomington and Indianapolis: Indiana University Press.

Ziegler, Charles E. (1986), 'Issue Creation and Interest Groups in Soviet Environmental Policy: The Applicability of the State Corporatist Model', *Comparative Politics*, Vol.18, No.2, pp.171–92.

PART IV

ENVIRONMENTAL OUTCOMES: FROM STATE SOCIALISM TO EU MEMBERSHIP

Environmental Pasts/Environmental Futures in Post-Socialist Europe

PETR PAVLÍNEK and JOHN PICKLES

One result of the social movements for ecological defence and political change that challenged the party throughout Central and Eastern European (CEE) states in 1989 has been the broad scope and rapid pace of environmental, economic and political reform. Throughout the region legal and regulatory frameworks have been rewritten and environmental, economic and political practices have been transformed in many important ways [*Pavlínek and Pickles, 2000*]. One result has been a shift in the issues people face on a daily basis as struggles over de-communisation have shifted to issues involved in the implementation of institutions and practices of representative democracy, and as the consequences of economic collapse have shifted to efforts to manage regionally and socially uneven growth and decline. At the heart of these new experiences is the now impending prospect of accession to the European Union (EU). This involves the challenges of meeting the Copenhagen political and economic requirements (stable democratic institutions, rule of law, protection of human and minority rights, and a functioning market economy) and the necessity of transferring and adapting to the regulations, norms and practices required by the many chapters and over 80,000 pages of the *acquis communautaire* to the specific circumstances of post-socialist states.

In assessing the integration of environmental policies and practices with those of the EU, a broader history of environmental struggles and post-socialist environmental transformations remains important. In particular, as the administrative tasks of aligning post-communist structures with those of the EU increasingly come to dominate discussions of environmental transitions in the region, and to the extent that EU enlargement reinforces and enhances existing structures of civil society and patterns of public participation, it remains necessary to understand the ways in which these realignments and integrations are shaped in fundamental ways by the complex legacies of struggles over environmental regulations and practices prior to 1989 and environmental transformations and problems emerging since 1989.

In this study, we focus on these legacies through an empirical assessment of the ongoing influence and importance of state socialist environmental conditions, practices and policies and their potential role in shaping the path to integration and accession. In particular, we focus on the ways in which state socialist and post-socialist environmental conditions, practices and policies are currently shaping the path to accession and integration. In focusing on the issues of geographical context and historical legacy we seek to render a pragmatic geo-history of actual practices of regulatory reform and ecological defence and, by implication, to put in question some of the millennial accounts of structural adjustment and the managerial optimism surrounding the process of EU enlargement.

Specifically, we seek to illustrate four specific propositions. First, environmentalists in CEE countries have welcomed EU accession in part because of the centrality of environmental requirements in the *acquis*. In this sense, the requirements of the *acquis* have served rhetorically and politically as an important instrument to maintain pressure on politicians to address environmental concerns and to accept the environmental question as a legitimate part of the political decision-making process. They have also provided direct support for environmental and other civil society groups to sustain and broaden public participation in this process. Second, while the governments of all accession countries have accepted environmental reforms as part of their preparations for membership of the EU, serious questions remain about their willingness and ability to implement the directives on the ground. The economic and political commitments to produce sustainable environmental outcomes remain highly contested elements of post-socialist transformations and EU realignment. Third, while the new environmental regulations provide common standards and procedures for applicant states, the minimum standards required by the *acquis* are, in some cases, also being used by neo-liberals in CEE countries to water down existing more comprehensive environmental legislation. In this sense, we point to the dialectical reception of accession instruments. Fourth, EU accession may have important indirect consequences on environmental conditions. For example, the accession process is creating enabling legislation and capacities for CEE governments to fight corruption. Legislation such as that dealing with conflict of interest, transparency and open government may, in time, diminish the ability of city councillors to pork barrel favourite projects or to benefit materially from their political decisions on issues such as new road-building programmes or opposition to restraints on urban sprawl.

State Socialist Environmental Legacies

It is now well known that post-socialist Europe inherited a complex legacy of environmental problems with their own distinct and important geographies (see, for example, Alcamo [*1992a*]; Carter and Turnock [*1993*]; Pavlínek and Pickles [*2000*]). By the late 1980s, large areas of the region suffered from excessive air pollution, water pollution and land degradation, particularly in the former East Germany, the Czech Republic and Poland. These northern regions had histories of more extensive industrialisation with higher levels of heavy industry, such as chemicals. Additionally, the northern region relied excessively on brown coal for energy, particularly compared to the rest of CEE countries.[1] East Germany and Czechoslovakia, in particular, used large amounts of low-quality brown coal (lignite) with low heating value but high ash and sulphur content to produce most of their electricity and heat. The consequence in the electricity and heat producing regions and the nearby areas affected by long-range transport of emissions was poor air quality and negative health consequences.[2] While power plants in these countries were generally equipped with scrubbers to remove most of the particulate matter from their emissions, they were not equipped to remove sulphur dioxide and other gases. Consequently, areas close to the Czech–German–Polish border, known locally as the 'black triangle', consistently recorded peak sulphur dioxide air pollution levels in CEE countries as a whole. Marquardt, Brüggemann and Heintzenberg [*1996*] have estimated that the area probably had the highest concentration of brown coal power plants in the world, emitting about 3 million tons of sulphur dioxide annually and accounting for 20 per cent of the total European sulphur dioxide emissions in the late 1980s [*Nowicki, 1993*].

Compared to this 'northern group' of CEE countries, Hungary had diversified its energy production away from coal towards nuclear energy, oil and natural gas in the early 1980s and derived only 31 per cent of its electricity from coal by the late 1980s [*Várkonyi and Kiss, 1990*]. Romania, Bulgaria and former Yugoslavia relied less on lignite for energy production and had lower concentrations of heavy industries. Consequently, they suffered lower levels of air pollution across smaller areas, with air pollution from point sources such as power plants and industrial enterprises affecting their immediate environs (Table 1). Compounding these background and point source polluters from large power plants and industrial enterprises were the highly concentrated effects of motor vehicle emissions especially in large urban areas[3] and the effects of the long-range transboundary air pollution [*see Pavlínek and Pickles, 2000*].

Water pollution had become the second most serious environmental problem under state socialism, and it affected all CEE countries. In Poland, Hungary, Bulgaria, Romania and FYR Macedonia water pollution was considered to be a more pressing environmental issue than air pollution [*REC, 1994a*]. Heavy industry releasing heavy metals and toxic chemicals, high biochemical oxygen demand levels that cause low dissolved oxygen

TABLE 1

SULPHUR DIOXIDE AND NITROGEN OXIDES EMISSIONS IN CEE IN 1989

Country	SO_2 emissions			NO_x emissions		
	Total (1,000 tons)	Per capita (kg)	Per km² (tons)	Total (1,000 tons)	Per capita (kg)	Per km² (tons)
Albania	50*	16	1.7	9*	2.8	0.3
Bulgaria**	2,180	243	19.6	411	45.7	3.7
Czechoslovakia	2,571	164	20.1	1,143	73.1	8.9
Czech R.	1,998	193	25.3	916	88.4	11.6
East Germany	5,250	316	48.5	604	36.4	5.6
Hungary	1,102	104	11.8	246	23.2	2.6
Poland	3,910	103	12.5	1,480	39.1	4.7
Romania**	1,647	71	6.9	813***	35.1	3.4
Slovakia	573	108	11.7	227	42.7	4.6
Yugoslavia	1,650	65	4.6	190*	8.1	0.7

* Estimated emissions

** Data for Bulgaria and Romania are not reliable. Based on different data sources, 1989 SO_2 emissions in Romania ranged from 200 to 1,647 thousand tons and from 390 to 1,753 in the case of NO_x emissions [*Livernash, 1992: 64–5; UN, 1995: 4–6*]. It is also hard to believe that NO_x emissions increased by 693 per cent in 1989 compared with 1988 and then dropped by 50 per cent in 1990 [*UN, 1995: 6*]. In the case of Bulgaria, previously accepted total 1989 SO_2 emissions of 1,030 thousand tons have been adjusted to double previous official estimates (to 2,180 thousand tons), putting Bulgaria in the same class with the Czech Republic and East Germany as among the most polluted countries of CEE [cf. *Livernash, 1992: 64*; *Kabala, 1991: 385; UN, 1995: 4*]. Similarly, 1989 NO_x emission figures have been almost tripled from 150 thousand tons to 411 thousand tons [cf. *Livernash, 1992: 65; UN, 1995: 6*].

*** 1987–91 average

Sources: UN [*1995: 4–6*]; Livernash [*1992: 64–5*].

concentrations, bacterial contamination, high nitrogen and phosphorus levels caused by agricultural practices, and high salinity all pose problems for water quality [*Novotny and Somlyódy, 1995*]. Heavily polluted rivers have also contributed significantly to increased pollution in the Baltic, Black and Adriatic Seas, and heavy metal pollution of the Polish Baltic Sea coast by zinc, cadmium lead, silver and phosphorus has been well documented [*Szefer et al., 1996*]. Throughout CEE countries untreated sewage, overuse of fertilisers and pesticides in rural areas, and petroleum leaks led to large-scale contamination of underground water resources. As with air pollution, large-scale pollution of primary rivers began with the industrialisation drives of the 1950s and 1960s when newly built factories began to discharge their wastewater directly into rivers without sufficient, if any, treatment. In the 1980s, the largest areas (relative to country size) with very poor and poor water quality were in Czechoslovakia, Poland and then Romania, and the smallest areas of polluted waters (relative to country size) were in Yugoslavia, Bulgaria and Hungary [*Alcamo, 1992b*]. Using the ratio of polluted river length to total river length, Novotny and Somlyódy [*1995*] have suggested a different ranking, with the worst situations being in Poland, Slovakia and Bulgaria, with these three countries each having more than half of their total monitored river lengths in the worst water quality class. However, despite this evidence of neglect, by the late 1980s only a few rivers were believed to be biologically 'dead' and general levels of river pollution were considered to be less severe than those of the most polluted rivers in industrial regions of Western Europe up to 20–30 years ago [*OECD, 1994*].

Soil degradation was a serious problem in all CEE countries except Poland and former Yugoslavia. These primarily resulted from a combination of factors related to the effects of collectivisation, especially resulting from the introduction of new large-scale farming methods, inadequate levels of investment in soil-conservation machinery and practices, and the systemic incentives for over-application of pesticides, herbicides and fertilisers under production for the 'plan' [*see Meurs, 2001*]. These environmental problems intensified during the period of state socialism have been carefully documented in the literature for the individual countries and their regions across CEE countries [*Alcamo, 1992a; Carter and Turnock, 1993, 2002; Pavlínek and Pickles, 2000*].

Contextualising the State Socialist Environmental Degradation

Despite popular Western generalisations about 'East European ecocide' arising from widespread and widely reported environmental problems [*McCuen and Swanson, 1993; Feshbach and Friendly, 1992*], the nature and extent of environmental degradation was actually highly uneven across CEE countries. Polluted regions and environmental hot spots coexisted with the areas of 'pristine' nature that covered an estimated 30 per cent of the region [*REC, 1994b*]. Such complexity and unevenness make sweeping generalisations about the state of the environment in CEE countries at the end of the state socialist period difficult, and this difficulty is compounded by the notoriously questionable nature of state socialist environmental data (Table 1). The nature and quality of environmental data collected before 1989 differed among the individual CEE countries, but in general data collection (where it existed) was focused on the most heavily polluted areas and during periods with the highest levels of pollution. Data over-represented the polluted areas both spatially and temporally and under-represented less polluted regions and periods.

Any assessment of post-socialist environmental change, and in particular any assessment of the role of the adoption of EU regulatory frameworks and practices, therefore needs to engage with the myths about state socialist environmental policies that circulate in Western and Eastern European policy arenas: the myth of 'ecocide', 'toxic nightmare' and 'ecological disaster' [*Feshbach, 1995; Feshbach and Friendly, 1992; McCuen and Swanson, 1993*], myths about the almost total ignorance of environmental problems by state socialist governments, and myths of state socialist environmental problems as being completely different from the situation and environmental challenges of Western capitalisms. These myths run deep, shaping the debates and science about the consequences of reform and remediation. Thus, for example, Klarer and Francis [*1997: 7–8*] have argued that:

> … with practically *no exception*, production processes were wasteful [under central planning] … there were *no incentives* to introduce efficient or environment-friendly technologies … the neglect of environmental problems was pervasive throughout the system [and] environmental pollution officially did not exist [emphasis added].

Similarly, Rowntree *et al.* [*2003: 324*] have recently maintained that:

> During the period of Soviet economic planning (1945–90) [in CEE countries], little attention was paid to environmental issues because of an overt emphasis on short-term industrial output. Additionally, since communist economics had no way to calculate environmental costs, they were not a concern. As a result, *the environment was not an issue* … [emphasis added].

In reality, state socialist environmental problems had a very distinct geography, the reality of state socialist environmental protection was complex, and the excessive environmental degradation in hot spots did not result merely from ignorance on the part of state socialist governments but from their failures to deal with environmental problems successfully within the limits of the state socialist system and the type of development model pursued. While the state formally acknowledged the importance of environment and nature protection, its environmental policies were strictly subordinated to overriding economic goals.

Formal recognition of the need for environmental management resulted in ever-changing institutional arrangements (the system of 'state nature protection', various environmental commissions, research institutes, and later ministries) and the introduction of environmental legislation (beginning in the 1950s and 1960s). The communist state had developed both the institutions and legal instruments to protect and manage the environment, and parts of this system (such as the extensive systems of nature preserves in all CEE countries) had achieved remarkable results well before 1989. The quality of the environment (particularly in Central Europe) suffered most during the 1970s and 1980s when 'forced industrialisation' expanded extensive production methods and increased the demand for, and reliance on, low-quality local coal for electricity generation.

There were important similarities between state socialism and capitalism in these production and consumption processes, with parallels in both environmental consequences and ideological understandings of nature and society. These parallels grew as CEE governments attempted to emulate Western production methods. In this sense, the environmental crisis of the 1970s and 1980s must be understood as an integral part of the crisis of state socialism that, in turn, was closely related to the broader global economic changes of the same period. Increasing indebtedness of state socialist countries in the 1970s and 1980s, for example, led directly to declines in environmental investment in countries such as Poland. More importantly, rapid increases in oil prices in the 1970s and reductions in oil deliveries from the Soviet Union in the early 1980s led to increased dependence on highly polluting domestic coal (typically low-quality lignite) to produce electricity and heat across CEE countries [see *Pavlínek and Pickles, 2000; DeBardeleben, 1991; Kramer, 1991, 1987*].

Environmental crises became symptoms of an economic crisis of state socialism. And it was these crises that gave rise to intense struggles within the apparatuses of the state, such as in Hungary and Czechoslovakia, as governments tried to deal with the issues of excessive environmental degradation in the 1980s. In some instances it was precisely these institutional struggles and the reforms they generated that produced the ideological, institutional and regulatory foundations for environmental change in the 1990s.

Environmental Effects of Post-Communist Transformations

The collapse of state socialism and the transformations that ensued had both positive and negative effects on the environment in CEE countries. In the early 1990s collapsing industrial production and gradual industrial restructuring away from heavy industries and from industries to services led to 'clean-up by default' in which lower emissions of pollutants were the by-product of economic and structural change. New environmental legislation was introduced with varying speeds across CEE countries, existing legislation was better enforced, and governments pursued energy policies for electricity and heat production aimed at switching permanently away from highly polluting coal and lignite to less polluting fuels such as nuclear power, hydroelectricity and natural gas. In some CEE countries, the largest polluters that survived the economic crisis were forced to cut their emissions drastically to comply with new emissions limits. Together these contributed to significant declines in air and water pollution across CEE countries, declines that have largely sustained as industrial production has stabilised and in some countries has begun to grow. However, while the region witnessed plummeting levels of air and water pollution, new environmental problems related to mass consumerism such as uncontrolled growth in the car traffic or rapidly increasing consumer waste emerged in the 1990s. In each of these cases of 'clean-up by default', re-capitalisation for more efficient machinery and pollution control equipment, legislative reform, or the emergence of new problems, it is important to consider the ways in which these trends were geographically and temporally uneven across the region in the 1990s (Tables 2 and 3, Figures 1, 2 and 3) [see also *Pavlínek and Pickles, 2000; Carter and Turnock, 2002*].

Air Pollution

Sulphur dioxide emissions illustrate such ambiguous temporal trends. Emissions had started to decline in the 1980s and it seems that the 1990s are the continuation of these trends (Figure 1). However, a closer examination of the 1980s emission data shows that while sulphur dioxide emissions generally declined in Central Europe (Poland, Czech R., Slovakia, Hungary and Slovenia) and less rapidly in the former Soviet Union (Russia, Ukraine, Belarus, Estonia, Moldova), they increased in South Eastern Europe (Romania, Bulgaria, Yugoslavia). This situation reflects different factors such as a greater emphasis on improving environmental quality in Central Europe in the second half of the 1980s, the diversification of energy resources (such as the introduction of nuclear power in Hungary and former Czechoslovakia and its rapid development in the former Soviet Union (FSU), and increased reliance on oil and natural gas in Hungary), and a series of milder winters in the second half of the 1980s demanding less heating and consequently lower emissions. However, the largest declines in sulphur dioxide emissions in the 1980s, at 35 per cent (see Table 2), were recorded in the European part of Russia in the areas covered by the EMEP programme.[4] This resulted from shifts from coal to oil, natural gas and nuclear power for electricity and heat production in the European portion of the FSU.[5]

In the 1990s, sulphur dioxide emissions declined further and rapidly across CEE countries. The extent of decline varied from 23.5 per cent in Yugoslavia to 95 per cent in Moldova. Only Yugoslavia, Croatia and Romania recorded below 50 per cent declines in sulphur dioxide emissions in the 1990s. The remaining CEE countries for which data is available

TABLE 2

PERCENTAGE CHANGE IN SULPHUR DIOXIDE EMISSIONS IN SELECTED CEE COUNTRIES, 1980–2000

	Belarus	Bulgaria	Croatia*	Czech R.	Estonia	Hungary	Latvia*	Lithuania
1980–85	–6.8	12.9	10.0	0.9	–11.5	–14.0	0.0	–2.3
1980–89	–9.7	6.3	18.0	–11.5	–11.5	–32.5	0.0	–4.2
1980–2000	–80.7	–52.1	–39.5**	–88.3	–66.7	–70.3	–84.8	–86.1
1989–2000	–78.6	–55.0	–48.8**	–86.8	–62.4	–56.0	–84.8	–85.5

	Moldova	Poland	Romania	Russia†	Slovenia	Slovakia	Ukraine	Yugoslavia
1980–85	–8.4	4.9	19.0	–13.5	3.0	–21.4	–10.0	17.7
1980–89	–22.7	–4.6	43.8	–34.7	–9.8	–26.5	–20.2	24.6
1980–2000	–96.1*	–63.1	–13.6**	–72.1	–59.0	–82.8	–70.6**	–4.7
1989–2000	–94.9*	–61.4	–39.9**	–57.3	–54.5	–76.6	–63.2**	–23.5

* 1980–89 emissions based on expert estimates
** Refers to 1999 data, 2000 data not available
† Data refer to the EMEP area only

Source: Developed from EMEP [*2002*].

TABLE 3

PERCENTAGE CHANGE IN NITROGEN DIOXIDE EMISSIONS IN SELECTED CEE COUNTRIES, 1980–2000

	Belarus	Bulgaria	Czech R.	Estonia*	Hungary	Lithuania	Moldova
1980–85	1.7	0.0	–11.3	0.0	–3.8	9.2	13.8
1980–89	12.4	–1.2	–1.8	–1.4	–9.6	13.8	20.7
1980–2000	–42.4	–55.7	–57.6	–40.9	–31.4	–68.8	–70.9**
1989–2000	–48.7	–55.1	–56.8	–40.0	–24.2	–72.5	–75.9**

	Poland	Romania	Russia†	Slovenia	Slovakia*	Ukraine	Yugoslavia
1980–85	22.1	3.6	9.7	3.9	0.0	–7.5	23.4
1980–89	20.4	10.7	47.2	13.7	15.2	–7.0	31.9
1980–2000	–31.8	–39.0**	35.9	13.7	–42.1	–60.2**	6.4
1989–2000	–43.4	–44.9**	–7.7	0.0	–49.8	–57.3**	–19.4

* 1980–86 emissions based on expert estimates
** Refers to 1999 data, 2000 data not available
† Data refer to the EMEP area only

Source: Developed from EMEP [*2002*].

FIGURE 1

TRENDS IN SO2 AND NOX EMISSIONS IN SELECTED CEE COUNTRIES, 1980–2000

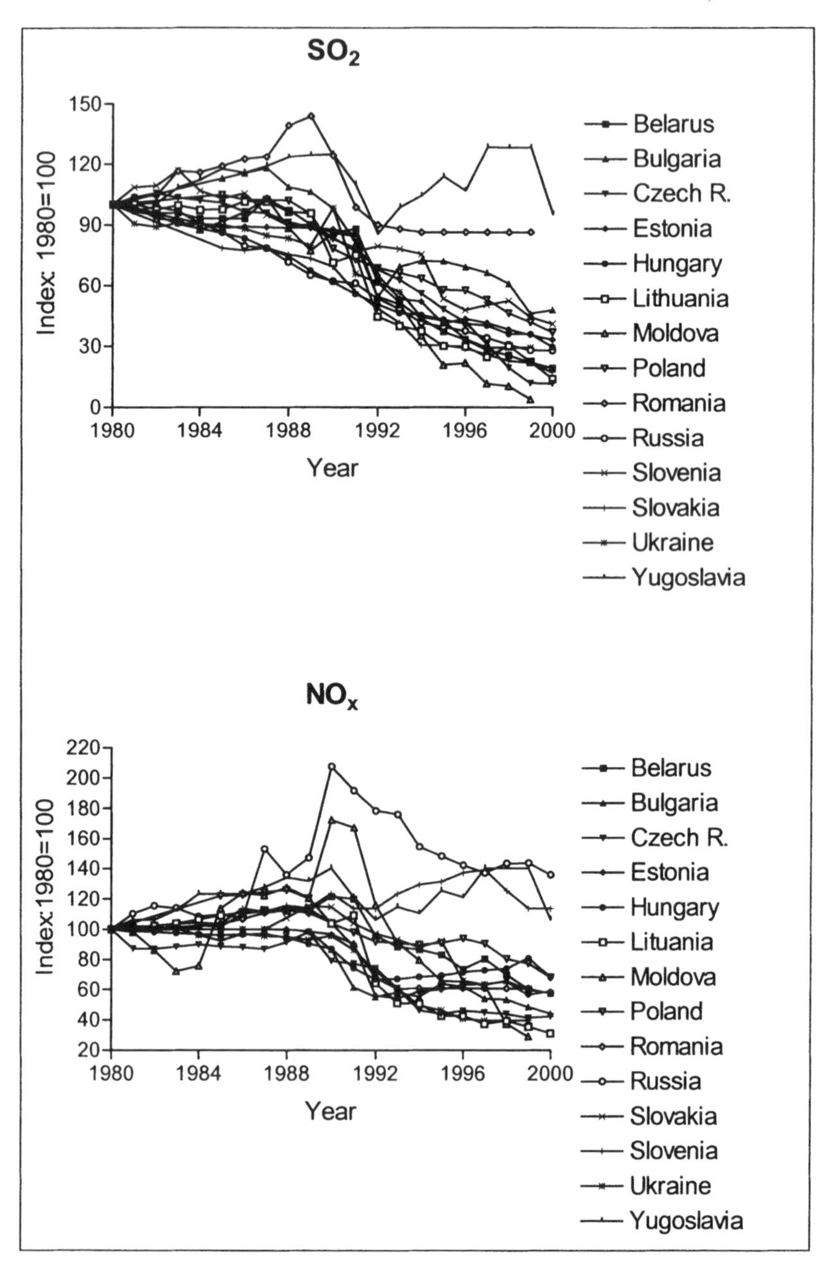

Source: Developed from EMEP [2002]

FIGURE 2

GDP AND INDUSTRIAL PRODUCTION TRENDS IN THE SELECTED CEE COUNTRIES

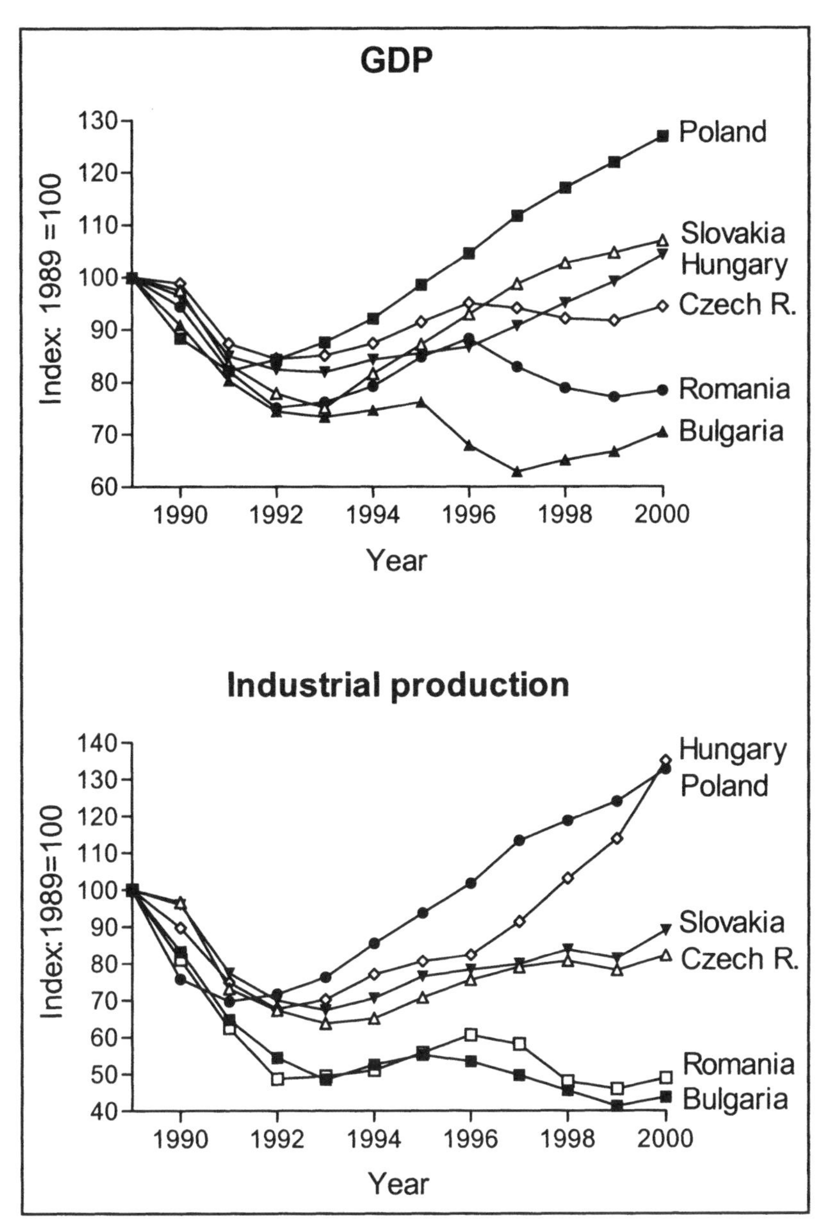

Source: Developed from BCE [*1999*] and SOSR [*2001*].

FIGURE 3

EMISSION TRENDS IN THE CZECH REPUBLIC, SLOVAKIA, HUNGARY AND POLAND, 1989=100

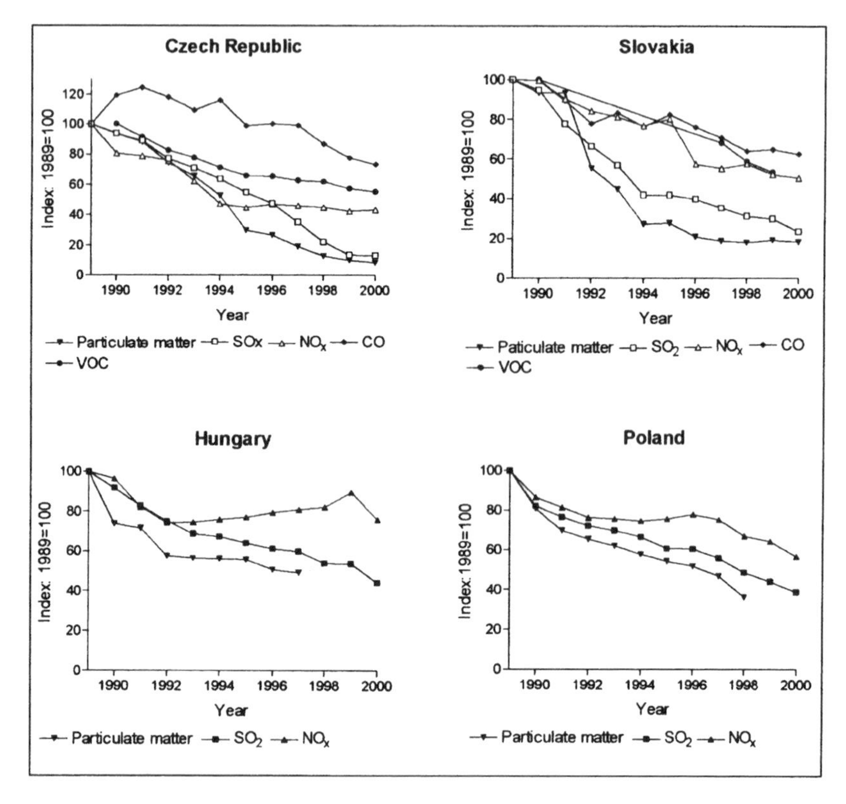

Source: Developed from EMEP [*2002*]; MoE and CSU [*2001*]; MoE [*1997*]; MoE SR [*2001, 1998*]; MERP [*1999*]; GUS [*1997, 1999, 2001*].

recorded declines in excess of 50 per cent. In Yugoslavia war and economic sanctions increased the reliance on domestic coal and delayed the introduction of policies of economic transformation compared to the rest of CEE countries. Elsewhere economic decline was certainly partially responsible for dramatic decreases in sulphur dioxide emissions,[6] but declines continued as economies stabilised and as they experienced some growth in the second half of the 1990s, suggesting the importance of structural economic changes, changes in the structure of fuel consumption, and the role of governmental environmental policies (Figures 2 and 3). Emissions continue to be high in several countries. In 2000, the Czech Republic recorded the fourth highest sulphur dioxide emissions per populated square kilometre of its territory in the world (7,980 tons), Slovakia ranked ninth (4,850 tons), Bulgaria tenth (4,610 tons) and Poland fourteenth (3,980 tons) [*The Economist, 2002*].

In most CEE countries nitrogen dioxide emissions continued to grow throughout the 1980s, although Bulgaria, Czech Republic, Estonia, Hungary and Ukraine experienced small declines in emission. In the 1990s, nitrogen dioxide emissions dropped across the entire region, albeit less rapidly than declines in sulphur dioxide emissions (Figure 1).

Data for particulate matter emissions across CEE countries in the 1990s are not readily available. However, in those countries where the data are available particulate matter emissions also declined substantially in the 1990s. For example, particulate matter emissions declined by 91.5 per cent in the Czech Republic between 1989 and 2000 and by 94.4 per cent between 1985 and 2000 [*MoE, 2002, 1997*]. They dropped by 81.8 per cent in Slovakia between 1989 and 2000 [*MoE SR, 1998, 2001*], by 64 per cent in Poland between 1989 and 1998 [*GUS, 1997, 1999, 2000*], by 50.9 per cent in Hungary between 1989 and 1997 (72.2 per cent between 1985 and 1997) (Figure 3) [*MERP, 1999; OECD, 1996*] and by 57.3 per cent in Russia between 1990 and 1997 [*GUS, 1999*]. Emissions of other air pollutants also declined significantly in the 1990s, including the emissions of carbon dioxide, volatile organic compounds (VOC) and heavy metals (Figure 3). It is important to keep in mind, however, that the data presented above mainly represent pollution from large stationary sources. Much less detailed and precise information is available about the small local point sources and mobile sources of air pollution.

Water Pollution

The lack of availability of data and its incompatibility between countries make any generalisations about water pollution difficult. Available evidence suggests that aggregate levels of water pollution have declined substantially since 1989, but that these declines have also been geographically uneven. Improvements in surface water quality were achieved because of decreases in the overall amount of discharged wastewater, construction and operation of new effluent treatment plants, refurbishment and increased efficiency of the existing ones, industrial restructuring, and decreases in the use of industrial fertilisers and pesticides by agriculture with corresponding decreases in their run-off (Figure 4).[7] As with air pollution, in some countries, such as the Czech Republic, levels of water pollution also began dropping in the mid-1980s under state socialism with the push to build more effluent treatment plants in order to treat more wastewater.[8] This trend has been strongly reinforced since 1989. In this period, overall withdrawals and use of water have declined considerably. In the Czech Republic, agriculture cut its water use by 83 per cent, the industrial sector by 36 per cent, the energy sector by 19 per cent, and drinking water withdrawals dropped by 28 per cent between 1990 and 1997 [*MoE, 1999*]. The volume of wastewater released by these sectors declined correspondingly. The sharp declines in water withdrawal by agriculture and industry originally resulted from production slumps in water consumptive sectors, and household water consumption declined after governmental subsidies on the price of drinking water were removed and prices for water rose rapidly.

In Slovakia, the number of water treatment plants almost doubled in the 1990s (179 in 1990, 344 in 2000) and their overall capacity increased by 48 per cent. The share of treated wastewater discharged into the public sewage system increased from 85 per cent in 1989 to 95 per cent in 2000 [*MoE SR, 2001*]. As in the Czech Republic, household water consumption was declining in the 1990s. Per capita daily household water consumption decreased by 28 per cent between 1991 and 1997 (from 183 to 132 litres) [*MoE SR, 1998*]. Between 1993 and 2000 surface water consumption by industry dropped by 20 per cent (from 700 million cubic meters to less than 600 million) and the underground water industrial withdrawal decreased by 50 per cent during the same period. Between 1990 and 2000 the application of industrial fertilisers per hectare plummeted by 80 per cent [*MoE SR, 2001*].

In Poland, the total amount of untreated industrial and household sewage discharged into rivers decreased by 70 per cent between 1990 and 1998, while the share of discharged sewage without any treatment in all discharged wastewater declined from 32 per cent to 17 per cent [*GUS, 1999*]. Only one per cent of effluent water discharged by industry was not treated in 1999 compared to 4.6 per cent in 1990. The total household and industrial production of sewage declined by 32 per cent as overall water consumption dropped by 20 per cent between 1990 and 1998 [*GUS, 1999*]. In the mid-1990s about 1,000 wastewater treatment plants were built in

FIGURE 4

TRENDS IN THE APPLICATION OF NPK FERTILISERS AND PESTICIDES IN THE CZECH REPUBLIC, POLAND AND SLOVAKIA IN THE 1990s

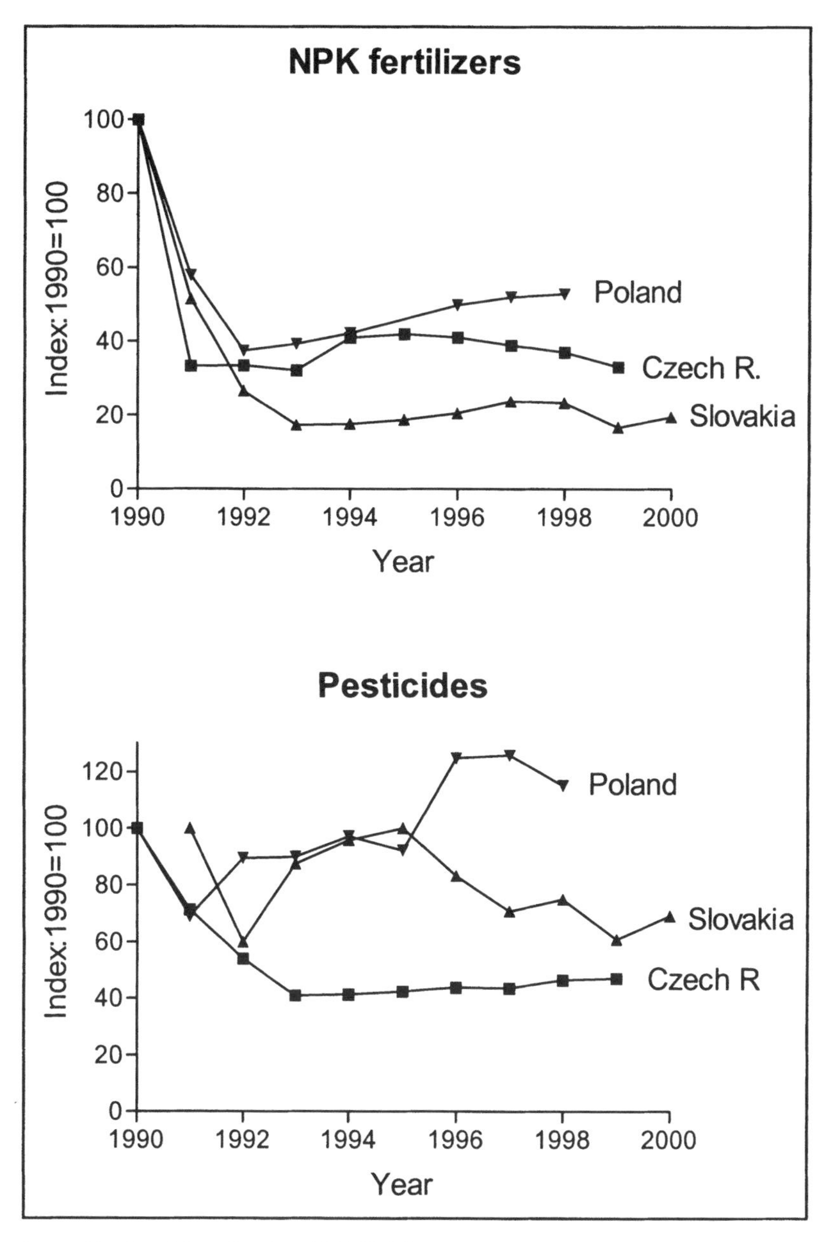

Source: Developed from data from MoE [*2000*]; MoE SR [*2002*]; GUS [*1999*].

Poland, at the rate of about 350–400 per year, although it is expected that it will take an additional 15–20 years before all sewage is properly treated [*Nowicki, 1997*].

In Hungary, water quality has also improved as a result of the post-1989 economic transformation. However, improvement has been slow and major 'changes in water quality were rare, only observable in a few sectors' [*MERP, 1994; Lehoczki and Balogh, 1997*]. Evaluating any changes in water quality is also difficult because of unreliability of data associated with changes in the statistical classification system in 1992. In 1998, more than 75 per cent of Budapest's sewage was still discharged directly into the Danube without any treatment and half of Hungarian households were not connected to a sewage system [*MF Dnes, 1998*]. In 1999, the Somes, Tisza and Danube Rivers were badly contaminated by a cyanide leak from the Aurul goldmine in Baia Mare, Romania, and threats to water quality from industrial accidents remain in many areas.

New Environmental Problems in CEE Countries

Socialist and post-socialist environmental regulations and economic changes have ameliorated many of the worst environmental legacies of forced industrialisation and lax state socialist environmental policies and enforcement. At the same time, post-communist transformations have themselves altered the nature of the environmental problems the region is facing. While post-socialist countries witnessed general declines in levels of air and water pollution in the 1980s and 1990s, they have not disappeared and new environmental problems typical of deregulated market capitalism have quickly emerged. These include, among others, the uncontrolled development of car transportation leading to worsening urban air quality, rapidly increasing waste production by households and the break-up of established systems of land use control and water resources and forestry management leading, for example, to uncontrolled urban development on precious agricultural land. In what ways, then, have post-socialist environmental and economic policies, along with the politics of *acquis* reform, not yet addressed the environmental problems of state socialism and set the region on a path to an environmentally sustainable future?

Private car ownership grew quickly across CEE countries in the 1990s largely fuelled by imports of millions of used cars from Western Europe (Figure 5).[9] Governments and the EU supported the development of car transportation through ambitious and expensive highway construction projects, occasionally leading to political struggles over environmentally valued landscapes and nature preserves. At the same time, governmental funds supporting public transport have dwindled. Not surprisingly, countries such as the Czech Republic experienced rapid declines in the amount of passengers and cargo transported by rail at the expense of more polluting road transport (Figure 6).[10] Uncontrolled development of road transportation has led to congestion in major urban areas, and consequently to the worsening of their air quality especially in terms of nitrogen oxide emissions. In Slovakia, for example, the transport sector accounted for 43.5 per cent of total emissions of carbon dioxide, 36.5 per cent of nitrogen oxides and 36.5 per cent of volatile organic compounds in 1999 [*MoE SR, 2001*]. These negative environmental developments were somewhat compensated for by the fact that the old (often two-stroke engine) and more polluting cars have been gradually replaced by less polluting automobiles and that leaded gasoline has gradually been phased out across the region.[11] However, the overall trend of increasing automobile dependence with attendant increases in energy consumption and air pollution, reductions in the use of public transportation, and increased landscape and wildlife habitat destruction through rapid highway construction has been environmentally damaging.

FIGURE 5

TRENDS IN THE NUMBER OF REGISTERED PASSENGER CARS IN POLAND, SLOVAKIA, THE CZECH REPUBLIC AND HUNGARY BETWEEN 1993 AND 1999

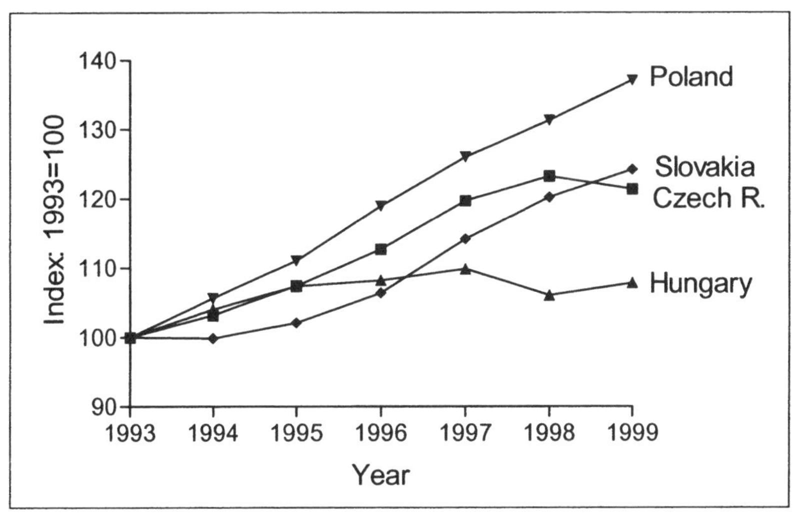

Source: Data from MTC [*2002, 1999*].

FIGURE 6

CARGO AND PASSENGERS TRANSPORTED BY CZECH RAILWAYS IN THE 1990s

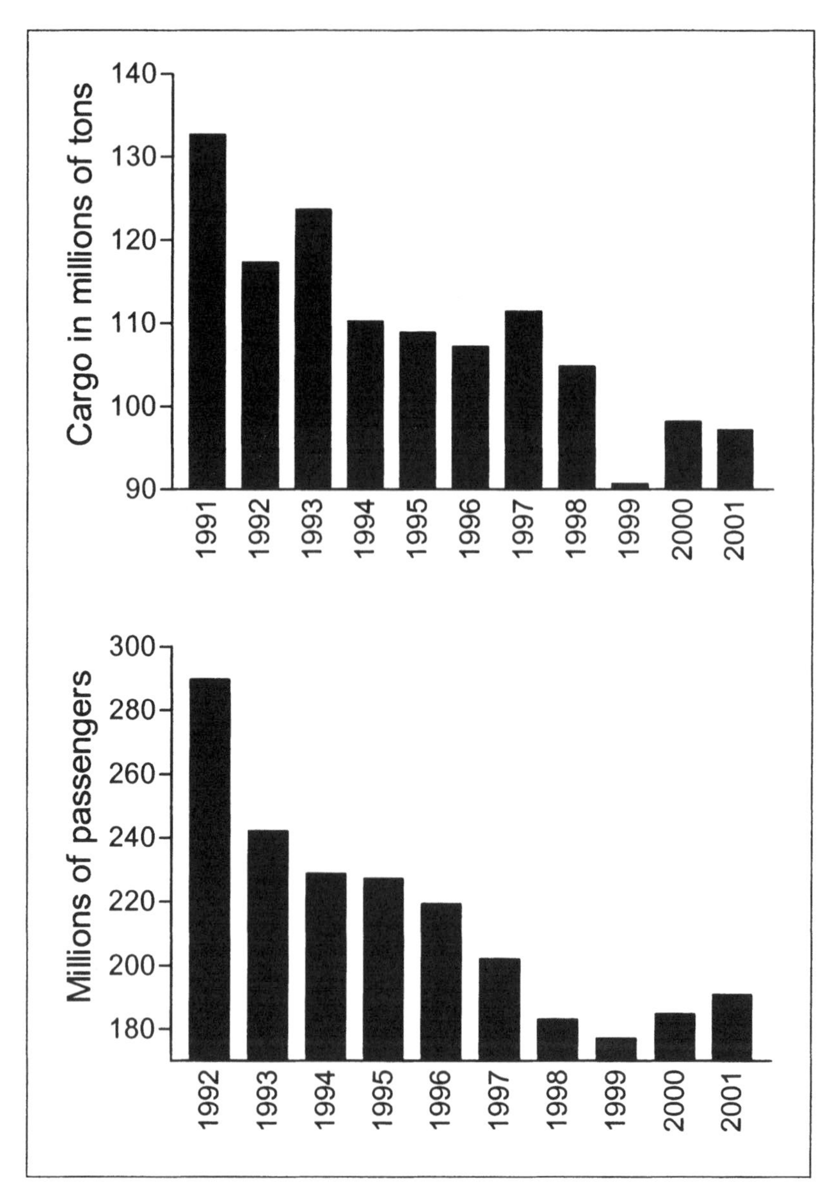

Source: Data from MTC [*2002, 1999*].

Western-style consumerism introduced in the region after 1989 also brought its own problems, particularly plastic packaging and increased domestic waste. Although CEE countries are producing less than half as much domestic waste as Western Europe, the region is quickly catching up [*Bisschop, 1996*].[12] The system for the collection of old paper and glass bottles collapsed after 1989 and no adequate system for recycling, composting and incineration of waste has been established. As a result, after 1989 municipal waste increased by 20–30 per cent and the number of illegal dumping sites grew rapidly. In Poland, for example, municipal waste generation more than doubled (+234 per cent) between 1990 and 1997 and the number of illegal dumps exceeded 10,000 in 1997 [*Nowicki, 1997; Carter and Kantowicz, 2002*]. Glass bottles, traditionally used many times over by beverage producers, were replaced by plastic bottles, but few efficient recycling programmes were put in place to deal with them. The result was a large increase in plastic waste across CEE.

Uncontrolled urban development also created its own problems, even as state policies and regulations are brought into line with the *acquis*. These included the rapid construction of large retail outlets by Western chain retailers at the outskirts of urban areas, rapid construction of houses for the newly rich, and the construction of industrial parks and green-field factories on previously protected high-quality agricultural land across CEE countries. Many foreign investors prefer green-field investments and build their factories on previously unused sites to avoid the potential environmental liability. For some Western investors CEE countries still offer lower environmental, health and safety standards (as they did before 1989) and governments have been known to ignore environmental issues while approving privatisation projects. In this context, in particular, the European Union is ambiguously placed. Committed as it is to environmental reform, it is nonetheless similarly committed to improving the economic climate for EU firms by providing opportunities for those firms with a pan-European economic region. The imperatives of lowering the cost of investments and production and of building environmentally sound policies do not always coincide.

Along with its broader requirements for environmental legislation and regulation, the EU does support environmental standards for producers seeking to export goods into the EU. Particularly since the Doha Round of trade negotiations, the EU is making progress on establishing criteria and gaining acceptance for eco-labels and ISO 14000 criteria. However, since the 1980s the EU has also operated beneficial customs arrangements for the outward processing to CEE countries of goods made by EU-based firms. Outward processing trade (OPT) arrangements permitted EU-based manufacturers (and later retailers) to carry out offshore assembly production by exporting raw materials for assembly if the finished goods were re-imported into the EU. Such OPT arrangements incurred no tariffs or taxes on the re-imported finished goods [see *Begg et al., 2003*]. In practice, OPT encouraged a system of sub-contracting assembly production to low wage regions under conditions of intense competition and extreme control by the manufacturers and retailers in Western Europe. Although the trade regime of OPT has been in place since the 1980s, the scale of its use has increased dramatically in recent years and the expansion of normal trade along similar lines has also been enormous. The health and environmental consequences for both workers and regions of such systems of sub-contracting are unknown at this time.

Privatisation in tandem with foreign direct investment (often from firms based in the EU) has itself generated uneven outcomes. Indeed, according to the former Czech environmental minister Ivan Dejmal, the Czech government was approving large privatisation projects without allowing experts from the MoE to assess potential environmental effects of privatisation and foreign investment. In some cases, MoE officials were never shown the proposals for privatisation [*Kolebaba and Petrlík, 1994*]. In Slovakia, codes of environmental conduct for foreign companies were not prepared, much less enforced, by the government as late as 1997 [*Podoba, 1998*]. In Hungary, environmental concerns and liability questions were also ignored during the initial phase of privatisation [*Lehoczki and Balogh, 1997*]. In Poland, the 1988 Law Governing Economic Activity Involving Participation of Foreign Companies provided special tax incentives for foreign investors who contributed to environmental protection and clean-up. Permission to invest in Poland could also be refused. However, both measures were removed when the 1988 Law was replaced by the Joint Venture Law in 1991 [*Kruszewska, 1993*].

Overall, any generalisations about the effects of influx of foreign capital to CEE countries are impossible to make because, while some Western companies investing in the region bring with them high environmental standards and behaviours, others have brought ‘dirty investments’. For example, the largest environmental accident in CEE countries in the 1990s was caused by a cyanide leak into the Tisza River. This originated at the Romanian–Australian joint venture gold mining company under conditions in which the Australian party had not transferred responsible environmental practices to their joint venture in Baia Mare.

Post-communist transformation has also resulted in new dangers of reckless exploitation. For example, privatisation of forests and uncertainties over ownership rights and cutting regulations have led new private owners to cut their newly acquired forests for profit, ignoring environmental consequences and, in some cases, established legal codes [*Pickles et al.,*

2002]. In another case, increased limestone mining in the Czech Karst Preserve associated with the German and Belgian investment in the Czech cement industry led to significant environmental damage [*Nika, 1994*].

Prior to 1989, large areas along the former Iron Curtain and numerous military exercise areas had been closed to the public and no economic activities had been allowed there; in effect Cold War border tensions had unintentionally created natural sanctuaries. When opened to the public after 1989, environmental quality was quickly compromised. In the 1990s, the management of existing natural parks was often negatively affected by efforts of private entrepreneurs supported by anti-environment-oriented politicians to reduce their size or remove restrictions on particular uses in order to allow economic activities such as logging and mining. Natural parks were also negatively affected by various forms of private construction activities designed to profit from the rapid expansion of domestic and foreign tourism.

Finally, it is also important to mention the negative environmental effects of war in the former Yugoslavia ranging from the environmental damage caused by large-scale refugee and troop movement to the environmental consequences of NATO bombardment. For example, bombings of chemical plants in Panèevo, the oil refinery at Novi Sad and the Zastava car plant at Kragujevac released clouds of toxic chemicals, contaminated the Danube and led to ground contamination with polychlorinated biphenyls and dioxins [*Clarke, 2002a*]. The use of depleted uranium weapons by NATO raised the fears of potentially health-threatening radioactive contamination and millions of mines remain a major environmental hazard particularly in Bosnia-Herzegovina, Kosovo and Krajina [*Clarke, 2002b; Jordan, 2002*].

Environmental Policies

Following the political changes of 1989, the general desire to clean up the worst cases of environmental degradation and control future pollution across CEE countries led initially to considerable efforts by new governments, the EU, the European Bank for Reconstruction and Development, and the US Environmental Protection Agency (among others) to introduce more effective environmental management and policies. While different CEE countries approached environmental reform differently and at different speeds, reflecting the urgency and diversity of environmental problems and the strength of local environmental movements, this change took four basic forms. First, existing environmental ministries and related environmental management institutions (such as environmental inspectorates and various institutes of environmental protection) were created or restructured, typically with much broader competencies compared to those of their state socialist predecessors. Second, the overhaul of existing environmental legislation resulted in the preparation and enactment of new environmental laws, and changes in the enforcement of existing environmental legislation (see Pavlínek and Pickles [*2000*]; Carter and Turnock [*2002*] for details). Third, the sense of urgency to deal with environmental problems in the early 1990s quickly evaporated as economic problems overwhelmed both the governments and the public. And fourth, the role of the EU in shaping environmental policies and legislation in the candidate countries increased in the second half of the 1990s (especially from 1999 onwards), particularly as a result of political requirements to meet the EU's environmental *acquis* and the full commitment this received in 1999 [*Turnock, 2002*].

Not all such changes were beneficial. Indeed, changes in environmental policies and management have been highly contested across the region. On several occasions politicians who were sceptical about environmental regulation used the process to water down existing environmental legislation to bring it into line with less demanding minimum standards required by the *acquis*.[13] Environmental interests were routinely overshadowed by economic imperatives and by anti-environment attitudes of leading politicians. 'Clean-up by default' was presented as a success of governmental economic and environmental policies, and was used politically to support neo-liberal views of environmental management in which the 'market will protect nature', as well as to silence the governments' most vocal critics [*Klaus, 2002*]. In the process, environmental movements were fragmented and marginalised as they suffered from the outflow of prominent figures into the governmental positions and private sector in the early 1990s, internal conflicts, increased public apathy towards the environmental issues, and attacks by anti-environmentalists in the government, high-level politics, private businesses and state administration. Indeed, many policies and strategies pursued to improve environmental quality have been questionable. For example, Slovakia forced through the controversial Gabèíkovo dam project on the Danube in an effort to increase hydroelectricity production and to regulate the Danube River. The country also completed the Mochovce nuclear power plant in 1998 despite strong protests from neighbouring Austria and long-standing opposition from environmentalists. The power plant is based on Soviet technology and is located less than 100 kilometres from Austria. Similarly, despite strong Austrian protests, the Czech government decided to complete construction of a nuclear power plant at Temelín at a site planned and started before 1989. Constructions of both power plants were justified by energy needs (false in the Czech case) and the fact that clean nuclear energy would contribute to further reductions in air pollution from coal-based power plants. Both power plants represent an important state socialist legacy and together with the Gabèíkovo dam point towards sometimes environmentally problematic origins of rapid declines in pollution across CEE countries.

The concrete ways in which these changes and processes took place in CEE countries varied considerably and have been analysed elsewhere [*Pavlínek and Pickles, 2000; Carter and Turnock, 2002*]. Detailed country analyses of environmental policies during the post-1989 period reveal a surprisingly strong influence of state socialist legacies in shaping the content and practices of post-1989 environmental reform. In some countries, such as the Czech Republic, the most important environmental laws enacted between 1990 and 1992 had already been prepared by environmentalists in the second half of the 1980s. The trend of environmental quality improvement also started in the 1980s, not *after* 1989, and new environmental legislation could not have been enacted as quickly as it was after 1989 without this preparation. While lack of investment in industry and agriculture shaped patterns of environmental degradation after 1989, investments made by the state before 1989 also resulted in beneficial environmental outcomes in the 1990s. Perhaps most important, it seems that attitudes towards the environment have changed little, and perhaps this is not surprising if one sees continuities rather than discontinuities in shaping policies and practices. The goal of state socialism to achieve prosperity and to be economically successful rendered the environment disposable, a limit to be overcome. For many in the ecological movements of the late 1980s and early 1990s, the 'transition to capitalism' offered a chance to achieve prosperity and ecological security. In practice, within only a few years, neo-liberals held sway, they were suspicious of state regulation especially when its origins lay in the *ancien regime*, and the environment was seen only as a luxury to be addressed after material prosperity has been achieved. That is, there is a surprising correspondence between the neo-liberals after 1989 and communist politicians before 1989 in the ways in which they understood and acted in regard to the environment.

Conclusion

There is no doubt that important changes in environmental regulation and management occurred in the decade of the 1990s and that the overall quality of the environment improved. These improvements resulted from a number of factors such as economic decline in heavy and polluting industries, a shift from an industrial to a more service-oriented economy, anti-pollution measures and policies introduced by CEE governments leading to reduced emissions (such as the desulphurisation of major coal-based power plants), and shifts in the structure of energy production away from highly polluting coal towards less polluting natural gas, oil, hydroelectricity and nuclear power.

At the same time some improvements have their origins in investments and regulations put in place before 1989 and in other cases the strategies pursued to achieve improvements were environmentally questionable (such as the reliance on nuclear power and large water hydroelectricity schemes in the Czech Republic and Slovakia). It is still too early to assess the overall consequences for environmental quality of new and potentially damaging developments associated with deregulation, industrial restructuring, contract outsourcing and marketisation. These include largely unregulated increases in automobile use, which are already having negative effects on environmental quality in urban areas and are undermining reductions in air pollution achieved from stationary sources. Additional problems are emerging as consumption patterns change.

The region has seen dramatic increases in plastic packaging and waste production by households, poorly regulated urban development on previously protected agricultural land, and efforts to undermine the protected natural areas [see also, *Gille, this volume; Beckmann and Dissing, this volume*]. Thus, while short-term environmental change certainly seems to be positive and represents a major improvement over the late 1980s, the longer-term environmental consequences of the 'transition to capitalism' remain less clear. However we interpret the past decade of environmental change, it is clear that the notion of a clear break with state socialist environmental practices in the early 1990s was more apparent than real. The economic and environmental crises of the 1980s produced both popular and anti-state movements for ecological defence *and* vigorous political and policy responses on the part of state institutions. The ecological defence movements that blossomed in 1989 had their social origins in the second half of the 1980s. Post-1989 environmental improvements were, in part, a result of pre-1989 policies and practices. Post-1989 environmental laws and regulations were – at least for a few years – heavily influenced by the reform planning prior to 1989.

After 1989, attitudes of the public and politicians towards the environment and its management were slow to change. Ingrained commitments to productivist practices were, of course, embedded in the political 'realities' of the day, particularly the deep and increasing deprivation experienced by increasing numbers of the population across the region. But it is precisely this continued commitment, and the consequences for human health and daily experience, that point to the open nature of environmental futures and environmental politics in the years ahead. Such contested legacies continue to shape both the natural and the political environments within which reforms aimed at EU alignment and accession are articulated and implemented.

How we interpret the environmental consequences of EU programmes and institutional reforms to meet the accession requirements for EU membership is then far from clear. Many of the improvements in air and water quality have been achieved as a result of political struggles waged in the 1980s and economic reforms arising in the 1990s. The long-term sustainability of these gains remains uncertain, particularly as long-standing production practices may not have altered, the ability to purchase new technologies may not be present in the commanding heights of severely weakened industries, and the 'drive to the bottom' remains a distinct possibility as sub-contracting and outsourcing become ever more important in a pan-European space economy. Where EU policies and commitments are having a direct and clear effect, and where functioning market economies and efficient production systems are emerging, they also must be understood in terms of the parallel adoption of patterns of consumption and mobility that bring with them new environmental challenges. As (and if) structural and cohesion funds begin to result in significant changes in the poorer CEE regions, we may begin to see significant impacts of EU accession on environmental conditions in industry and agriculture. As (and if) eco-labels and other product standards begin to become effective regulators of industrial systems, we may begin to see upgrading in technical efficiencies, work conditions and environmental externalities. For the moment, the dialectic of post-socialist transformation continues to keep open the environmental futures for regions and places throughout CEE countries.

NOTES

Petr Pavlínek would like to acknowledge the International Research & Exchanges Board (IREX), the National Endowment for the Humanities, the United States Department of State Title VIII Program, and the IREX Scholar Support Fund for supporting research for this contribution. John Pickles would like to acknowledge the Geography and Regional Science Program at the National Science Foundation, award No. BCS/SBE 0225088. None of these organisations is responsible for the views expressed.

1. East Germany, Czechoslovakia and Poland derived 69 per cent of their energy from coal in 1989 while Hungary, Bulgaria and Romania only 24 per cent [*Russell, 1990: 8*].
2. Approximately 85 per cent of East German electricity was derived from lignite in 1989 [*Elsom, 1992*]. In the case of former Czechoslovakia, coal provided 55 per cent of energy needs in 1989 and 78 per cent of electricity [*Russell, 1990; Statistická roèenka Èeskoslovenské socialistické republiky, 1989*]. Lignite accounted for 70–78 per cent of coal used in the thermal production of electricity between 1970 and 1985 [*World Bank, 1992*]. In Poland in 1989, 76 per cent of energy requirements was derived from coal [*Russell, 1990*], most of which was hard coal.
3. Automobiles accounted for less than 5 per cent of total sulphur dioxide emissions and less than 10 per cent of particulate matter, but 30–60 per cent of nitrogen oxides, 40–90 per cent of carbon monoxide emissions, and 35–95 per cent of lead emissions in CEE countries [*OECD, 1994*].

4. The EMEP programme (Cooperative Programme for Monitoring and Evaluation of the Long-Range Transmission of Air pollutants in Europe) was set up by The Convention on Long Range Transboundary Air Pollution (LRTAP), signed in 1979.
5. Nuclear power accounted for 0.5 per cent of Soviet electricity production in 1970, 5.6 per cent in 1980 and 12.4 per cent in 1989. All new nuclear power plants launched in Russia in the 1980s were located in its European part [see *Mounfield, 1991*]. While coal's share as the fuel supply for thermal power plants declined from 70 per cent in 1960 to less than 50 per cent by the mid-1980s in FSU as a whole, oil and natural gas dominated in the European part of the FSU because of insufficient local coal supplies. Outside the European part of the FSU (Siberia, the Far East and Kazakhstan) there was a shift back to coal for electricity production at the end of the Soviet period as the Soviet planners strove to sell more oil on the world market for hard currency [*Bater, 1996*].
6. For example, sulphur dioxide emissions declined by 40 per cent in heavily polluted northern Bohemia of the Czech Republic between 1990 and the middle of 1993 before any desulphurisation equipment was installed on local coal-based power plants. The decline occurred because of sharply lower electricity demand resulting from economic crisis and the closure of a 500MW output capacity at the Tušimice I power plant [*Pisinger, 1993*].
7. While the use of pesticides actually increased in the case of Poland between 1990 and 1998 by 15 per cent (measured in tons of active ingredients), it was still 30 per cent lower in 1998 compared to 1985. This case raises the question of sustainability of the decreases that occurred in the early 1990s (Figure 4).
8. In the second half of the 1980s, the Czechoslovak communist government launched the construction of 300 wastewater treatment plants in the Czech Republic. These water treatment plants were completed in the early 1990s and contributed significantly to the improvements in water quality in the 1990s [*Dejmal, 2001*]. Overall, 333 new municipal wastewater treatment plants were completed between 1990 and 1999 [*MoE, 2000*].
9. The total number of used vehicles imported to CEE in the 1990s is not available. The Czech Republic imported more than 1.1 million used cars between 1991 and 2001, 1.5 million were imported to Poland in 1991 alone, and 810,000 to Russia in 1998. In Ukraine, used car imports accounted for 84 per cent of car sales between January and August 1998 [see *Pavlínek, 2002*].
10. Car use increased by 50 per cent while urban public transport decreased by 13 per cent and passenger rail transport by 35 per cent in the Czech Republic between 1990 and 1999 [*Moldan and Hak, 2000*].
11. In 1999 only 26.6 per cent of cars (981,000 cars) were equipped with catalytic converters in the Czech Republic, up from 0.8 per cent in 1990. The corresponding number for Slovakia was only ten per cent in 1996 [*MoE, 2000; Závodský and Zuzula, 1997*].
12. Municipal waste per capita increased by 58.9 per cent in the Czech Republic, 25.8 per cent in Bulgaria, and by 18.9 per cent in Poland between 1985 and 1992. At the same time it decreased by 17.5 per cent in Slovakia and by 8.9 per cent in Hungary. The Western European OECD members recorded an average increase by 20.6 per cent [*OECD, 1996*]. There are several possible reasons why municipal solid waste per capita has declined in Slovakia and Hungary: households have become poorer and produce less waste; rapidly rising prices of waste collection, previously provided free of charge, have reduced the number of households willing to subscribe; and there is less waste collection because of the overall difficulties associated with transformation [*Lehoczki and Balogh, 1997*].
13. This has been the case of environmental impact assessment legislation in the Czech Republic, for example.

REFERENCES

Alcamo, J. (ed.) (1992a), *Coping with Crisis in Eastern Europe's Environment*, London and New York: The Parthenon Publishing Group.

Alcamo, J. (1992b), 'A Geographic Overview of Environmental Problem areas in Central and Eastern Europe', in J. Alcamo (ed.), *Coping with Crisis in Eastern Europe's Environment*, London and New York: The Parthenon Publishing Group, pp.27–45.

Bater, J.H. (1996), *Russia and the Post-Soviet Scene: A Geographical Perspective*, London: Arnold.

BCE (1999), 'Key Data 1990–98', *Business Central Europe*, June.

Begg, B., Pickler, J., and Smith, A. (2003) Cutting it: European integration, trade regimes, and the reconfiguration of East-Central European apparel production. Environment and Planning A Vol.35(12), in press.

Bisschop, G. (1996), 'Optimism Wanes for A Prompt Cleanup', *Transition*, Vol.2 No.1, pp.42–5.

Carter, F.W. and E. Kantowicz (2002), 'Poland', in F.W. Carter and David Turnock (eds.), *Environmental Problems of East Central Europe,* London and New York: Routledge, pp.181–206.

Carter, F.W. and D. Turnock (2002), *Environmental Problems of East Central Europe*, Second Edition, London and New York: Routledge.

Carter, F.W. and D. Turnock (eds.) (1993), *Environmental Problems in Eastern Europe*, London and New York: Routledge.

Clarke, R. (2002a), 'Yugoslavia', in F.W. Carter and David Turnock (eds.), *Environmental Problems of East Central Europe*, London and New York: Routledge, pp.396–416.

Clarke, R. (2002b), 'Bosnia and Herzegovina', in F.W. Carter and David Turnock (eds.), *Environmental Problems of East Central Europe*, London and New York: Routledge, pp.283–304.

DeBardeleben, J. (1991), 'The Future Has Already Begun: Environmental Damage and Protection in the GDR', in J. DeBardeleben (ed.), *To Breath Free: Eastern Europe's Environmental Crisis*, Washington, DC, Baltimore, London: The Woodrow Wilson Center Press and The Johns Hopkins University Press, pp.175–96.

Dejmal, Ivan (2001), Former Czech Minister of the Environment, interview conducted in Prague, Czech Republic, 26 Feb.

The Economist (2002), *The Economist Pocket World in Figures, 2003 Edition*, London: Profile Books Ltd.

Elsom, D. (1992), *Atmospheric Pollution: A Global Problem*, Oxford and Cambridge, MA: Blackwell Publishers.

EMEP (2002), http://www.emep.int/index_data.html.

Feshbach, M. (1995), *Ecological Disaster: Cleaning up the Hidden Legacy of the Soviet Regime*, New York: The Twentieth Century Fund Press.

Feshbach, M. and A. Friendly, Jr. (1992), *Ecocide in the USSR: Health and Nature Under Siege*, New York: Basic Books.

GUS (1997), *Ochrona Środowiska 1997: Informacje i opracowania statystyczne* (Environment 1997: Information and statistical papers), Warszawa: Główny Urząd Statystyczny.

GUS (1999), *Ochrona Środowiska 1999: Informacje i opracowania statystyczne* (Environment 1997: Information and statistical papers), Warszawa: Główny Urząd Statystyczny.

GUS (2001), *Mały Rocznik Statystyczny 2000* (Small Statistical Yearbook 2000), Warszawa: Główny Urząd Statystyczny.

Hann, C.M. (ed.) (2002), *Postsocialism: Ideals, Ideologies, and Practices in Eurasia*, London: Routledge.

Jordan, P. (2002), 'Croatia', in F.W. Carter and David Turnock (eds.), *Environmental Problems of East Central Europe*, London and New York: Routledge, pp.330–46.

Kabala, S.J. (1991) 'The Environmental Morass in Eastern Europe', *Current History*, November, pp.384–389.

Klarer, J. and P. Francis (1997), 'Regional Overview', in J. Klarer and B. Moldan (eds.), *The Environmental Challenge for Central European Economies in Transition*, Chichester: John Wiley & Sons, pp.1–66.

Klaus, V. (2002), 'Přírodu ochrání trh' (Market will protect the nature), *Lidové noviny*, Vol.15, No.34, pp.13–14.

Kolebaba, I. and J. Petrlík (1994), 'Klíčem k řešení je zákon' (Law is the key to solution), *Nika*, Vol.15, No.10, pp.21–2.

Kramer, J.M. (1987), 'The Environmental Crisis in Poland', in F. Singleton (ed.), *Environmental Problems in the Soviet Union & Eastern Europe*, Boulder & London: Lynne Rienner Publishers, pp.149–67.

Kramer, J.M. (1991), 'Energy and the Environment in Eastern Europe', in J. DeBardeleben (ed.), *To Breathe Free: Eastern Europe's Environmental Crisis*, Washington, DC, Baltimore, London: The Woodrow Wilson Center Press and the Johns Hopkins University Press, pp.57–79.

Kruszewska, I. (1993), *Open Borders, Broken Promises: Privatization and Foreign Investment: Protecting the Environment Throught Contractual Clauses*, Amsterdam: Greenpeace International.

Lehoczki, Z. and Z. Balogh (1997), 'Hungary', in J. Klarer and B. Moldan (eds.), *The Environmental Challenge for Central European Economies in Transition*, Chichester: John Wiley & Sons, pp.131–68.

Livernash, R. (1992), 'Central Europe', in *World Resources 1992–93*, New York and Oxford: Oxford University Press, pp.57–74.

Marquardt, W., E. Brüggemann and J. Heintzenberg (1996), 'Cleaning Eastern Germany with an Acid Bath', *Ambio*, Vol.25, No.3, pp.215–16.

McCuen, G. and R.P. Swanson (eds.) (1993), *Toxic Nightmare: Ecocide in the USSR & Eastern Europe*, Hudson, WI: Gary McCuen Publications, Inc.

MERP (Ministry for Environment and Regional Policy) (1994), *Environmental Indicators of Hungary*, Budapest: Ministry for Environment and Regional Policy.

MERP (Ministry for Environment and Regional Policy) (1999), *State of the Environment in Hungary*, Budapest: Ministry for Environment and Regional Policy.

Meurs, M. (2001), *The Evolution of Agrarian Institutions: A Comparative Study of Post-Socialist Hungary and Bulgaria*. Ann Arbour: The University of Michigan Press.

MF Dnes (1998), 'Madarsko se hospodářsky vzchopilo, ale musí ještě mnohé dohnat' (Hungary's economy improved but the country still needs to do a lot), *Mladá Fronta Dnes*, Vol.9, No.78, p.8.

MoE (1997), *Zpráva o stavu životního prostředí České Republiky v roce 1996* (Report on the state of the environment in the Czech Republic in 1996), Prague: Ministry of Environment of the Czech Republic.

MoE (1999) *Zpráva o stavu životního prostředí České Republiky v roce 1997*, (Report on the state of the environment in the Czech Republic in 1997) Prague: Ministry of Environment of the Czech Republic.

MoE (2000) *Zpráva o stavu životního prostředí České Republiky v roce 1999*, (Report on the state of the environment in the Czech Republic in 1999) Prague: Ministry of Environment of the Czech Republic.

MoE (2002) *Zpráva o stavu životního prostředí České Republiky v roce 1999*, (Report on the state of the environment in the Czech Republic in 1999) Prague: Ministry of Environment of the Czech Republic.

MoE and CSU (2001), *Statistická Ročenka životního prostředí České Republiky 2001* (Statistical Environmental Yearbook of the Czech Republic 2000), Prague: Ministry of Environment of the Czech Republic and the Czech Statistical Office.

MoE SR (1998), *Správa of stave životného prostredia Slovenskej republiky v roku 1997* (Report on the environment of the Slovak Republic in 1997), Bratislava: Slovak Ministry of Environment and Slovak Environmental Agency.

MoE SR (2000) *Správa of stave životného prostredia Slovenskej republiky v roku 2000* (Report on the environment of the Slovak Republic in 1997), Bratislava: Slovak Ministry of Environment and Slovak Environmental Agency.

Moldan, B. and T. Hak (2000), *Czech Republic 2000, Ten Years On: Environment and Quality of Life after Ten Years of Transition*, Prague: Charles University.

Mounfield, P.J. (1991), *World Nuclear Power*, London and New York: Routledge.

MTC (Ministry of Transport and Communications) (1999), *1998 Czech Republic Transport Yearbook*, Prague: Ministry of Transport and Communications.

MTC (Ministry of Transport and Communications) (2002), *2001 Czech Republic Transport Yearbook*, Prague: Ministry of Transport and Communications.

Nika (1994), 'Úvodem' (editorial), *Nika,* Vol.15, pp.10–11.

Novotny, V. and L. Somlyódy (1995), 'Water Quality Management: Western Experiences and Challenges for Central and Eastern European Countries', in V. Novotny and L. Somlyódy (eds.), *Remediation and Management of Degraded River Basins with Emphasis on Central and Eastern Europe*, Berlin, Heidelberg, New York: Springer, pp.1–34.

Nowicki, M. (1993), *Environment in Poland: Issues and Solutions*, Dordrecht: Kluwer Academic Publishers.

Nowicki, M. (1997), 'Poland', in J. Klarer and B. Moldan (eds.), *The Environmental Challenge for Central European Economies in Transition*, Chichester: John Wiley & Sons, pp.193–227.

OECD (Organisation for Economic Cooperation and Development) (1994), *Environment for Europe: Environmental Action Programme for Central and Eastern Europe*, the Document approved by the Ministerial Conference Lucerne, Switzerland, 28–30 Apr. 1993, OECD and World Bank.

OECD (Organisation for Economic Cooperation and Development) (1996), *Environmental Indicators: A Review of Selected Central and Eastern European Countries*, Paris: Organisation for Economic Cooperation and Development, Center for Cooperation with the European Economies in Transition.

Pavlínek, P. (2002), 'Restructuring the Central and Eastern European Automobile Industry: Legacies, Trends and Effects of Foreign Direct Investment', *Post-Soviet Geography and Economics*, Vol.43, No.1, pp.41–77.

Pavlínek, P. and J. Pickles (2000), *Environmental Transitions: Transformation and Ecological Defence in Central and Eastern Europe*, London and New York: Routledge.

Pickles, J., M. Nikolova, C. Staddon, S. Velev, Z. Mateeva and A. Popov (2002), 'Bulgaria', in F.W. Carter and D. Turnock (eds.), *Environmental Problems in East Central Europe*, London and New York: Routledge, pp.305–29.

Pisinger, René (1993), Advisor of the Czech minister of the environment, interview conducted in Teplice, Czech Republic, 4 Aug.

Podoba, J. (1998), 'Rejecting Green Velvet: Transition, Environment and Nationalism in Slovakia', *Environmental Politics*, Vol.7, No.1, pp.129–44.

REC (Regional Environmental Center for Central and Eastern Europe) (1994a), *Strategic Environmental Issues in Central and Eastern Europe: Volume 2, Environmental Needs Assessment in Ten Countries*, Budapest: REC.

REC (Regional Environmental Center for Central and Eastern Europe) (1994b), *Strategic Environmental Issues in Central and Eastern Europe: Volume 1, Regional Report*, Budapest: REC.

Rowntree, L., M. Lewis, M. Price and W. Wyckoff (2003), *Diversity Amid Globalization: World Regions, Environment, Development*, 2nd edition, Upper Saddle River: Prentice Hall.

Russell, J. (1990), *Environmental Issues in Eastern Europe: Setting an Agenda*, London: Royal Institute of International Affairs.

SOSR (Statistical Office of the Slovak Republic) (2001), *CESTAT Statistical Bulletin 2000/4*, Bratislava: SOSR.

Statistická ročenka Československé socialistické republiky, (1989), *Statistická ročenka Československé socialistické republiky* (Statistical Yearbook of the Czechoslovak Socialist Republic), Prague: SNTL.

Szefer, P., K. Szefer, G.P. Galby, J. Pempkowiak and R. Kaliszan (1996), 'Heavy-Metal Pollution in Surficial Sediments from the Southern Baltic Sea off Poland', *Journal of Environmental Science and Health*, Vol.A31, No.10, pp.2723–54.

Turnock, D. (2002) The Central importance of the EU. In Carter, F. and Turnock, D. *Environmental Problems of East Central Europe*. London: Routledge, pp.56–91.

UN (United Nations) (1995), *Strategies and Policies for Air Pollution Abatement: 1994 Major Review Prepared under the Convention on Long-Range Transboundary Air Pollution*, Geneva: Economic Commission for Europe.

Várkonyi, T. and G. Kiss (1990), 'Air Quality and Pollution Control', in D. Hinrichsen and G. Enyedi (eds.), *State of the Hungarian Environment*, Budapest: The Hungarian Academy of Sciences, the Ministry for Environment and Water Management and the Hungarian Central Statistical Office, pp.49–65.

World Bank (1992), *Czech and Slovak Federal Republic: Joint Environmental Study*, A joint report of the Governments of Czechoslovakia, the Czech and Slovak Republics, the European Community, the United States Government, and the World Bank, Nov. 1991. Washington, D.C., World Bank.

Závodský, D. and I. Zuzula (1997), 'Vývoj znečistenia ovzdušia Slovenskej republiky' (Development of air pollution in Slovakia), *Ochrana ovzduší*, Vol.9, No.4, pp.5–9.

Market Liberalisation and Sustainability in Transition: Turning Points and Trends in Central and Eastern Europe

SANDRA O. ARCHIBALD, LUANA E. BANU
AND ZBIGNIEW BOCHNIARZ

There is a growing body of literature assessing the impact of the historical transformation that has taken place in the Central and Eastern European (CEE) countries over the past 12 years. This is a particularly timely issue on the brink of the accession of ten of these CEE[1] countries to the European Union (EU). The purpose of this contribution is to assess the impact of radical reforms on the sustainability of the transformation of these CEE countries to civic societies with market economies. The radical reforms encompass three main components – market liberalisation, macro-economic stabilisation and institutional reforms [*Balcerowicz, 2000/2001*].

This study focuses on the impact of market liberalisation – its speed, scope and type – and other institutional reforms on selected indicators measuring progress towards more sustainable development. Thus, sustainability of transformation is measured by proxy by the achieved progress in sustainable development, defined as development that secures meeting the basic – economic, social and environmental – needs of contemporary generations without jeopardising the needs of future generations. Specifically, we first investigate, for the CEE countries, the impact of liberalisation and other reforms on environmental pressure holding incomes constant utilising an augmented environmental Kuznets curve (EKC) framework. Second, using a broader data set that includes additional CEE countries and the new independent states (NIS), we examine the effects of reforms focusing on liberalisation.

Market Liberalisation

Market liberalisation is one of the more controversial reforms and also a symbol of radical change. We are aware that even the most radical market liberalisation will not produce expected sustainability benefits without other well-designed components. There is, however, a great deal of evidence that liberalisation significantly influenced other institutional reforms, particularly in the environmental area. For example, market-based environmental policies, including environmental charges, funds, emissions trading and assignment of environmental liability for damage or voluntary agreements are examples of solutions taken from liberal economic systems. These institutions and policies were implemented relatively early in the transition process in Central European countries before the EU accession processes began to dominate the direction of institutional and policy reforms [*Zylicz, 1995*]. It is noteworthy that most of the ten CEE countries have introduced national environmental strategies and/or policies in the first half of the 1990s and thus have influenced the broader development strategies that followed.[2] It was not the case of the majority of the NIS where economic and political changes were not accompanied by comprehensive environmental reforms.

Our hypothesis is that the reform process, including the degree, speed and type of liberalisation, has had significant positive implications for sustainability in the CEE10. Understanding this will help assess whether or not we can expect the performance of the ten CEE countries admitted to the European Union to be replicated by the rest of the countries in transition including the NIS, given the different reform models present among the set of countries vying for entrance to the EU. We examine impacts of reforms incorporating liberalisation specifically on key environmental pollution indicators, to determine whether reforms have led to improvements in environmental quality and quality of life as a measure of promoting a more sustainable development path in the CEE region.

Previous work [*Archibald and Bochniarz, 1998, 1999, 2000/2001, 2002*] has shown that reforms introduced in early 1990s have had positive effects on the environment and explain a significant share of the observed declines in key air pollution emissions that took place through that time period. We also observed significant improvement in economic and social performance, particularly in those countries that took more radical steps in liberalising their economy, stabilising their markets, and introducing deep reforms in their financial institutions.

Debates over the Impact of Liberalisation

Liberalisation and its consequences in CEE countries is one of the most analysed aspects of transition. Balcerowicz [*2000/2001: 52*] maintains that the main features of liberalisation are the:

> removal of the remaining restriction on private activity, liquidation of the remnants of the central allocation of inputs, price liberalisation, the removal of most quantitative restriction on exports and imports,

unification of the exchange rate, and the introduction of convertibility of the Polish currency for current account operations.

While studies [*Dabrowski et al., 2001; Archibald and Bochniarz, 2002*] tend to agree on this definition, there is a continuing debate over the effects of liberalisation on sustainability. For example, based on an analysis of transitional reforms and economic policies in China and Russia [*Stiglitz 1999: 3; Stiglitz and Ellerman 2001*] argues that liberalisation has not led to sustainable development, claiming:

> ... the countries of Eastern Europe, whose history, geography and prospects are at least more similar to each other than they are to the Central Asian or Baltic states, early liberalisation and overall average growth exhibit no positive relationship; if anything, they appear to have a negative correlation, ...

Polish economists including Dabrowski, Gomulka and Rostowski [*2001*] disagree with Stiglitz. They argue that by considering only two models of transition, Chinese and Russian, Stiglitz over-generalises the transition experience. They conclude that CEE countries' transition experience shows features of economic and environmental sustainability that do not appear in the dichotomous China–Russia comparison. In addition, Japanese economist Gemma's research indicates that the transition experience of CEE countries, including Baltic countries, differs compared to that of the NIS [*Gemma, 2000*]. This recent polemic inspired us to examine the question of the impact of liberalisation over a broader set of countries, focusing on the most advanced ten CEE countries as well as on other CEE nations and the NIS.

Assessing the Impact of Market Liberalisation on Sustainability

This contribution analyses the impact of rapid liberalisation, as defined by the European Bank for Reconstruction and Development (EBRD),[3] on economic, social and environmental performance for the CEE countries and is then expanded to include the NIS. The EBRD measures liberalisation as a function of small-scale privatisation, price liberalisation, and trade and foreign exchange system, using a 'measurement scale where 1 represents little or no change from rigidly planned economy and 4+ represents the standard of an industrialised market economy'. Based on this scale, we developed an index classifying a country as fully liberalised when it has achieved a score of at least three on price liberalisation and at least four on trade and foreign exchange. For each of the analysed years, the countries received a score of one when fully liberalised and zero when not.

Using the notion of cumulative liberalisation[4] as a criterion, we divided the CEE10 into two groups: Early and Late Liberalisers. When the 15 additional CEE and NIS countries are added to the analysis, the group of Late Liberalisers expands and a new group of Non-Liberalisers appears. The countries, clustered by degree of liberalisation, are summarised in Table 1.

TABLE 1

CUMULATIVE LIBERALISATION INDEX FOR CEE AND NIS COUNTRIES, 1990–2000[5]

Early Liberalisers CEE6	Score	Late Liberalisers CEE4	Score	Additional Late Liberalisers	Score	Non-Liberalisers	Score
Czech Rep.	10	Bulgaria	6	Albania	9	Azerbaijan	0
Estonia	8	Latvia	8	Armenia	6	Belarus	0
Hungary	11	Lithuania	8	Croatia	8	Tajikistan	0
Poland	9	Romania	7	FYR Macedonia	8	Turkmenistan	0
Slovak Rep.	10	*Average*	*7.25*	Georgia	5	Ukraine	0
Slovenia	9			Kazakhstan	3	Uzbekistan	0
Average	*9.5*			Kyrgyzstan	7	*Average*	*0*
				Moldova	7		
				Russia	2		
				Average	*6.5*		

Source: Developed from EBRD [*2001*].

The Environmental Kuznets Curve (EKC)

The environmental Kuznets curve (EKC) hypothesises that environmental damages increase, at first, with rising income as consumption increases and then declines as demand for environmental improvements increase with rising income [*Shafik, 1994; Selden and Song, 1994; Grossman and Krueger, 1995; Holtz-Eakin and Selden, 1995; Andreoni and Levinson, 2001; Hill and Magnani, 2002*]. This hypothesised relationship between environmental degradation and income per capita thus takes the form of an inverted U. There are several key factors other than income that can play a role in environmental emissions. These include technology, composition of the economy (structure), environmental expenditures/investments, openness and other policy reforms [*Stern and Common, 1996*].

There have been several empirical analyses of the 'augmented' EKC hypothesis with mixed results [*Lucas et al., 1996; Archibald and Bochniarz, 1998*]. For example, Archibald and Bochniarz' [*1998*] analysis of the EKC in CEE countries provides support for the hypothesis, suggesting that rising incomes in the CEE region initially contributed to higher levels of emissions of key air quality indicators. However, beyond a specific level of income, a 'turning point', higher incomes led to reduced pressure on the environment. These turning points were significantly lower than turning points identified for developed countries, mostly ranging from $1,600 to $13,3796[6] [*Archibald and Bochniarz, 1998*]. For example, Lucas et al. [*1992*] estimated that for CO_2 the turning point was US$24,586 (1987) and for VOC $20,000. NO_2 at first declines with income and then rises with a turning point beyond the range of observed data. Additionally, we found some evidence that policy reform with regard to the openness of the economy, environmental investments, and the degree of privatisation was effective in reducing environmental pressure for some selected pollutants [*Archibald and Bochniarz, 1998*].

Economic and Policy Factors Critical to Environmental Quality

The beginning of the CEE transition is marked with the landslide victory of *Solidarnosc* movement in June 1989 when the Polish parliamentary election shifted the balance of power overnight. A similar peaceful transition of power took also place in Hungary, East Germany, Czechoslovakia, Bulgaria and Romania in the following months of 1989. These political changes were followed by dramatic changes in economic policies and institutional reforms introducing the foundation for a market economy from the very beginning of 1990.

These critical changes in policy and institutions occurred during the time when the structure of the economy was dominated by state-owned industries. In addition to fundamental changes in economic policy to push towards a more market-based structure, new regulations and policies were imposed and new investments were made to reduce and mitigate the significant damage done to the environment under previous regimes. These reforms did impact on the environment, each in different ways. Table 2 provides the expected effects on the environment for a set of selected critical policies (factors). We provide a definition of the critical factor, the expected effect on the environment, define an empirical measure and select a useful indicator. The effects of each of these factors on pollution emissions must be sorted out in order to determine the contribution of each to the observed reductions in environmental pressure and enhancing sustainability under our definition. Using statistical methods, we can identify how each of the critical factors contributes to observed changes in pollution emissions and make some preliminary determination of the significance and size of these separate effects.

TABLE 2

SELECTED CRITICAL POLICY, ECONOMIC AND INSTITUTIONALFACTORS OF ENVIRONMENTAL SIGNIFICANCE

Factor	Effect on the Environment	Definition	Indicator
Consumption Patterns	Rising incomes expected to have a negative effect (increase pressure)	Per capita income; income growth rate	GDP per capita (GDP)
Structure of Economy	Faster privatisation is expected to have a positive effect (reduce pressure)	Contribution of small-scale private sector to the economy	EBRD Index of small-scale privatisation (SMPV)
Regulatory Structure	Institutions to regulate the environment expected to have a positive effect (reduce pressure)	Extent of environmental regulation	Ratio of environmental expenditures to GDP (ENVEXP)
Liberalisation	Liberalisation is expected to have a positive effect (reduce pressure) due to efficiency gains	Complete price liberalisation Full current convertibility	Cumulative EBRD Index of liberalisation (LIB)
	No a priori expectations about the effects of foreign direct investment	Contribution of foreign direct investment to the economy	Ratio of foreign direct investment to GDP (FDI)
Country-specific Effects	Larger countries are expected to have a negative effect (increase pressure)	Country size	Population in millions (POP)
Efficiency Gains	Efficiency gains are expected to have positive effects (reduce pressure)	Change over time	Time trend (t)

Consumption Patterns

A key objective of the EKC analysis is to examine consumption patterns, as measured by income per capita, to determine the extent to which rising (declining) per capita incomes can be expected to contribute to pollutant emission levels and to test the inverse U-shaped relationship. If rising incomes due to transition result in environmental degradation, the claim of sustainable development cannot be made. If, on the other hand, positive income growth and progress towards market reforms lead to a decline in environmental quality, development would be considered to be more sustainable.

Increased consumption that results from rising incomes is generally expected to increase environmental pressures, *ceteris paribus*. We expect to find that declining incomes have contributed to a reduction in observed environmental pressure. As incomes begin to rise again with economic recovery an increase in pollutant emissions is expected to follow. This increase in incomes is expected to continue well into the middle of the next century, although debate continues as to whether and at what point rising incomes might result in a slowing of pollution increases. The measure of income adopted is gross domestic product (GDP) per capita measured at constant prices in local currency units and expressed in US dollars using the World Bank 1998 conversion factor [*Bochniarz and Toft, 1995*]. It was necessary to use World Bank GDP data to obtain a sufficient time series for all ten countries. To test for inverse U-shaped patterns, income per capita is included in quadratic form. In order to isolate and measure the effect of income on environmental change, our analysis includes measures of other critical economic and policy factors believed to affect the environment including importantly the degree of liberalisation.

Structure of the Economy

Because we are testing the hypotheses that economic reforms have not had a deleterious impact on the economy, measures of structural change are included. Changes in the structure of the economy include mainly the development of a private sector, as well as declining manufacturing and fast growing service sectors. These are critical factors predicted both to reduce and to increase environmental pressures. The present research uses only privatisation of small-scale enterprises, since previous research [*Archibald and Bochniarz, 2002*] showed that privatisation of smaller firms was both more successful and also a better economic indicator than was privatisation of medium and large businesses. One major reason for their better performance was that these enterprises were forced by tight budget constraints to restructure and to attain efficiency from the start. In the case of medium and large enterprises, privatisation was introduced mostly through a variety of mass privatisation schemes, including vouchers, which did not lead either to rapid restructuring of management or to raising new capital. Declining manufacturing and fast rising services, or a more general shift from industrial to post-industrial economy has also had a positive impact on reduction of environmental pressure [*Archibald and Bochniarz, 2002; Panayotou et al., 2000*].

The degree to which market reforms have been achieved is also expected to have a significant effect on pollutant levels although there is debate about the observed direction of its effect. Some argue that privatisation and industrial restructuring have been promoted with little awareness of the potentially negative environmental impacts [*Pearce and Warford, 1993*]. Others believe that market reform has a positive effect at least through gains in efficiency of resource use if nothing else.

Also important in the transition process in these countries is the role of foreign direct investment (FDI). Here it is initially measured as net inflows of foreign direct investment as a percentage of GDP. There are no a priori expectations about the direction of the effects of FDI on emissions levels, although for sustainable development it is hoped that investments would be made to reduce pressure on the environment.

Regulatory Structure

The forces of economic transformation alone are not likely to reduce environmental pollution. Regulatory structures and investments are also needed to adjust for externalities and provide more environmentally friendly technologies. Environmental regulation should reduce pressure on the environment. With appropriate regulatory institutions, it is possible that economic growth can continue without undue harm to the environment. Environmental expenditures (percentage of GDP) are used as a proxy for regulatory intensity. This captures investments in environmental improvement as well.

Liberalisation

Liberalisation, or increased openness of these economies to the global economy, is also predicted to impact on the environment in a positive way, although some argue the opposite effect is possible. The increased openness of the economy is hypothesised by some economists to increase resource efficiency and should thus result in lower pollution levels. Others argue that trade increases environmental pressures as resources are exploited for hard currency gains. There is little empirical analysis of these hypotheses. Therefore, liberalisation may or may not affect sustainability, and must be taken into account in an empirical model.

Environmental Quality Indicators

The initial empirical analysis of the effects on economic growth and transition on the environment focuses on air quality in the region. While it is preferable to rely on indicators of environmental quality that measure the state of the environment directly (for example, ambient concentration of air pollution), such data are not consistently available for these countries. Typically, pressure indicators are relied upon as a proxy for environmental quality in such cases. Changes in pressure indicators imply changes in environmental quality. The three dependent variables (for instance the key indicators measuring air pollution pressures) are defined in Table 3. A regional analysis is critical given that these key air pollutants have global and as well as local impacts. Furthermore, the indicators developed for the

TABLE 3

KEY AIR QUALITY PRESSURE INDICATORS

Pollutant	Indicator Definition and Unit of Measurement	State Indicator Indicator Definition	Alternative	Scope of Effect
Emissions of carbon dioxide (CO_2)	Emissions of CO_2 in kt per capita	CO_2 concentrations, global temperature	None	Global climate change
Emissions of sulphur dioxide (SO_2)	Emissions of SO_2 in kt per capita	Concentration of sulphates (SO_2), global temperature	An aggregated indicator for aerosol particles	Global, regional and local climate change Acidification
Emissions of nitrogen oxides (NO_2)	Anthropogenic emissions of NO_2 in kt per capita	Atmospheric nitrous oxide concentrations, global temperature	Emissions of NO_2 per unit of area; aggregate indicator for CO_2, CH_4 and NO_2 emissions based on global warming potential	Global and regional climate change Ozone depletion

Source: Developed from Eurostat (1999) *Eurostat's Environmental Pressure Indicators Project Methodology Sheets*, Eurostat

Pressure-State-Response[7] (PSR) system are most appropriate to regional analysis [*Eurostat, 2001*].

Model Specification

An econometric analysis is conducted using a pooled-time series data set for the ten countries covering the period 1990–99. We specified a reduced form of the general linear fixed effects regression model. The reduced form assumes that environmental outcomes are related to predetermined endowments and economic measures. Clearly some interdependence in environmental indicators is probable. The model specification was estimated (Table 4) using the EKC policy augmented framework. The model applied to all three pollutants (CO_2, SO_2 and NO_2) tested for the effects on emissions of income, openness, structure of the economy, regulatory structure, time effects and country size. The EKC allows us to test the hypothesis that the relationship between income and environmental pressure follows an inverted-U curve with rising incomes initially leading to increased environmental pressures but after some point declines with further increases in income. The model is specified as:

$$AP_{it} = \alpha_i + \gamma_t + \beta_1 (GDP) + \beta_2 (GDP_2) + \beta_3 (ENVEXP) + \beta_4 (FDI) + \beta_5 (SMPV) + \beta_6 (LIB) + \beta_7 (POP) + \varepsilon_{it}$$

where: i denotes country and t represents time; AP is the vector of air pollution indicators outlined in Table 3 measured in kilograms per capita. The parameter α_i reflects specific country effects and γ_t the time effect. The three dependent variables measuring pollution pressures are as defined in Table 3. Critical factors are as defined in Figure 2: GDP is gross domestic product per capita (constant 1995 US$); ENVEXP is environmental expenditures (percentage of GDP), to reflect regulatory intensity, FDI represents net inflows of foreign direct investment as percentage of GDP, SMPV measures the contribution of the small-scale private sector to the economy. LIB represents liberalisation as defined by the cumulative EBRD index, POP is population in millions and å is the error term. Some interdependence in environmental indicators is probable. A summary of the econometric results is shown in Table 4.

Results of the Econometric Analysis

The power of the economic and policy variables to predict changes in the air quality indicators was quite strong; adjusted R^2 for each model ranged from 0.79 to 0.96. The signs – direction of impact on the environment – were generally as expected. Incomes at first increase emissions pressure and then reduce it. For two out of the pollutants these estimated parameters were statistically different from zero.

Environmental expenditures as a measure of regulation generally showed a positive effect on the environment (decreasing pressures) although for CO_2 and NO_2 the results were statistically weak. As regulatory intensity increases the CEE10 can expect environmental pressures to

TABLE 4

SUMMARY OF ECONOMETRIC RESULTS OF EKC ANALYSIS

Variables	SO_2	NO_2	CO_2
Time	–*	–	–
GDP	+	+*	+*
GDP_2	–	–*	–*
ENVEXP	–**	–	–
FDI	–	–	–
SMPV	–*	–	–
LIB	–	–	–
POP	+*	+*	+*
	R2=0.74	*R2=0.89*	*R2=0.96*

*Significant at .01 level ** at .05

decrease. These results were uniformly strong with relatively weaker confirmation for CO_2 and NO_2. The population variable indicates that more heavily populated countries are reducing pressures on the environment to a greater extent. Time effects indicated a decreasing relationship between population and pollutant emissions, especially in the case of SO_2. Economic structure indicators were highly significant; however, they were mixed in their influence on pressure indicators. Although small-scale privatisation has a positive influence by reducing SO_2 emissions, at the same time CO_2 and NO_2 emissions have increased. These mixed results might be the consequence of different models of privatisation implemented in CEE countries.

The degree of liberalisation had a positive effect on the environment for all these pressure indications although the results were not statistically significant in this case. Turning points are the level of per capita at which further increases in income can be expected to result in reduced environmental pressure. Consumers demand greater environmental quality. Results for these models were quite robust with respect to the turning points derived. For CO_2, the level of income at which emissions begin to decline is US$6,727; for SO_2 is $5,826, and for NO_2 the level is $6,108. These turning points are below those observed in developed countries. For example, in developed countries, the average income at which CO_2 emissions stop rising and begin to fall is approximately US$8,000 [*Grossman and Krueger, 1995*]. Furthermore, these turning points are in the range of incomes observed among these countries during the decade under consideration, indicating that they are indeed achievable and have already been experienced in some places. The lowest income per capita observed was US$1,317, which occurred in Bulgaria in 1997; the highest was US$11,160, observed for Slovenia in 1999. The average annual income across all countries included in the model was US$4,092.

Analysis of Trends in Key Sustainability Indicators

The results of the EKC model clearly indicate that the CEE10 have made steps towards sustainable development during their transition process. This provides some support to the claim that policy, economic and institutional reforms, including liberalisation, have had a positive impact on both incomes and environmental pressure. These positive conclusions should be further examined, taking into account other sustainability indicators and their development over the last 12 years, and whether they can be replicated by other CEE countries and NIS.

Economic Sustainability

Despite well-known deficiencies, GDP was selected an indicator of economic sustainability, owing to its availability for all 25 countries under consideration here. The GDP indicator shows striking differences in economic performance between Early, Late and Non-Liberalising states (see Figure 1). Among the CEE10, only Early Liberalisers noticed significant gains in per capita incomes, achieving an average increase of about 25 per cent. In addition, only the economic improvements among Early Liberalisers proved to be sustainable, experiencing continuing positive economic growth; all of them passed the pre-transition income level and stabilised gains. Despite a one-year decrease and roughly two years of stagnation at around $3,500 GDP per capita, since 1994 Early Liberalisers experienced sustained annual increases in income. In 1999, these countries reached a GDP of over $4,500 per capita measured by the commercial exchange rate (CER).

Late Liberalisers experienced negative growth until 1993, with their average GDP per capita falling from $2,000 to $1,500. After a short period of a weak growth (1994 to 1996) they again lost ground, falling in 1999 to a level slightly above $1,500. This constitutes an average decrease of almost 25 per cent. Since 1999 they have stagnated at this low level. Thus, early

FIGURE 1

GROSS DOMESTIC PRODUCT: LIBERALISERS CEE10 (CER)

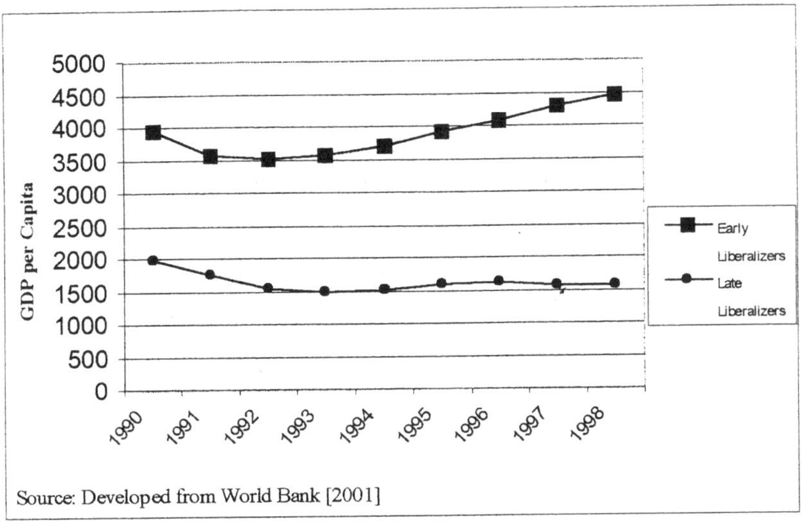

Source: Developed from World Bank [2001]

Source: Developed from *World Bank Development Indicator, 2001.*

liberalisation in the leading six countries produced GDP increases of over 25 per cent while delayed liberalisation in the remaining CEE4 contributed to a decline in GDP of almost 25 per cent.

The results for the other countries in transition – the Late Liberalisers – are similar to those of the CEE4 (see Figure 2). Late Liberalisers experienced a loss in per capita income over the same period, falling 37 per cent from about $3,000 to about $1,900. Losses in relative terms were even larger for Non-Liberalisers, falling by 47 per cent from $1,700 to $900 per capita. Unlike the CEE6, the GDP for the rest of the countries in transition, considered together, decreased from about $2,500 to about $1,600 between 1990 and 1996, and then remained stagnant until the end of the period of analysis. In general, Late Liberalisers and Non-Liberalisers did not make any progress towards reaching economic stability and sustainability, particularly in reducing inflation, until 1999. As a consequence they have not yet experienced any significant economic gains from transition, except that of stopping economic decline in most cases.

According to our previous study [*Archibald and Bochniarz, 2002*] analysing GDP expressed in purchasing power parity (PPP), the income gap between the six CEE Early Liberalisers and the four CEE Late Liberalisers ($7,800 and $6,400, respectively) was less dramatic, as well as smaller relative losses and gains during that period. Early Liberalisers gained from about $7,800 to over $10,000 (about 13 per cent). Late Liberalisers lost about nine per cent, falling from about $6,400 to a level slightly below $6,000, and have since stagnated there. These indicators confirm again the importance of well-designed liberalisation policies introduced early.

Economic and social sustainability does not depend only on the level of GDP but also, equally important, on income distribution. Economists

FIGURE 2

GROSS DOMESTIC PRODUCT LIBERALISERS CEE10 AND CEE15 AND NIS

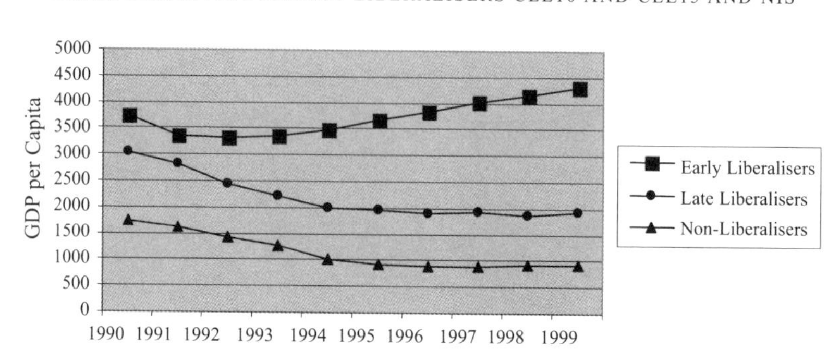

Source: Developed from *World Bank Development Indicator, 2001.*

developed measurements of income distribution such as the Gini Ratio derived from the concept of the Lorenz Curve representing the ideally equal distribution. The Gini Ratio can vary from zero – which represents a perfectly equal distribution of income, to one (or 100 per cent) – for a completely unequal distribution. It is worth mentioning that CEE countries entered the transition with relatively high apparent equality (without taking into account special access to goods and services and other privileges enjoyed by the communist elites). The 'planned' equality occurred at very low average income levels. However, during the process of transition inequality has significantly increased. It has been also visible partially because of greater transparency, freedom of media to publicise the issues and easy access to information. This inequality has provoked widespread public outcry in some of the 25 countries.

Contrary to some hypotheses, including that of Stiglitz, the most liberalised CEE countries – Hungary, Czech Republic, Poland, Estonia, Slovenia and Slovakia – have not suffered the highest inequality (see Table 5). For these countries, the Gini Ratio rose on average by about five or six percentage points during the ten-year period (1990–99). On the other hand, Late Liberalisers, including Bulgaria, Latvia, Lithuania, Romania and Russia, reached a level of inequality almost three times higher at 15 percentage points [*EBRD, 2001*]. The worst inequality was experienced by Russia at 21 percentage points (30 percentage points according to Milanovic [*1998*]), followed by Romania (18 percentage points) and Latvia (13 percentage points). Non-liberalised Ukraine reached 10 percentage points, the same level as deeply liberalised Poland.

These differences in income inequality are probably due not to the rate of liberalisation, but to policy strategies implemented in each country. Early Liberalisers followed the Western pattern, building a civic society based on

TABLE 5

CHANGE IN GINI RATIO

Country	Gini Coefficient 1990	Gini Coefficient 1999	Overall Gini Change
Bulgaria	22	34	12
Czech Republic	21	28	7
Estonia	30	38	8
Hungary	30	32	2
Latvia	22	35	13
Poland	21	31	10
Romania	18	36	18
Slovakia	18	23	5
Slovenia	24	30	6
Russia	24	48	21
Ukraine	25	35	10

Source: Developed from UNICEF (2001) *A Decade of Transition: The MONEE Project: CEE/CIS/Baltics*. Florence, Italy: Innocenti Research Centre, UNICEF.

a strong middle class. Some of the Late Liberalisers (like Russia) or Non-Liberalisers (for example Ukraine) developed a polar structure with a small upper class of oligarchs and a large lower class [*Kakwani, 1995; Morvant, 1996*].

The unemployment rate is another indicator of economic and social sustainability (see Appendix for details). Under communist rule, there was no recognised unemployment. However, state enterprises were typically heavily overstaffed. Liberalisation of the economy and the introduction of hard budget constraints as a crucial element of stabilisation led to massive bankruptcy of non-competitive firms. This wave of bankruptcy served to move labour and capital from negative-value-added to positive-value-added firms and industries [*Dabrowski et al., 2001*]. The changes in unemployment rates and the success of employment strategies implemented are visible in Hungary, Latvia, Lithuania and Slovenia over the last 11 years. One of the most important factors in the reduction of unemployment was achieving positive economic growth. Another factor was the ability of countries to attract foreign direct investment (FDI).

This was particularly important in Estonia, Hungary, Czech Republic and Latvia (see Appendix for details) where liberalisation was the major engine behind economic growth and inflows of FDI. The six most liberalised CEE states have attracted about 80 per cent of all FDI for the whole region, including both CEE and NIS countries, during the transition period [*Bochniarz et al., 1998*].

Environmental Sustainability

CEE countries, particularly the CEE6, made significant progress in reducing air pollution emissions during the last decade, while improving their well-being. Unfortunately, the same improvement cannot be found in the NIS. The level of CO_2 emissions decreased from over 10,000 kg per capita to about 8,500 kg for Early Liberalisers over the decade (Figure 3) while their GDP per capita increased significantly (Figure 2).

This might indicate an efficiency gain achieved mostly in energy use and implies a significant structural change towards a lower ratio of emission per dollar of GDP. In the meantime, the Non-Liberalisers reduced their level of pollution but simultaneously suffered a decreasing GDP per capita. Their percentage decrease in both CO_2 levels and GDP per capita was approximately 54 per cent. Thus, this pollution pressure indicator improved, but did so at the expense of economic well-being. In the case of SO_2 (Figure 4), Early Liberalisers greatly reduced emissions per capita from about 105 kg to about 45 kg while maintaining robust economic growth during that period. This is a remarkable achievement. Late Liberalisers also reduced

FIGURE 3

CARBON DIOXIDE EMISSIONS OF CEE10, AND CEE15 COUNTRIES AND NIS

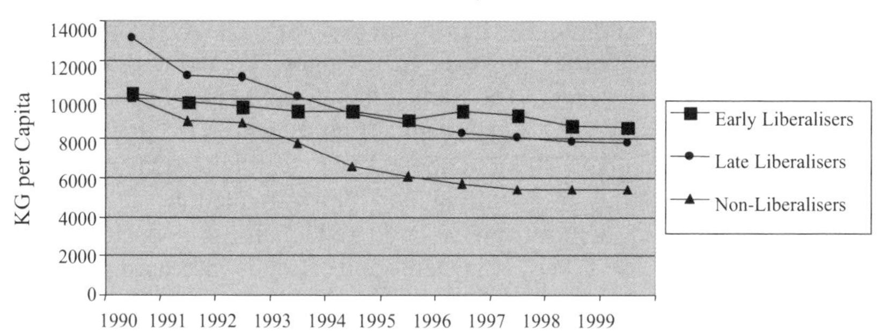

Source: Developed from EMEP [*2001*].

FIGURE 4

SULPHUR DIOXIDE EMISSIONS OF CEE10

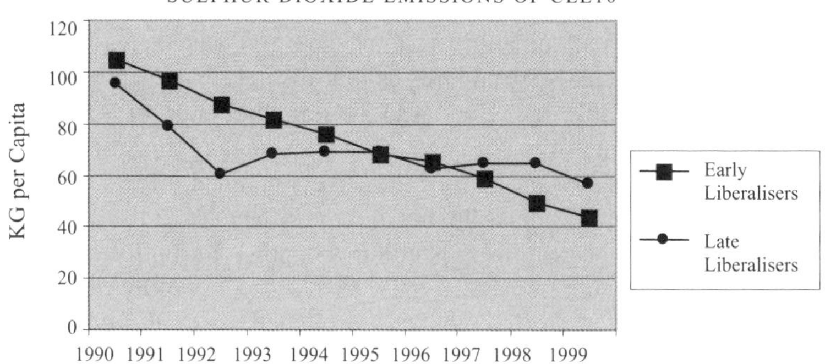

Source: Developed from EMEP [*2001*].

FIGURE 5

NITROGEN DIOXIDE EMISSIONS OF CEE10

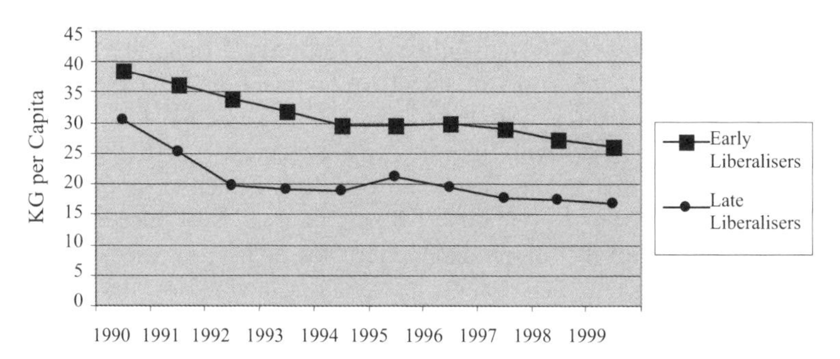

Source: Developed from EMEP [*2001*].

their emissions from about 95 kg in 1990 to about 55 kg, but they did not achieve economic growth during that period.

Regarding NO_2 (Figure 5), significant reductions in emissions were made in all CEE10 countries; however, the changes were not as dramatic as in the case of SO_2 emissions. The Early Liberalisers reduced emissions of NO_2 from about 38 kg to 27 kg per capita. Late Liberalisers cut emissions almost in half, from about 30 kg to about 17 kg per capita. This reduction originated in industrial restructuring and the introduction of pollution control technologies. Unfortunately, large reductions in industrial NO_2 emissions among Early Liberalisers were soon offset by increased pollution from transportation, mostly consumers' cars. Higher income levels among CEE6 countries led to much greater private automobile ownership and its associated pollution.

Finally, we will move to analysis of policy response indicators such as 'expenditures for environmental protection as a share in GDP'. As we have mentioned earlier, the CEE states experienced significant environmental and policy reforms. Those initiatives resulted in significant increases (two–five times) in environmental expenditures in both absolute and relative terms to the achieved GDP (Table 6). They also shifted the bulk of environmental expenditure financing from public to industrial sources [*Eurostat, 2003*].

Table 6 reveals that the Early Liberalisers attained a level close to that of OECD countries. While at the beginning of the decade the average level of environmental protection expenditures among Early Liberalisers was about 0.5–0.7 per cent of GDP, it grew to 1.5–2 per cent by the end of the 1990s. Here again, early liberalisation produced positive outcomes; these positive results should be further strengthened and sustained in the process of accession to the European Union.

Broader Trends in Sustainability

To understand the degree to which countries in transition have achieved progress towards social sustainability, it is important to examine some of the most fundamental indicators of social welfare such as life expectancy and infant mortality rates. Analysis of the data clearly indicates that liberalisation has positively affected life expectancy in the CEE states. Figure 6 shows that among the CEE6 countries life expectancy rose from 70.8 years in 1990 to 73 years in 1999 and the trend continues upwards. By comparison, Late Liberalisers experienced losses in life expectancy during the first six years and then change the trend reaching almost the initial level of 1990 at the end of the period.

The infant mortality rate provides a related but different measure of improvement in both quality of life and social sustainability. Analysis

TABLE 6

ENVIRONMENTAL EXPENDITURES AS A SHARE OF GDP

Country	1990	1999
Bulgaria	1.0	1.2[2]
Czech Republic	1.1	1.5[3]
Estonia	n.a.	1.2
Hungary	0.6	1.7[4]
Latvia	n.a.	0.8
Lithuania	n.a.	0.5[2]
Poland	0.7	2.4
Romania	0.7	1.8[2]
Slovakia	1.1	1.7[5]
Slovenia	0.2[1]	1.3
US	1.6	1.6
EU15	n.a.	1.0

Sources: Developed from *OECD Environmental Indicators, 1994*; *World Bank Development Indicators CD-ROM, 2001*; *Eurostat Environmental Statistics Yearbook, 2001*; *Environment 2002 GUS, Warsaw.*

[1] 1992; [2] 1998l; [3] Investment only; [4] Public – 0.9; private – 0.8 (CES-Budapest); [5] 1996

FIGURE 6

LIFE EXPECTANCY AT BIRTH OF CEE10

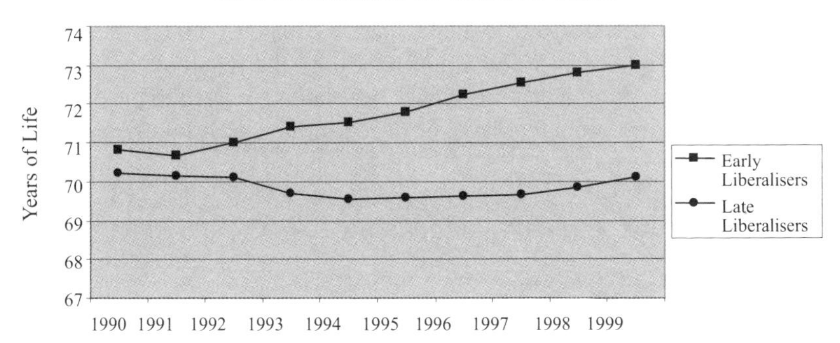

Source: Developed from *World Bank Development Indicator, 2001.*

suggests significant progress among CEE6 countries. Figure 7 shows that in the CEE6 infant mortality rates declined from 17 deaths per 1,000 live births in 1990 to eight deaths per 1,000 live births in 1999. Despite little progress earlier in the decade, the non-CEE6 countries did reach an infant mortality level of 17 deaths per 1,000 live births in 1999. This was the level at which the CEE6 began the decade of transition.

Conclusions

The analysis presented above – both the assessment in trends of sustainability indicators and the econometric EKC analysis – provide some solid indication that market liberalisation did not undermine the sustainability of transformation. In fact, the analyses show that market liberalisation in Central and Eastern Europe countries significantly contributed to strengthening the sustainability of the leading CEE10 invited to join EU. These results provide quantitative support for our hypothesis that early liberalisation contributed significantly to a much smoother, more successful and more sustainable transformation of the CEE10. In addition, they support our previous research, which also showed that the three major processes of systemic transformation – liberalisation, stabilisation and institutional reforms – could positively influence sustainability in transitional countries. However, this was not indicated with most of the other 15 CEE countries and the NIS.

This initial analysis using available data for the CEE countries and NIS provides some important information regarding the relationships between key pollutant emission indicators and critical factors related to the economy and institutional reforms that have occurred in the CEE countries over the

FIGURE 7

INFANT MORTALITY RATE OF CEE10, CEE15 COUNTRIES AND NIS

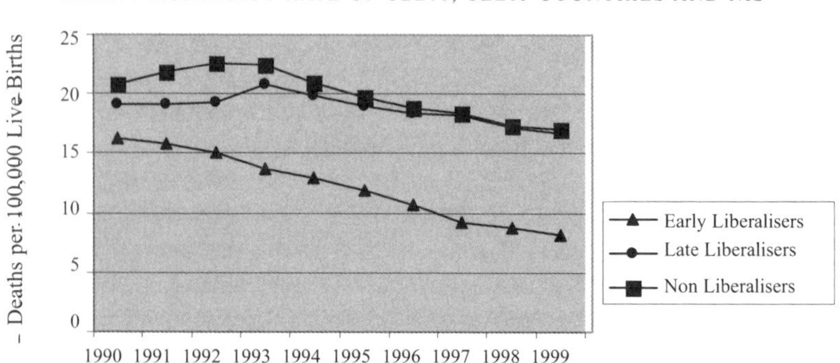

Source: Developed from *World Bank Development Indicator, 2001.*

past decade. Taking into account both the trends and econometric results we can conclude that there is already clear evidence that emissions have been reduced during growth in these countries. This is a result of several factors including the design of the transition processes, to include not only economic but also environmental institutions and policies, much stronger and consequent enforcement of the designed model of transition, and finally a much more liberal, market-based approach to resolve not only economic and social but also environmental problems. Given these results we can argue that both market liberalisation and environmental policy and institutional design can and should be done together. The early liberalising countries adopted the strategy of relatively aggressive environmental policy, facilitating the attainment of both sustained economic growth and significant reduction of all major air pollutants emissions.

Fast liberalisation implemented in six of the CEE10 not only produced significant economic and social gains but facilitated their progress towards environmental sustainability. Upon entry into the EU, the critical question is whether or not these positive trends in their development linked with liberal policies will be sustained and strengthened. Several of these ten countries selected for admission rely more heavily on liberal market-based mechanisms (MBM) than the majority of EU countries (Peszko and Zylicz, 1998). Despite many declarations of the need to apply more market-based mechanisms and instruments, the EU is moving very slowly in this direction. So far, only CO_2 emission trading has received a green light from the European Commission with rather discouragements for CEE countries to apply the tradable permits to other pollutants [*Zylicz, 1999*]. Recently completed negotiations for accession with eight of the CEE countries to be admitted in 2004 clearly revealed that the command-and-control (CAC) approach was preferred by EU negotiators, including costly, troublesome and bureaucratic best-available-technology-approach (BATA). At the same time, the originally designed and implemented institutions and policies in parts of Central and Eastern Europe did not receive positive evaluation or encouragement for further development. This contradicts the evaluations made by many international organisations such as the World Bank, OECD or UNDP and experts.

Wise policy choices are clearly needed to speed up and continue the observed trends in sustainability indicators. They are also critical to future gains in environmental quality. The strategic question is how much imposition of costly CAC is necessary to comply with EU policies and how much those CEE states can retain their effective MBM approaches being members of the EU? The answer to this question will have a significant impact on the costs of adjustment to EU environmental requirements to the CEE nations and the competitive position of their industries. This is a particularly important issue to secure economic sustainability, since most of the environmental expenditures are spent in the CEE10 by industry, contrary to EU countries, where the public sector finances most of the expenditures with exception of the UK [*Eurostat, 2003*]. Imposing high and often unnecessary expenditures on emerging industries might be questioned not only in terms of the impact on their sustainability but also of fairness, effectiveness, and efficiency. It could influence the scope and types of environmentally sound restructuring of CEE industries. Given this analysis, the key question that future developments in the region will have to respond to is whether the observed progress will be sustained when the CEE10 join the EU? This key question provokes one to consider the question as to whether decision makers in EU and beyond can learn lessons from the effective liberal market approach taken by several of CEE countries.

APPENDIX

FIGURE 8

CHANGES IN UNEMPLOYMENT RATES, CEE10

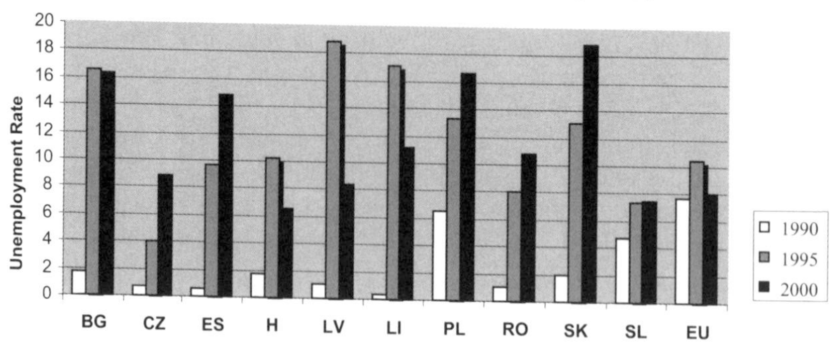

Source: Developed from *World Bank Development Indicators, 2001* and *Economic Survey of Europe, 2002*.

FIGURE 9

CUMULATIVE FOREIGN DIRECT INVESTMENT IN THE CEE10 AS A SHARE OF GDP

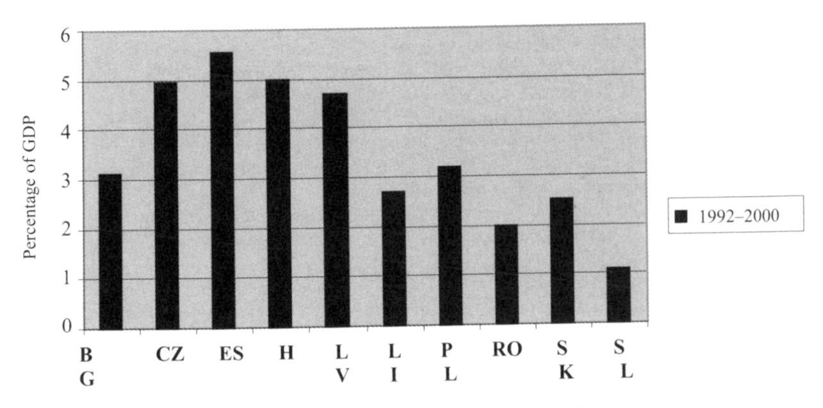

Source: Developed from *World Bank Developoment Indicators, 2001*.

NOTES

The authors would like to thank Andrea Cutting for her significant contribution to this study.

1. These ten countries are the Czech Republic, Estonia, Hungary, Latvia, Lithuania, Poland, Slovak Republic, Slovenia (in 2004), and Bulgaria and Romania (in 2007).
2. There are several good examples of such strategies: in Bulgaria, the Ministry of the Environment influenced by *Ecoglasnost* elaborated a national environmental strategy in 1990 that, although did not last too long but produced a Basic Environmental Law passed by Parliament in October 1991. In Czechoslovakia, the federal government approved a strategic document called *The Concept of the State Environmental Protection Policy* in 1990 that was later published and popularised as *Rainbow Book*. In Poland, the parliament passed *National Environmental Policy* in May 1991 that set directions of environmental policy and institutional changes until the year 2000.
3. The European Bank for Reconstruction and Development was established in 1991 to assist countries from Central Europe to Central Asia 'to nurture a new private sector in a democratic environment'. The Bank works not only as a financial institution, but also as think-tank oriented on economic policy research. EBRD publishes annually, starting from 1994, a *Transition Report*, that contains critical information on the progress of transition concerning markets and trade, enterprises, infrastructure and financial institutions in 28 former communist countries.
4. Cumulative liberalisation was obtained by adding the number of years in which a country has achieved a score of at least three on price liberalisation and at least four on trade and foreign exchange liberalisation and the relevant EBRD transition indicators.
5. The authors' decision to include countries with the same number of liberalised years in different categories, for example Estonia as an Early Liberaliser and Albania as a Late Liberaliser, was made on the basis of the country's political and social stability. For similar reason Croatia and FRY Macedonia were not listed in the first group.
6. This variance is a function of the differences between the indicators analysed (for instance SO_2, CO_2, etc.) and the statistical model applied
7. The PSR system was introduced first by OECD and modified later by the World Bank in a form of environmental performance indicators. See: *OECD Core Set of Indicators for Environmental Performance Reviews. A Synthesis Report by the Group of the State of the Environment*, 1993, OECD, Environment Monographs, No.83, Paris.

REFERENCES

Andreoni, James and Arik Levinson (2001), 'The Simple Analytics of the Environmental Kuznets Curve', *Journal of Public Economics*, Vol.80, No.2, pp.269–86.

Archibald, Sandra and Zbigniew Bochniarz (1998), 'Environmental Outcomes Assessment: Using Sustainability Indicators for Central and Eastern Europe to Estimate Effects of Transition on the Environment', World Congress of Environmental and Resource Economics: Book of Abstracts and website, Venice. June 1998.

Archibald, Sandra and Zbigniew Bochniarz (1999), 'An Empirical Analysis of the Environmental Kuzntes Curve in Central and Eastern European Transitional Economies', Ninth Annual Conference of the European Association of Environmental and Resource Economics: Book of Abstracts and website, pp.36. Oslo, June 1999.

Archibald, Sandra and Zbigniew Bochniarz (2000/2001), 'Assessing Environmental Impacts of the Transition in Central and Eastern European Countries', *Periphery: Journal of Polish Affairs*, Vol.6/7, pp.86–92.

Archibald, Sandra and Zbigniew Bochniarz (2002), 'Assessing Sustainability of the Transition in Central European Countries: A Comparative Analysis', University of Minnesota: Center for Austrian Studies.

Balcerowicz, Leszek (2000/2001), 'Polish Reform and its Results', *Periphery: Journal of Polish Affairs*, Vol.6/7, pp.52–6.

Bochniarz, Zbigniew and David Toft (1995), 'Free Trade and the Environment in Central Europe', *European Environment*, Vol.5, pp.52–7.

Bochniarz, Zbigniew and Jermakowicz Wladyslaw (1998), 'Foreign Direct Investments and Their Environmental Implications in Central and Eastern Europe', Institute of the Future Studies, Stockholm.

Dabrowski, Marek, Stanislaw Gomulka and Jacek Rostowski (2001), 'Whence Reform? A Critique of the Stiglitz Perspective', *Journal of Policy Reform*, Vol.4, No.4, pp.291–325.

EBRD (European Bank for Reconstruction and Development) (2001), *Transition Report*, London: EBRD.

European Commission Eurostat (2001) *Environment Statistics Pocketbook*, Office for Official Publications of the European Community, Luxembourg.

European Commission Eurostat (2003), *Energy, Transport and Environment Indicators – Pocketbook*, Luxembourg, Office for Official Publications of the European Communities.

Gemma, Masahiko (2000), 'Industrial Development in Transition Economies: Lessons and Implications', *Waseda Studies in Social Science*, Vol.1., No.1, pp.19–31.

Grossman, Gene. M. and Alan B. Krueger (1995), 'Economic Growth and the Environment', *The Quarterly Journal of Economics*, Vol.110, No.2, pp.353–77.

Gus Warsaw, *Ochrona Srodowska/Environment 2002*, pp.498.

Hill, Robert J. and Elizabeth Magnani (2002), 'An Exploration of the Conceptual and Empirical Basis of the Environmental Kuznets Curve', *Australian Economic Papers*, June, pp.239–54.

Holtz-Eakin, Douglas and Thomas M. Selden (1995), 'Stoking the Fires? CO_2 Emissions and Economic Growth', *Journal of Public Economics*, Vol.57, No.1, pp.85–101.

Kakwani, Nanak (1995), 'Income Inequality, Welfare, and Poverty: An Illustration Using Ukrainian Data', *World Bank Policy Research Working Papers*, http://www.worldbank.org/html/dec/Publications/Workpapers.

Lucas, Robert, David Wheeler and Hemamala Hettige (1992), 'Economic Development, Environmental Regulation and the International Migration of Toxic Industrial Pollution: 1960–1988', *World Development Report*, Background Paper No.33.

Milanovic, Branko (1998), 'Income, Inequality and Poverty during the Transition from Planned to Market Economy,' Policy Research Working Paper, The World Bank Development Research Group Poverty and Human Resources.

Morvant, Penney (1996), 'Russia: The Changing Face of Poverty', *Transition*, Vol.2, No.1, pp.56–61.

OECD, Organisation for Economic Co-operation and Development (1994), *Environmental Indicators: OECD Core Set*, Paris.

Panayotou, Theodore, Alix Peterson and Jeffrey Sachs (2000), 'Is the Environmental Kuznets Curve Driven by Structural Change? What Extended Time Series May Imply for Developing Countries', CAER II Discussion Paper No.80, Cambridge, MA: HIID.

Pearce, David W. and Jeremy J. Warford (1993), 'World Without End: Economics, Environment and Development', New York, Oxford University Press.

Peszko, Grzegorz and Tomasz Zylicz (1998), 'Environmental Financing in European Economies in Transition', *Environmental and Resource Economics*, Vol.11, Nos.3–4, pp.521–38.

Selden, Thomas and Daqing Song (1994), 'Environmental Quality and Development: Is There a Kuznets Curve for Air Pollution Emissions?', *Journal of Environmental Economics and Management*, Vol.27, No.2, pp.147–62.

Shafik, Nemat (1994), 'Economic Development and Environmental Quality: An Econometric Analysis', *Oxford Economic Papers*, Vol.46, pp.757–73.

Stern, David I and Michels Common (1996), 'Economic Growth and Environmental Degradation: the Environmental Kuznets Curve and Sustainable Development', *World Development*, Vol.24, No.7, pp.1151–60.

Stiglitz, Joseph (1999), 'Quis Custodiet Ipsos Custodes? Corporate Governance Failures in the Transition', keynote address at the Annual Bank Conference on Development Economics in Europe, World Bank, April 1999.

Stiglitz, Joseph and David Ellerman (2001), 'Not Poles Apart: "Whither Reform?" and "Whence Reform?"', *Journal of Policy Reform*,Vol.4, No.4, pp.325–39.

UNECE/EMEP (United Nations Economic Commission for Europe/Co-operative Programme for Monitoring and Evaluation of Long Range Transmissions of Air Pollutants in Europe) 2001.

UNECE/EMEP emission database. Webdata whttp://webdata.emep.int/

United Nations Economic Commission for Europe (UNECE) (2001). *Economic Survey of Europe No.2*, Geneva.

World Bank Group (2001), *World Development Indicators*, World Development Indicators 2001 CD-ROM.

Zylicz, Tomasz (1994), 'Taxation and Environment in Poland', in *Taxation and the Environment in European Economies in Transition*, Paris: OECD, OCDE/GD(94)42, pp.36–56.

Zylicz, Tomasz (1995), 'Pollution Taxes as a Source of Budgetary Revenues in Economies in Transition', in L. Bovenberg and S. Cnossen (eds.), *Public Economics and the Environment in an Imperfect World*, Boston: Kluwer, pp.203–18.

Zylicz, Tomasz (1999), 'Obstacles to Implementing Tradable Pollution Permits. The Case of Poland', in OECD, *Implementing Domestic Tradable Permits for Environmental Protection*, Paris: OECD, pp.147–65.

EU Enlargement and the Environment: Six Challenges

JOHN M. KRAMER

> Environment will be one of the most difficult chapters to close in the accession negotiations. Taking on board the environmental EU legislation is a tremendous challenge for any national parliament and administration, not to mention the financial, administrative and technical aspects of putting it into practice.
>
> Margot Wallström, European Commissioner for Environment, 20 June 2000

The prospective enlargement of the European Union (EU) constitutes the pre-eminent force shaping policies towards the environment in ten Central and Eastern European (CEE) countries that have opened formal negotiations to join its ranks (Bulgaria, the Czech Republic, Estonia, Hungary, Latvia, Lithuania, Poland, Romania, Slovakia and Slovenia). This circumstance derives from the EU mandate that a prospective entrant before admission must adopt the *acquis communautaire* (*acquis*) 'the common body of EU legislation' of which the environmental *acquis* comprises an integral component. In the legal sense in which the EU understands this dictum, 'it means the complete alignment of national legislation so that it complies 100 percent with the requirements of EU legislation. And not just on paper, but' of course 'also in fact' [*Commission, 1997b: 3*]. Adoption entails three distinct elements with respect to the *acquis*: transposition (incorporation into national legislation), implementation and enforcement. That fulfilment of this desideratum presents profound political, economic and social challenges extending far beyond the environment constitutes a truism. Indeed, by consensus, adoption of the environmental *acquis* because of the stringent deadline entailed therein, requisite costs for its implementation and the breadth and complexity of the issues involved is among the most challenging components of the accession process. The challenge is especially acute given that the candidate countries must rely primarily on their own financial and other resources to meet it – resources already severely strained in meeting numerous other demands including those entailed in the overall accession process. In meeting this challenge, the applicant countries have had to battle against the pernicious legacy of communist misrule where political indifference left massive degradation of the environment and a relative paucity of resources (human, technical, financial, institutional and legal) available to improve environmental quality [*Kramer, 1983*].

To what degree accession countries meet these challenges remains an open question given that, as EU officials themselves candidly admit, all of them attach a far lower priority to protecting the environment than their attachment to entering the EU as quickly as possible and in addressing what they consider much more pressing problems of economic revitalisation and growth. Speaking before the European Parliament, Environment Commissioner Margot Wallström acknowledged that there is 'sometimes a discrepancy between the ambitious target dates for accession and the level of priority with which the environmental chapter of the *acquis* is dealt with', a circumstance that makes her 'impatient' [*Wallström, 2000*]. Privately, some EU officials also worry that member states will offer the applicant countries a *quid pro quo* by 'letting them off' on strict and timely fulfilment of the environmental *acquis* to compensate for being especially tough with them on such politically charged issues as the free movement of labour and refugees (personal interview with Commission staff analyst, Brussels, 1999). In contrast, as Pavlínek and Pickles argue in this volume, the EU has performed an important political function by allowing CEE environmentalists to exploit the imperatives of accession to place (and keep) environmental issues on the political agenda and pressure sometimes indifferent and reluctant officials to act upon them.

Six Challenges

Eight of the ten CEE applicant countries – all except Bulgaria and Romania – closed the environmental chapter of the *acquis* in their accession negotiations. The European Commission (the 'Commission') defines 'closed' to mean that 'schedules for transposition and implementation of the environmental *acquis* have been fully clarified, including plans for further strengthening of the administrative capacity' [*Commission, 2002a*]. Bulgaria and Romania began negotiations, in July 2001 and March 2002 respectively, on the environmental chapter. Consequently in late 2001, Jean François Verstrynge, head of the European Commission's enlargement directorate, contended that 'we are almost home on environment' given the advanced state of negotiations on the environmental *acquis* in the accession talks (*Reuters*, 23 November 2001). However, both the authors in the present volume and the Commission itself in numerous reports make it clear that this is an overly sanguine assessment, as all candidate countries confront pressing challenges as they seek to fulfil the environmental commitments they made to the EU. These challenges increasingly involve what Environment Minister Wallström calls the 'much more difficult nut to crack' of effective implementation and enforcement of the environmental *acquis* [*Wallström, 2003*].

This study provides an overview of six challenges – most addressed at discrete points in these pages – with which the candidate countries must successfully cope if they are to fulfil the environmental *acquis* and do so in such a way that they 'crack' the even more fundamentally important 'nut' of building a sustainable environment in the region: (1) the *fiscal* challenge of providing requisite monies; (2) the *administrative* challenge of building both institutional and staffing capacity; (3) the *environmental* challenge of promoting a sustainable environment while fulfilling the *acquis*; (4) the *democratic deficit* challenge of ensuring substantive input for *Vox Populi*; (5) the *energy* challenge of reducing the excessive consumption of environmentally threatening liquid and, especially, solid fuels and coping with the dangers of obsolete nuclear power stations built in the Soviet era; (6) the *political* challenge of mobilising the support necessary to respond effectively to these foregoing challenges.

Several considerations should be borne in mind before proceeding with this analysis. First, assessing the challenges confronting the candidate countries should not obscure the substantive progress they have made, to varying degrees, in the process of adopting the environmental *acquis*. That they have done so in little more than a decade, having emerged from communist misrule with a legacy of profound neglect and indifference towards the environment, and being forced to rely mostly on their own human and financial resources is indeed an impressive achievement. The contributions to this volume speak this, as well.

Even more important, one person's challenge is another's opportunity. This truism manifestly applies to environmental accession. The candidate countries have a historic opportunity to meet the challenges of this accession in ways consistent with realising the EUs own professed goal of ultimately achieving a sustainable environment in Europe. Hence, fulfilment of the environmental *acquis* constitutes a necessary, but not in itself sufficient, step in realising this latter end. If the candidate countries successfully seize this opportunity, they will benefit themselves and prospective candidates in future waves of EU enlargement by providing the latter with a 'road map' of how 'to get it right' in meeting the challenges of environmental accession. The EU itself can also profit from such a 'road map' since it confronts many of the same challenges besetting the candidate countries, including those related to financing and administration and, most fundamentally, of transforming its rhetorical commitment to building a sustainable environment into a substantive reality.

The Fiscal Challenge

The daunting fiscal challenge of complying with the environmental *acquis* is a pervasive theme in the present volume. This challenge has sparked a contentious debate entailing both the obvious question of how these expenditures will be funded as well as whether the projected costs somehow can be reduced and/or 'stretched out' through sundry transitional periods. One must treat cautiously all estimated costs of compliance, but the requisite expenditures undoubtedly will be huge and impose a heavy additional burden on economies already severely strained in the transition from communist rule. In one widely cited study published in 1997, the Commission estimated that it would cost the CEE candidate countries approximately 120 billion Euro to comply with EU requirements for drinking water supply, wastewater management, large combustion plants and waste management [*Commission, 1997a*]. Subsequently, the Commission revised this estimate now calculating that costs would range between 80 billion Euro and 110 billion Euro, although the costs for individual candidate countries – both absolutely and per capita – differ substantially (see Table 1).

This revised estimate – like its predecessor – should not be accepted literally: at best, it is suggestive of the enormous sums that will be required to fulfil the environmental *acquis*. First, the estimate excludes the investment needs of important new and forthcoming EU environmental legislation such as the Water Framework Directive. Second, for some investment-heavy directives – for example, the Air Quality Framework Directive – the necessary expenditures will only become known after completion of an initial assessment of the scope of the problem and the

TABLE 1

ESTIMATED ENVIRONMENTAL FINANCING NEEDS IN APPLICANT COUNTRIES

Country	Estimated Total (Millions of Euro)	Euro Per Capita
Bulgaria	8610	1117
Czech Republic	6600–9400	643–915
Estonia	4406	3095
Hungary	4118–10000	497–989
Latvia	1480–2360	620–989
Lithuania	1600	443
Poland	22100–42800	572–1107
Romania	22000	983
Slovakia	4809	888
Slovenia	2430	125
Total	79260–11001	–

Source: Developed from Commission [*2001b: Annex 2*].

adoption of a strategy to combat it. Finally, the required expenditures will vary considerably depending on how strictly EU directives are interpreted and applied. For example, according to a World Bank study, this variable accounts for the enormous differences in the estimated costs of compliance for Poland contained in Table 1 [*Hughes and Bucknall, 1999*].

The nature and extent of any transitional periods granted by the EU for the applicant countries to ease the burden of complying fully with the environmental *acquis* has been an especially contentious issue in accession. Applicant countries themselves pointed to both the magnitude of the task in adopting this *acquis* and the limited financial and institutional resources they bring to this effort to justify the need for such transitional periods. In principle, the EU endorsed the need to grant transitional periods to fulfil those areas of the *acquis*, explicitly including the environment, 'where *considerable adaptations* are necessary and which require *substantial effort*, including *important financial outlays* ...' [*Commission, 1999*]. Behind this general statement of principle, EU officials made it clear that considerable restrictions would apply to any transitional periods granted. First, applicant countries needed to provide a substantively persuasive rationale for any transitional period they requested: Commission officials reportedly looked with particular disfavour on requests they considered a disguised form of protectionism for domestic industrialists anxious to escape the onerous financial burden entailed in fulfilling sundry EU environmental regulations (personal interview with Commission staff analyst, Brussels, 1999). Second, applicant countries would receive transitional periods only for relatively short, and clearly defined, periods of time and would need to submit a detailed strategy and timetable for complying with the regulation before they received the grace period. Third, applicant countries would receive transitional periods primarily for the fulfilment of regulations necessitating substantial financial investments rather than for the much less costly transposition of the environmental *acquis*.

Applicant countries at times sought to pressure the EU to make these desiderata less stringent. For example, Poland's chief negotiator with the EU argued that Poland might consider postponing its target date of 2003 for joining the EU unless the latter compromised over its proposed timetable for adopting sundry components of the *acquis*, explicitly including the chapter on environment (*PAP*, 7 November 2000). Hungary pursued a variant of this strategy by asking why it should adopt politically and economically onerous provisions in the *acquis* when it remained unclear when or if the EU actually would enlarge its ranks. 'The government's view is that the measures disadvantageous to Hungary should be taken at the time of EU accession', not before, as originally planned, by 2002, according to an official governmental spokesman (*Reuters*, 26 April 2000). Yet the overriding desire to enter the EU as rapidly as possible ultimately undermined this stratagem. This was most obvious in Poland where, in April 2001, its chief EU negotiator announced that Poland would rescind sundry requests for transitional periods to meet EU environmental standards to keep its entry talks on track: 'To stay in the rhythm of negotiations, we had to change our position' he explained in providing the rationale for the decision (*Reuters*, 8 April 2001). In the end, the longest transitional periods granted to meet EU standards typically were for urban wastewater and the quality of drinking water – components of the *acquis* whose fulfilment entails especially onerous fiscal burdens. Environment Commissioner Wallström promised to ensure strict enforcement of these transition deadlines explaining that 'existing member states will not allow environmental dumping' – a reference to concerns that exporters in CEE countries would have a competitive cost advantage over them if they were not subject to the same environmental regulations (*Reuters*, 23 January 2003). The EU also mandated that transitional periods do not apply to new investments, which must conform to the environmental *acquis* thereby preventing applicant countries from luring investors with the enticement of less stringent environmental standards and, concomitantly, lower costs of production [*Commission, 2002a*].

Overall, the EU estimates that candidate countries on average must spend between two per cent and three per cent of gross domestic product (GDP) to ensure implementation of the environmental *acquis*. To place this task in perspective, consider what the following candidate countries spent on environment as a percentage of GDP in 2001: Czech Republic (1.04), Hungary (1.0–1.1), Lithuania (0.22), Poland (1.7), Romania (0.40), Slovakia (1.5) [*Commission, 2001c*]. In the EU itself, such expenditures now average about one percent of GDP. These data raise the pressing question of where the candidate countries will acquire the monies to fulfil their obligations.

EU officials have made it clear that whatever actual sums the candidate countries ultimately spend on environmental compliance must come primarily from their own resources. At best, EU aid for this 'purpose projected at about five per cent of the estimated requisite monies overall' (personal interview with Commission staff analyst, Brussels, 1999) will target selected high priority projects. The 'Instrument for Structural Policies for Pre-Accession' (ISPA) has been the principal vehicle for EU environmental aid to the CEE candidate countries, although upon accession these monies will come primarily from the EU's regular structural and cohesion funds [see also, *Schreurs, this volume*]. Between 2000 and 2006, the Commission plans to allocate approximately 500 million Euro annually through ISPA for environmental investments [*Commission, 2001b*]. ISPA funds are also designed to serve as a catalyst to promote aid from other external donors, including the European Bank for Reconstruction and Development (EBRD), the World Bank, and private capital as well. The first fruits of these efforts are now appearing. In June 2000, the EU, the World Bank and the government of Denmark announced they would help co-finance a $97 million loan to Poland to build small geothermal and gas-fired power plants to produce environmentally clean energy (*Reuters*, 29 June 2000).

Yet total foreign assistance, including ISPA funds, still accounts for a relatively limited share of environmental investments in the candidate countries. For example, in 2001 external assistance comprised no more than ten percent of the environmental investments in Poland [*Commission, 2001d: 82*]. In Romania, external donors have pledged just $1 billion towards the projected cost of $22 billion to fulfil the environmental *acquis* (*Mediafax*, 11 March 2003). Not surprisingly in these circumstances, some advocate a substantial increase in EU assistance, contending that it is simply unrealistic to expect the applicant countries to marshal the requisite monies for environmental compliance primarily from their own already heavily burdened domestic economies. In this vein, the European Parliament has asked the EU to double the amount of its environment-related assistance to the applicant countries by 2006 [*European Parliament, 2000*]. Similar arguments are often voiced in the applicant countries themselves. Thus, in representative arguments from Poland, the Minister of Environment contends that 'our firms cannot take on more burdens on environmental issues and stay competitive' while a local governmental official simply states 'without EU funds, it would take us 20 years to become environmentally friendly, with Europe's help, three' (*Reuters*, 20 June 2001). In contrast, a World Bank study sharply challenges these arguments, contending that companies themselves 'should, in almost all cases, be expected to finance any necessary environmental improvements through regular channels as part of their overall programs of capital investment'. The report dismisses concerns about higher environmental outlays undermining industrial competitiveness as 'seldom necessary', citing surveys from the United States on this subject to substantiate its position [*Hughes and Bucknall, 1999: 10, 11*].

Whatever the merits of these respective arguments, it seems clear that the private sector – both producers and consumers – will shoulder a heavy load in financing EU-related environmental investments. To this end, it becomes critical that candidate countries vigorously pursue the privatisation of environmental services such as water and power supply and waste removal and the concomitant establishment of so-called full cost recovery pricing – in plain English, the elimination of subsides and the establishment of market-based prices – for them (*Environment for Europeans*, October 2000). While no panacea, these initiatives potentially have the salutary effect of: (1) making the provision of environmental services a profitable business wherein environmental-related investments are included in the costs of production and reflected in the retail price of the service; (2) attracting investment to these now profitable businesses which can be utilised to promote technological modernisation and the more efficient use of resources as a means of reducing costs and increasing profits; (3) forcing consumers to be more thrifty in the utilisation of resources for which they now pay market-based prices; (4) generating a predictable stream of revenue for government from taxes and user fees which can be used to finance environmental projects of national significance, including in partnership with the private sector [*Commission, 2001b: 14*]. This position is consistent with the overall argument in the Archibald, *et al.* contribution [*this volume*], which strongly stresses the interlocking environmental and economic benefits of market liberalisation. It stands in contrast to those authors [e.g. *Pavlínek and Pickles, this volume*] who express reservations with aspects of this process.

The Administrative Challenge

Strengthening their administrative capacity to transpose and, even more importantly, implement and enforce the environmental *acquis* is rapidly emerging as one of the key challenges confronting the applicant countries. Calling this challenge a 'central plank of environmental protection', Environment Commissioner Wallström in January 2003 told CEE environment ministers meeting in Brussels that requisite administrative capacity has been 'identified from the start as one of your weak points' [*Wallström, 2003*]. A recent EU-sponsored study on this subject found 'specific capacity problems in almost every candidate country', but singled out Bulgaria, Poland and Romania for where 'the most pressing capacity problems occur' [*Commission, 2001a: 258*]. Reportedly, 'quite a lot of tensions' have existed over ISPA projects because of the Commission's view that at times candidate countries have lacked the requisite administrative capacity to implement them effectively (personal interview with Regional Environmental Center (REC) staff analyst, Szentendre, Hungary, 2001). Yet as Schreurs [*this volume*] reminds us, capacity problems are hardly unique to the applicant countries but often bedevil the EU itself, especially its less developed states, in the effective implementation and enforcement of environmental legislation. That said, the pernicious legacy of communism with its profound neglect of the environment meant that the applicant countries, to varying degrees, began accession with exceedingly limited administrative capacities to cope with the highly complex, interconnected, and substantively challenging imperatives of the environmental *acquis*.

The Commission has identified many administrative problems in Poland that are found to varying degrees in all applicant countries [*Commission, 2001d: 82*]:

> Poland's administrative capacity for EC environmental directives remains a matter of concern. … Staff resources are limited and the awareness about the requirements of EU environmental directives needs to be improved. Significant training in EU environmental policy is still necessary. Poland's division of tasks over numerous agencies and administrative levels has, in some cases, caused unclear responsibilities. Different bodies are responsible for setting objectives, permitting monitoring, inspection and financial instruments. This risks diminishing the accountability for achieving environmental standards.

The challenge of staffing is multifaceted encompassing the need to increase the number of personnel engaged in environmentally related accession issues, to provide them with requisite training and resources to execute their duties competently, and to retain qualified staff by compensating them adequately so that they do not, as have so many of their erstwhile colleagues, leave for better paying employment in the private sector. The problem, according to the Minister of Environment in Slovakia, whose ministry has only one-third the level of staff that the Commission considers necessary, involves money – or the lack thereof: 'It's all about money – if we had money, we would have administration, too' (*The Slovak Spectator*, 27 January 2003). That a lack of photocopying facilities prevented staff in the Bulgarian Ministry of Environment from receiving copies of EU legislation illustrates – perhaps in caricature form – the simple and basic constraints under which environmental personnel often labour in the candidate countries [*Commission, 2001a*].

The EU has sought to mitigate such problems through the 'twinning' mechanism under its PHARE programme, which overall seeks to enhance the administrative capacity of applicant countries to implement and enforce the *acquis*. Environment has been designated as one of the priority areas for twinning, which primarily entails the long-term secondment of highly qualified civil servants from EU member states to assist their colleagues in the applicant countries. 'Twinning' has assisted Bulgaria in developing a task force to participate in ISPA, Hungary to transpose environmental legislation and establish and manage a Central Environmental Protection Fund, and Slovakia to enhance its administrative capacity to implement measures on air pollution.

The respective Ministries of the Environment in the applicant countries, which under communism were typically politically and administratively powerless, also require considerable strengthening if they are to fulfil their designated role as the principal agency responsible for the preparation and execution of environmental legislation. Besides the aforementioned problem of staffing hindering them in this regard, these ministries are often reduced to a 'coordinating' or 'consultative' role since they lack administrative power and competence over key sectors of the environment. Problems of coordination and clarification of administrative responsibilities are common in such a decentralised system. Further, sectoral agencies tasked with many responsibilities besides environment may not always accord requisite priority to environmental protection. Exacerbating this problem, these ministries are relatively weak politically compared to many other sectoral agencies and often lack the political muscle to overcome this sectoral bias in policy formulation and execution. Such sectoral bias is becoming especially pernicious in the accession process given the imperative to integrate environmental considerations into other policy sectors and the concomitant need to devise the most cost-effective strategies to realise environmental objectives (personal interview with REC staff analyst, Szentendre, Hungary, 2001).

The capacity challenge will be especially pressing at the regional and local levels where governments – in sharp contrast to their mostly moribund state under communism – will increasingly become key players, especially in the implementation of EU legislation. As a ranking EU official explains:

> Local authorities will, for example, be responsible for issuing permits for industrial plants. They have to be able to evaluate requests, and this often requires highly technical knowledge. Another task is monitoring air and water quality. To begin with, monitoring equipment has to be developed and certain procedures followed to make sure that everybody is measuring the same emissions. In the end, results have to be reported to the Commission, and also made available to the public (*Environment for Europeans*, February 2002, p.4).

The challenge is reflected in Poland where officials at the regional ('voivodship') level will be responsible for issuing permits to 70 per cent of the installations covered under the key Directive on Integrated Pollution Prevention Control (IPPC). In its 2002 report on Poland, the Commission noted that no environmental departments were yet 'fully operational' at the voivodship level and that Poland must 'urgently' begin training their staff to implement the IPPC Directive [*Commission, 2002b*].

This analysis hardly exhausts the challenge of administrative capacity. In the present volume, for example, both Bell and Kružíková – with their substantive focus on, respectively, civil society and law – address issues that bear directly on this subject while Pavlínek and Pickles make the useful point that accession may have important 'indirect consequences' for the environment by, *inter alia*, strengthening administrative capacities through combating corruption.

The Environmental Challenge

This contribution argues that fulfilment of the environmental *acquis* constitutes only a necessary first step towards the realisation of a fundamentally more important goal: the creation of a sustainable environment in Europe. It would be a tragedy if EU accession, despite the claims of proponents of the *acquis*, actually led to policies that degraded, rather than enhanced, the CEE environment. Yet many environmentalists contend this is precisely what is happening because the EU is imposing on the applicant countries its own fundamentally flawed developmental 'model', which stresses the imperatives of large-scale economic development to the detriment of building a sustainable environment. In this volume, for example, Gille bluntly states this argument charging that the EU 'stands for unsustainable development' by placing economic, over environmental, interests.

CEE ministers of environment reportedly echoed this argument in meeting with Environment Commissioner Wallström where they told her the EU needs to take 'one hard look in the mirror' about the pressing need to revise EU policies – specifically the massive subsidisation of agriculture and motorisation – that promote environmental degradation and would make an enlarged EU 'no more sustainable than the present EU' [*Wallström, 2000*]. Bedrich Moldan, the Czech Republic's chief negotiator with the EU on environment, was even more graphic on this subject, calling the extension of subsidies for agriculture and motorisation to the applicant countries 'disastrous for the environment, biodiversity, social structures, and economies of Europe as a whole' [*Moldan, 2000*].

Environmentalists especially target EU pre-accession assistance for its alleged failure to promote the principles of sustainable environmental development. Thus, in two prominent reports, issued in 2000 and 2001, respectively, CEE Bankwatch Network and Friends of the Earth contended that the EU is actually promoting environmental degradation in CEE countries by failing to integrate the principle of sustainable development throughout its assistance programmes. They were especially critical of ISPA, charging that it favoured the creation of a 'car dependent' society through its funding of highway infrastructural projects while 'not at all' promoting a much more 'environmentally friendly' system of reliable and comprehensive urban public transport and rail service. A related point is that the reports charge that priorities in EU aid reflect the preferences of elites, not the public, since there is little meaningful opportunity for citizen input into the selection and preparation of accession projects [*CEE Bankwatch Network and Friends of the Earth Europe, 2000, 2001*]. Here, however, the culprit may more often be CEE governments themselves and the Commission part of the solution, according to the executive director of CEE Bankwatch Network: 'We, the people living in the accession countries have the right to decide over our future. Unfortunately, our Governments are not making it possible. We are drawing the attention of the European Commission to that problem and hope for their support to make the entire process more open and transparent' [*Friends of the Earth Europe, 2002*].

That the EU will soon require applicant countries to subject all EU pre-accession investment projects before their implementation to a rigorous environmental impact assessment (EIA), including mandatory public participation therein, may mitigate some of these aforementioned concerns. Separately, Environment Minister Wallström has convened a series of 'informal ministerial' meetings with the ministers of environment of the candidate countries to pursue a dialogue among them on environmental issues of common concern, including how to promote the principle of sustainable development. One of the substantive fruits of this dialogue has been an agreement to waive the requirement of a 5 million Euro threshold before a project is eligible for ISPA funds – a requirement that favoured larger projects that critics contend were often threatening to the environment – for worthy smaller scale less costly projects that concomitantly are more 'environmentally friendly' [*Informal Ministerial Meeting, 2001*].

Despite such initiatives, it is obvious that the challenge of promoting sustainable development remains a work in progress. If this challenge is to be met successfully, it will require policymakers in both Brussels and the applicant countries to transform their professed commitment to the EUs own sustainable development strategy, adopted by EU leaders at the 2001 Gothenburg Summit, into a substantively meaningful one that views accession holistically with environmental concerns permeating the *acquis*, not being compartmentalised only into the chapter on environment.

The 'Democratic Deficit' Challenge

An especially critical challenge entails overcoming the so-called democratic deficit wherein *Vox Populi* in general, and private sector environmental groups (NGOs) in particular, have been insufficiently heard in shaping public policy towards the environmental *acquis*. As a report issued in 2002 by the influential World Wildlife Fund bluntly contends, 'civil society access to information and decision-making regarding the accession process and programming for use of EU funds is still largely missing in the candidate countries' [*World Wildlife Fund, 2002*]. Persistence of the democratic deficit makes it difficult to mobilise political support and will to implement often fiscally onerous environmental policies and promote their sustainability by actively involving environmental NGOs in their formulation. The origins of the democratic deficit are rooted both in CEE societies themselves, where the legacy of communism still inhibits the development of vibrant civil societies, and in, as several authors in these pages have demonstrated, the *modus operandi* of EU institutions and how they are managing accession.

In CEE countries, as former President Vaclav Havel of the Czech Republic has observed, strengthening *Vox Populi* has been a 'difficult process' with many public officials retaining the communist era view of the citizenry as an adversary, not a partner, in the exercise of power (*Die Welt*, 18 February 2000). That Bulgaria's Ambassador to the EU recently chastised 'self-styled NGOs' who 'interfere into the negotiations for membership in the EU' by communicating directly with Brussels 'rather than through the national government' is emblematic of such attitudes [*CEE Bankwatch Network and Friends of the Earth Europe, 2001*]. Much work especially needs to be done to make NGOs in the region forceful and effective advocates for the environment, according to a comprehensive report on the subject. This report, based on a survey of approximately 3,000 CEE NGOs, found that the 'vast majority' of them are in an 'unstable, poor, or very poor financial state' that cooperation between them and national and international governmental authorities is 'for the most part poor', and that over two-thirds of them are concentrated in only four countries, the Czech Republic, Hungary, Poland and Slovakia [*Regional Environmental Center, 1998*]. Even in the Czech Republic, with its relatively well-developed NGO sector compared to many of its CEE counterparts, a July 2001 poll found that 58 per cent of the respondents could not name any environmental NGO (although 57 per cent of the sample said they would support the work of such organisations) (*CTK*, 9 January 2001). In an often Faustian bargain, local NGOs may mitigate many of their problems by affiliating with well-known international organisations such as Friends of the Earth or World Wildlife Fund, but at the risk of being, as Bell observes in this volume, 'overwhelmed and marginalised' by their far larger and more powerful Western patrons.

Yet the EU itself, even if unintentionally, has managed environmental accession in such a way largely to exclude CEE environmentalists from substantively meaningful participation in it. This circumstance arises most obviously because, as Bell notes herein, Brussels mostly has determined the agenda of environmental accession with the applicant countries playing the role of implementers of policies predetermined by the Commission. Then, too, Hallstrom effectively demonstrates in this volume how the Commission, including DG Environment, informally has a strong 'technocratic' bias which prizes the technical expertise of scientists and other professionals and sees the role of grassroots environmental organisations – typically lacking requisite expertise – as transmitters to their own publics of expert-determined policies made in Brussels.

However, initiatives are underway to mitigate this bleak situation. In conjunction with the REC, the internationally funded Hungarian-based organisation charged with facilitating environmental cooperation in the region, the EU is now conducting 'NGO Dialogue' conferences to facilitate discussion on environmental-related issues of accession and to enhance the public stature of these groups as spokesmen for the environment. While Gille [*this volume*] has dismissed such meetings as 'futile', Hicks [*this volume*] has noted more positively that recent sessions have exhibited more dialogue and substantive interchange about broader international programmes. 'Putting its money where its mouth is', the Commission in June 2001 announced that it was approximately tripling the financial support that it provides European environmental NGOs (from 10.6 million Euro to 32 million Euro over a five-year period) so that it could include CEE NGOs in the programme (*Green Horizon*, 29 June 2001). To promote these same ends, the European Parliament officially has invited environmental NGOs to submit annual reports to it identifying 'real or potential deficiencies' among applicant countries in complying with the environmental *acquis* [*European Parliament, 2000*].

The EU also correctly has recognised that if environmental NGOs are to perform their functions effectively, they must have largely unfettered access to reliable and contemporaneous data about the environment. Responding to the imperatives of EU accession, CEE governments increasingly are compiling these data, but, often claiming a need to protect commercial secrets, have been less forthcoming in disseminating them publicly, even to EU officials who request them (personal interview with Commission staff analyst, Brussels, 1999). Consequently, the EU has established a 'Public Right to Know Project' that works closely with environmental NGOs and private individuals to pressure CEE governments to establish minimum standards for public access to information regarding the environment. This promising initiative performs the dual function of simultaneously strengthening the capacity of NGOs to serve as effective advocates for the environment and arming the citizenry with the requisite information to make informed choices about environmental policy. EU legislation mandating public participation, such as the Environmental Impact Assessment Directive and the Access to Environmental Information Directive, similarly promotes these functions.

The Energy Challenge

Under communism, CEE governments pursued an energy policy enormously deleterious to the environment. It stressed the forced draft development of energy-intensive industries, including steel, chemicals and mining [*Kramer, 1990*]. Consequently, the economies of the region expended between 30 percent and 50 percent more energy than their counterparts in Western Europe to produce the same unit of national income. CEE countries also produce the wrong kinds of energy to promote environmental quality. Low grade soft coal of high sulphur, ash and cinder content predominates in the production of primary energy in CEE states, except in Poland, where hard coal predominates.

This environmentally threatening legacy persists in the region. Thus, in 1998, energy intensity – that is, tons of oil equivalent consumed per $1,000,000 of GDP – in the applicant countries was upwards of four to nine times higher than the average of Western European members of the Organisation for Economic Cooperation and Development (OECD). Comparing the energy intensity of the United States – the most energy-intensive user in the West – to those of the applicant countries in 1998, only Estonia had comparable energy intensity whereas all others had levels at least double that of the United States. In some CEE countries, energy efficiency has improved very little in the transition years [*Andonova, 2002*]. The consequences of this for the environment are predictable: for example, in 1998, the applicant countries, relative to output, produced anywhere from three to six times more greenhouse gases than did their counterparts on average in the EU. Simply stated, a decisive improvement in the CEE environment requires that these countries reduce substantially their energy intensities [*EBRD, 2001: 91–3*].

The highly controversial status of nuclear power in the CEE region also has its roots in the communist era where intensive exploitation of the atom comprised an integral component of energy policies in most states [*Kramer, 1995*]. In 1998, an EU-sponsored study of nuclear safety in CEE countries and the former Soviet Union concluded that six Soviet-era nuclear power stations (NPS) still operating within EU aspirant states – each of the two oldest units at, respectively, the Kozloduy NPS in Bulgaria, the Jaslovske Bohunice NPS in Slovakia and the Ignalina NPS in Lithuania 'were both impractical' to retrofit 'to cope with accidents which are normally safeguarded against in Western designs' and could only 'be operated for a short time without excessive risk' [*Commission, 1998: 37–41*]. The concerns are especially acute regarding Ignalina whose reactors (unlike those at its counterparts in Bulgaria and Slovakia) are of the same design and technology as the reactor at the Chernobyl NPS in Soviet Ukraine which caused the world's worst nuclear accident when it exploded on 26 April 1986.

After long and arduous negotiations beset with charges that it was not pursuing the issues with sufficient vigour, the EU finally secured from these states timetables for the closure of the controversial plants. In each instance, it is clear that pressure from the EU which explicitly made the opening of accession talks with each state contingent upon resolution of the issue proved decisive in affirming the decision for closure. In February 2002, Energy Commissioner Loyola De Palacio reaffirmed the link between accession and nuclear safety in the applicant countries, contending that the latter 'should be one of the preconditions for EU entry' (*Reuters*, 18 February 2002). To mitigate the onerous financial burden on the affected countries of decommissioning these reactors – for example, it will cost an estimated 3 billion Euro to close the Ignalina NPS – the EU and other Western donors in 2000 established Dedicated Nuclear Decommissioning Funds for Bulgaria, Lithuania and the Slovak Republic.

Yet these initiatives have failed to assuage critics. For example, the EU has angered many anti-nuclear activists in CEE countries and in EU member states by making it clear that it remains the sovereign prerogative of CEE states themselves to determine whether NPS in the region that the EU deems meeting prevailing standards of safety – a total of 15 reactors – should remain operative or be closed. Indeed, Energy Commissioner De Palacio has argued forcefully for expanded utilisation of nuclear power in the EU as a critical component of its strategy to reduce reliance on importation of energy, especially petroleum (*Reuters*, 18 February 2002). As Axelrod details in this volume, the latest nuclear controversy embroiling the EU involves the opening of the Temelín NPS in the Czech Republic which houses two giant Soviet-designed 1,000 MW reactors upgraded with Western-supplied safety and control equipment.

The EU also has many critics in those countries where NPS will be closed who argue that: (1) they are only closing the plants under duress since these now meet prevailing standards of safety after considerable sums were spent modernising them; (2) the funds Western donors are providing to decommission the plants fall far short of what is needed for this task; (3) the premature closure of these reactors will inflict considerable hardship since the affected countries depend so heavily on the atom as a source of power (in Lithuania, for example, Ignalina supplies upwards of 80 per cent of its electricity and the figure in Bulgaria for Kozloduy is approximately 50 per cent).

While these arguments have merit – for example, regarding Ignalina, the EU has pledged less than 10 per cent of the estimated 3 billion Euro needed to decommission the plant (*Reuters*, 21 March 2002) – they obscure the much more fundamental challenge confronting these countries and, indeed, other applicant countries as well: the imperative to reduce substantially their excessive consumption of energy. Tellingly, in 1999, compared to the United States, energy intensity in those candidate countries where NPS will be closed was approximately five times higher in Bulgaria, three times higher in Lithuania, and twice as high in Slovakia. In the Czech Republic, which operates the controversial Temelín NPS, energy intensity was also approximately double that of the United States. Further, each of these states has higher energy intensity than the average for the region [*EBRD, 2001*].

A detailed discussion of how best to achieve substantial reductions in energy intensity lies beyond the purview of the present study. Suffice it to say that an in-depth study of this subject found that the 'most significant' way to achieve this end entailed market liberalisation by eliminating the substantial under-pricing of energy in all transition economies, including those of the applicant countries, which hitherto has provided little economic incentive to their consumers to utilise energy resources prudently [*EBRD, 2001*]. Establishing energy prices approximating actual cost-recovery levels would provide a compelling economic *disincentive* to consume energy profligately and concomitantly would represent a 'win win' situation for the environment by both lessening the rationale for relying on dangerous NPS to produce power increasingly in excess of demand and reducing energy-related air and water emissions, which are among the most threatening sources of environmental pollution.

The Political Challenge

The critical challenge confronting applicant countries may well be marshalling the political will to enact requisite policies to fulfil the environmental *acquis* that are inevitably highly controversial, fiscally onerous, and disadvantageous to key groups in society. Stated succinctly, without sufficient political will, none of the foregoing challenges analysed herein is likely to be met successfully. So far in the accession process, the imperative to enter the EU as quickly as possible, rather than any substantive commitment to improve the environment *per se*, has been the overriding motivator compelling the applicant countries to fulfil whatever demands the EU has made regarding the environmental *acquis*.

However, after accession, the status of political will may become more problematic given that, as Jehlička and Tickle argue in this volume, the EU inevitably will have diminished leverage over the former applicant countries and the latter will have more opportunity to set their own agendas and priorities, including those towards the environment. If the environment is to be a priority under these circumstances, it becomes critical to eliminate the aforementioned democratic deficit in accession. Absent that, *Homo Politicus* is hardly likely to render support and shoulder the burden of complying with the environmental *acquis* if he is largely excluded from its formulation and execution.

To this end, proponents must make a convincing case that complying with the environmental *acquis* – despite its costs and burdens – overall is in the best interests of the applicant countries. As Environment Commissioner Wallström informs, 'In my meetings with political leaders from the Accession Countries, I have stressed over and over again that environment should not be seen as a cost but rather as an opportunity in the accession process' [*Wallström, 2000*]. Proponents can, for example, argue that this end materially enhances their prospects for EU membership since a symbiotic relationship exists between many requisite policies to fulfil the environmental *acquis* and policies the EU demands to promote marketisation and democratisation. From establishing market-based prices for energy resources that provide economically compelling incentives to utilise them prudently to overcoming the democratic deficit in environmental policy making, what is 'good' for the environment typically is similarly 'good' for fostering vibrant market economies and democratic polities.

Yet more fundamentally, proponents must drive home much more forcefully than hitherto the substantive environmental, health-related and economic benefits likely to ensue from compliance with the environmental *acquis*. A 2001 independent study on this subject, commissioned by the EU, identified many of these benefits, including better public health by reducing environmentally related respiratory diseases and premature mortality; less damage to commercially related natural resources and buildings; reduced risk of permanent damage to critical natural resources such as groundwater aquifers; promotion of eco-tourism; and increased economic efficiency, higher productivity, and enhanced economic competitiveness through the use of modern 'environmentally friendly' technologies and higher levels of reuse and recycling of primary materials [*Ecotec Research and Consulting, 2001*].

Overall, the study estimated that implementing the environmental *acquis* completely by 2010 would bring the candidate countries cumulative benefits during 1999–2020 amounting to between 134 billion Euro and 681 billion Euro. Given the obvious methodological uncertainties associated with these computations, the study correctly stresses the need to utilise the most conservative estimates of anticipated benefits when analysing this issue. Nevertheless, it is striking that even when following this dictum, the lowest estimate of anticipated benefits (134 billion Euro) is upwards of 18 percent greater than the *highest* estimated cost (110 billion Euro) of fully implementing the environmental *acquis*. Lower levels of air pollution – for example, full compliance with EU directives through 2020 is expected to reduce emissions of particulate matter by 1.8–3.3 million tons while sulphur dioxide (SO_2) emissions should be some 2–3 millions lower than if the EU directives are not implemented – account for upwards of 55 per cent of the total estimated value of these benefits through improved public health, preventive damage to buildings and reduced damage to crops. Particularly striking, the study estimates that fully implementing EU air quality directives can lead annually to between 15,000 and 34,000 fewer cases of premature deaths from exposure to air pollution – Poland alone could have between 7,000 and 10,000 fewer premature deaths by 2010 – and between 43,000 and 180,000 fewer cases of chronic bronchitis.

Naturally, these estimated benefits of full compliance will vary among the candidate countries both per capita and as a percentage of GDP (see Table 2). This compliance will also reduce transboundary air pollution originating in applicant countries thereby providing annually lowest estimated benefits of 1.7 billion Euro to other applicant countries, 6.5 billion Euro to EU member states, and 9.5 billion Euro to non-EU member countries (notably Belarus, Russia and Ukraine). If utilised effectively, even these most conservative estimates of the benefits of complying with the environmental *acquis* provide potent ammunition to its proponents in meeting the political challenge of convincing the public, in the words of Bedrich Moldan of the Czech Republic, that 'what we are doing is not because we want to satisfy Brussels clerks but because we, of course, want to have a better environment' (*CTK*, 27 October 1999).

If this effort is to succeed, it also becomes critical that the EU eschew the mixed messages that it too often sends on the environment – messages that in word typically say all the right things about environment and the need for sustainable development but in deed frequently entail policies such as the stress on large-scale intensive agricultural development that directly conflict with its rhetorical commitment to sustainability [see *Beckmann and Dissing, this volume*]. Such mixed messages only weaken those environmentalists in CEE countries pressing their polities for a substantive transformation in environmental policy to promote sustainable development in the face of considerable political indifference, at times, even overt opposition, to this end.

TABLE 2

ESTIMATED AVERAGE ANNUAL BENEFITS OF FULL COMPLIANCE WITH THE ENVIRONMENTAL *ACQUIS*

Country	Average Benefits Per Capita (Euro)	Average Benefits as % of GDP
Bulgaria	154.5	10.9
Czech Republic	467	9.65
Estonia	196.5	6.2
Hungary	400.5	8.9
Latvia	136	5.85
Lithuania	216	8
Poland	331	8.85
Romania	246.5	17.35
Slovakia	376	11.45
Slovenia	343.5	3.65

Source: Developed from Ecotec Research and Consulting [*2001: 24*].

Conclusion

Whether the bottle is half full or half empty distinguishes conflicting assessments – as reflected in this volume – about the status of environmental accession. The more positive ‘half full’ assessment argues that in the face of formidable obstacles – the daunting legacy of communism, exceedingly limited financial resources drawn primarily from their own reserves, and stringent EU-imposed deadlines – the applicant countries have made remarkable progress in a relatively short period of time in fulfilling key components of environmental accession. As this volume documents, these countries, albeit to different degrees, mostly have completed the transposition of the environmental *acquis*, adopted EU-approved strategies to fulfil their commitments towards the environment, engaged in substantial administrative capacity building both to implement and to enforce the *acquis*, and in many instances are spending as much as, if not substantially more on environment as a percentage of GDP than their counterparts in the present EU. In this perspective, where environmental accession remains ‘half empty’, it is often in areas where present EU members are themselves struggling to overcome similar challenges.

The contrasting, more negative, ‘half empty’ assessment argues that whatever commitments the applicant countries have made towards the environment are purely formalistic and meant only to serve their overriding goal of entering the EU as quickly as possible. This assessment does not spare the EU either, viewing it as, in effect, a hypocritical organisation that virtuously proclaims its commitment to building a sustainable environment but substantively pursues few policies to realise this end. In short, in both the applicant countries and the EU, political will is largely absent to move much beyond rhetorical affirmations about the pressing need to implement environmentally sustainable policies.

In reality, of course, both assessments – as most of their respective adherents would concede – contain critically important elements of truth. Similarly, all can agree that the challenges analysed in this contribution will persist – and, perhaps, become even more acute – after most of the applicant countries formally enter the EU in 2004.

This author is cautiously optimistic that the EU is evolving in ways – albeit at times hesitantly, erratically, and perhaps overly slowly – that will make it a much more 'environmentally friendly' institution than it is now. The clear thrust of this evolution is towards more openness, transparency, accountability and a greater utilisation of market-based solutions to environmental challenges. Critics, as evidenced in this volume, would especially express reservations about viewing the market as a panacea for the environment. This is a valid criticism that justifies prudent public policy to regulate the 'excesses' of the market towards the environment, not an argument against how market liberalisation can benefit the environment *per se*.

If this analysis proves largely correct and the applicant countries successfully seize the opportunities available to them, then environmental accession will promote meaningful progress towards the ultimate – and far more challenging – goal of creating a sustainable environment in Europe and, concomitantly, serve as a reasonably accurate 'road map' for future aspirants in EU enlargement about the best ways to arrive at that destination.

REFERENCES

Andonova, Liliana (2002), 'The Challenge and Opportunities for Reforming Bulgaria's Energy Sector', *Environment*, Vol.44, No.10, pp.8–19.

CEE Bankwatch Network and Friends of the Earth Europe (2000), 'Billions for Sustainability? The Use of EU Pre-Accession funds and their Environmental and Social Implication – First Briefing', CEE Bankwatch Network and Friends of the Earth Europe, Brussels. October. www.bankwatch.org/publications/index.html

CEE Bankwatch Network and Friends of the Earth Europe (2001), 'Sustainable Theory – Unsustainable Practice (Billions for Sustainability? The Use of EU Accession Funds and their Environmental and Social Implications) – Second Briefing', CEE Bankwatch Network and Friends of the Earth Europe, Brussels. June. www.bankwatch.org/publications.index.html

Commission (Commission of the European Communities) (1997a), 'Compliance Costing for Approximation of EU Environmental Legislation in the CEEC,' Brussels. April. www.europa.eu.int/comm/enlarge/pdf/compcos.pdf

Commission (Commission of the European Communities) (1997b), 'Guide to the Approximation of European Union Environmental Legislation,' Brussels. May. www.europa.eu.int/comm/environment/guide/contents/htm

Commission (Commission of the European Communities) (1999), 'Composite Paper: Reports on progress towards accession by each of the candidate countries,' Brussels. 13 October 1999. www.europa.eu.int/comm/enlargement/report/_10_99/pdf/en/composit_en.pdf

Commission (Commission of the European Communities) (2001a), 'Administrative Capacity for Implementation and Enforcement of EU Environmental Policy in the Thirteen Candidate Countries,' Brussels. March. www.europa.eu.int/comm/enlarge/pdfadministrative_capacity.pdf

Commission (Commission of the European Communities) (2001b), 'The Challenge of Environmental Financing in the Candidate Countries,' Brussels. 6 June 2001. COM(2001) 304. www.europa.eu.int/comm/environment/docum/01304_en.htm

Commission (Commission of the European Communities) (2001c), 'Regular Report on Progress Towards Accession for Each Candidate Country,' Brussels. 13 November 2001. www.europa.eu.int/comm/enlargement/report2001

Commission (Commission of the European Communities) (2001d), 'Regular Report on Poland's Progress Towards Accession,' Brussels. 13 November 2001, SEC(2001) 1752. www.europa.eu.int/comm/enlargement/report2001/pl_en.pdf

Commission (Commission of the European Communities) (2002a), 'Chapter 22 – Environment,' Brussels. June 2002. www.europa.eu.int/comm/enlargement/negotiations/chapters/chap22/index.htm

Commission (Commission of the European Communities) (2002b), 'Regular Report on Poland's Progress Towards Accession,' Brussels. 9 October 2002, SEC(2002) 1408. www.europa.eu.int/comm/enlargement/report2002/pl.en.pdf

EBRD (European Bank for Reconstruction and Development) (2001), *Transition Report*, London, EBRD.

Ecotec Research and Consulting (2001), 'The Benefits of Compliance with the Environmental Acquis,' Brussels. Contract B7-8110/2000/159960/MAR/HI. www.europa.eu.int/comm/environment/enlarge/pdf/benefit__long.pdf

European Parliament (2000), 'Enlargement of the European Union,' Committee on Foreign Affairs, Human Rights, Common Security, and Defense Policy. Strasbourg. Report A5-0250. 10 April 2000. www.europarl.eu.int/enlargement/positionep/default_en.htm

Friends of the Earth Europe (2002), Press Release, 'NGOs present results of three years of monitoring EU pre-accession funds "Billions for Sustainability?" – How to take the question mark away?' Brussels. 26 November 2002.

Hughes, Gordon, and Julia Bucknall (1999), 'Poland: Complying with EU Environmental Legislation,' World Bank Technical Paper, No.454, Washington, D.C.

Informal Ministerial Meeting (2001), Meeting Between Ms. Wallström and the Ministers of the Environment of the Candidate Countries, Brussels. 27 November 2001.

Kramer, John M. (1983), 'Environmental Problems in East Europe: The Price for Progress', *Slavic Review*, Vol.42, No.2. pp.204–20.

Kramer, John M. (1990), *The Energy Gap in Eastern Europe*, Lexington: Lexington Books.

Kramer, John M. (1995), 'Nuclear Power in Central and Eastern Europe', *Problems of Post-Communism*, July/Aug., Vol.42, No.4, pp. 37–41.

Moldan, Bedrich (2000), 'Environment After 10 Years of Transition,' Regional Environmental Center, Szentendre, Hungary. 18 June 2000. www.rec.org/rec/programs/10th_anniversary/speech.html

Regional Environmental Center (1998), 'Doors to Democracy: Current Trends and Practices in Public Participation in Environmental Decision-making in Central and Eastern Europe,' Regional Environmental Center, Szentendre, Hungary. June. www.rec.org/rec/publications/ppdoors/cee/cover.html

Wallström, Margot (2000), 'The Commission's Perspective,' European Parliament, Committee on Environment, Public Health, and Consumer Policy. Strasbourg. 20 June 2000. www.europa.eu.int/meetdocs/committees/envi/default_en.htm

Walström, Margot (2003), 'Welcome and Opening Speech to the Environmental Ministers of Candidate Countries,' The Commission, Brussels, 21 January 2003. www.europa.eu.int/comm/environment/index

World Wildlife Fund (2002), *WWF Agenda for Accession: An Update*, World Wildlife Fund, Brussels, 29 April 2002. www.panda.org/downloads/europe/agenda_summary.pdf

CONCLUSION

Assessing Conventional Wisdom: Environmental Challenges and Opportunities beyond Eastern Accession

STACY D. VANDEVEER AND JOANN CARMIN

Fifteen years into the transitions away from state socialism, the enormity of the changes and challenges in Central and Eastern Europe are clearer, but no less impressive. CEE states are implementing dramatic environmental policy reforms as they gain EU membership. Over the last decade, new institutions have been developed, new policies formed and implemented, and reductions in many pollutants achieved. Collectively, the contributors to this volume concur on these points. The authors express varied and more critical views as they consider the implications of the CEE transition towards EU accession. Each of the contributions to this volume makes unique arguments and offers important insights into EU enlargement to the east. At the same time, four themes emerge across the contributions: the importance of capacity limitations and efforts to address them, the presence of mixed messages and conflicting EU priorities, the importance of non-state actors, and the need for multi-directional exchange of ideas and information. These themes highlight the ways that the contributors to this volume enrich conventional notions about Europeanisation and challenge prevailing views about CEE impacts on environmental quality in an enlarged EU.

Capacity Limitations and CDE

In the transition towards EU accession, CEE countries are striving to meet the demands and requisites of the *acquis*. However, as noted by many of the authors in this volume, there are numerous instances of lagging capacity in governmental, non-governmental and private sectors across CEE countries. Effective and efficient governance requires a host of capabilities associated with human resources and with organisations and institutions at multiple levels of governance and across civil society [*Grindle, 1997*]. Although notable gaps in capacity levels still exist across the region, it is also clear that there is more public, NGO and private sector environmental expertise and policy capacity in CEE states and societies than was present ten or fifteen years ago.

One area where gaps in capacity remain is in CEE environmental ministries. Being poorly staffed and under funded, their organisational, financial and human resource capabilities cannot meet their mandates. Further, their domestic political importance remains low compared to the more traditional and powerful ministries, such as those associated with defence, economic development and labour. Environmental officials are struggling to keep pace with the growing volume of international and domestic laws and regulations that they must implement. This pressure at the national level may be even greater at the regional and local levels where environmental agencies and offices are tasked with monitoring, enforcing and implementing a host of complex and expensive policies. Often the limited capacity for innovation and implementation of new policies and practices results in CEE countries relying on readily available and time-honoured approaches to environmental protection. Many of these challenges are long-standing complaints among environmental policy analysts and advocates in other countries as well [*Holzinger and Knoepfel, 2000*]. While critical in CEE states, they also give environmental policymakers and advocates in these states interests in common with their counterparts across and beyond Europe.

Just as states are wrestling with capacity limitations, so too are civil society actors and organisations. Since the fall of communism, NGOs have gained a presence in the region and have the potential to play critical roles in the implementation and enforcement of environmental policies. However, many of these organisations are not only struggling to meet their basic expenses, they are also faced with shortages of new talent to take the place of leaders who are retiring [*Beckmann, Carmin and Hicks, 2002*]. In addition, historical reliance on government bodies to make decisions and policies that place restrictions on participation further limits civil society's role in environmental policy. Concern about particular issues provides a rationale for becoming involved in environmental politics, but civil society participation in institutional forums and the use of more expressive forms of activism require both access and adequate resources [*Carmin, 2003*].

If implementation of EU environmental policy requires an engaged civil society, as a number of the contributions suggest, then resource constraints seen among CEE environmental groups may pose a challenge for implementation, as well. While the volume's contributors focus attention on the important roles played by NGOs and these groups' tremendous potential, CEE environmental NGOs also face an important domestic challenge beyond the availability of human and financial resources. In view of the general lack of environmental mobilisation and activism among CEE publics, CEE NGOs must overcome a kind of mobilisation deficit if they are to realise their potential influence on policy development and implementation [*Auer, 2004; Börzel, 2002*].

Despite the gaps in capacity across the EU and general critiques raised about the limits and impacts of external funding on CEE governments and non-governmental organisations [e.g. *Mendelson and Glenn, 2002; Gutner, 2002; Wedel, 1998*], programmes aimed at capacity development for the environment (CDE) can work. Well-tailored programmes have helped to improve CEE environmental capacities [*VanDeveer, 1997*]. As a result of these efforts, governmental and non-governmental actors are engaged in the policy process and new environmental laws, regulations, modes of assessment and evaluation have been put into place. Furthermore, even limited international financial support has led to the formation of new policy instruments and to investments and improvements in areas such as sewage treatment, drinking water quality, nuclear power plant safety, air quality and habitat protection [*Gutner, 2002*]. At the same time, training programmes, such as those organised under the auspices of the Regional Environment Center for Central and Eastern Europe (REC) or built through long-term international partnerships with NGOs such as Friends of the Earth, have enhanced the technical and administrative expertise of environmental organisations.

The success of some multilateral and bilateral assistance programmes, often sponsored by EU bodies, development banks, EU member states and international NGOs, demonstrates that international assistance can build domestic governance capacities [*Carius, 2002; KPMG, 1998; VanDeveer, 2000; VanDeveer and Dabelko, 1999, 2001*]. To improve their chances of success, such programmes must assess and target underlying causes of incapacity – not just symptoms – and avoid simply replicating donor-driven programmes across countries in 'cookie cutter' fashion [*Grindle, 1997; Sagar 2000; VanDeveer and Sagar, forthcoming*]. If an 'era of implementation' is to characterise EU environmental policy in the future, public sector actors, both in the East and in the West, will need to draw lessons from previous capacity-building experiences.

Mixed Messages in the Quest for Sustainability

As the *acquis* is adopted, CEE states generally acquire stronger environmental policies that, if implemented, have the potential for promoting environmental quality. At the same time, as the contributions demonstrate, CEE states and publics find themselves sifting through mixed messages and conflicting EU priorities. For instance, some of the environmental values promoted in the environmental *acquis* are undermined by the consumerist values and development projects inherent in other areas of EU policy. Such mixed messages may occur within a single policy arena, as they do around waste management issues. They also appear in the form of divergent purposes across different issue and funding arenas, such as when investments and requirements in the energy and transport sectors conflict with or undermine those in environmental action plans. These differential purposes and messages pose challenges to state officials as they find themselves promoting divergent types of behaviour.

The effects of mixed messages and conflicting priorities certainly are not restricted to states and state decision-making. For example, a number of this volume's contributions suggest that what the EU espouses regarding the importance of civil society actors often diverges from the types of participatory practices that are being promoted and implemented. Participatory frameworks are articulated and codified in agreements such as the Aarhus Convention. However, even as the EU seeks to provide access, it also serves as a transnational force that constrains activist actions and agendas. Although the EU has an opportunity to foster the credibility of NGOs by promoting broad participation in decision making, officials in Brussels seem to prefer to work more closely with organisations that have specialised technical expertise. Under the former regime, a culture of expertise prevailed in CEE states [*Wolchik, 1991*]. By giving preference to technical experts, it appears that the EU may be replicating patterns of reliance that promote societal complacency rather than fostering the credibility of non-governmental actors and promoting civil society development overall.

Just as the EU is exporting the strengths of its environmental policies, it is also exporting its limitations. CEE accession and assumption of the environmental *acquis* reveal some of the limits of generally stringent EU environmental policy. In areas such as nuclear power and nuclear waste disposal, pollution emissions from traffic and transport, urban sprawl and other types of development patterns, continental biodiversity protection, and increased 'Western-style' consumption, EU policies and practices are not environmentally sustainable [*Carius, Homeyer and Bär, 2000; Princen, Maniates and Conca, 2002; Legro and Aver, 2004*]. The contributions therefore suggest that attempts to make CEE states and societies more like those of the EU – to 'Europeanise' them – will not produce environmental sustainability. In short, this experiment has already been run in Western countries and it did not produce sustainability. Particularly problematic from an environmental standpoint are the ways in which 'non-environmental' EU policies such as those associated with transport and agriculture often contradict or undermine many of the goals promoted by environmental policy. While EU bodies and treaties have stepped up their efforts in recent years to 'integrate' environmental factors across other policy areas, the contributors to this volume suggest that such efforts have met with only very limited success.

Importance of Civil Society Actors and NGOs

Environmental movements and organisations played important roles in the overthrow of the communist regimes. Although there were expectations that environmental and other types of NGOs would retain a strong presence, the visibility of these groups has not been maintained over time. As many of the authors note, civil society actors in CEE states must navigate domestic political and resource constraints as well as EU policies that provide access to decision-making processes, but do not seem to promote genuine opportunities to influence outcomes.

The criticisms raised in this volume about the constraints placed on NGOs and the limited involvement in environmental policymaking that they are afforded appear against a backdrop of expectation that these groups can play a wide range of roles and make important contributions in environmental policy processes and outcomes in an enlarged EU. The import and potential of these groups is perhaps most often seen in local communities where they are able to engage in direct forms of action designed to promote rural sustainable development that can have far-reaching implications and impacts [*Carmin, Hicks and Beckmann, 2003*]. For example, as discussed in the contribution by Beckmann and Dissing, NGOs in the White Carpathian region are pressing forward with their vision of how to foster environmentally friendly agriculture, tourism and local development. They are doing so in the face of EU policies designed to reward less sustainable forms of development.

Although NGO impact may be most noticeable at the local level, the authors in this volume assert that the potential exists for NGOs to play significant roles at the national and international levels of environmental policymaking, implementation and enforcement. At present, they maintain that the EU is using direct and indirect means to shape the actions and agendas of environmental activists, generally keeping them at a distance from critical policy processes. The one exception is the inclusion of non-governmental experts associated with research institutes or providing technical support. Other groups that have limited technical capacities are finding that they are largely excluded from the discussion. These barriers to participation and input are resulting in the EU and CEE states missing opportunities and benefits that NGOs can offer.

EU and CEE officials need not regard NGOs solely as sources of contestation. The contributions in this volume suggest that fostering NGO participation and the input of environmental activists can help to generate new ideas to ensure that environmental decisions are relevant and realistic within their national contexts. Promoting collaboration between government agencies and a wide range of NGOs can generate sources of support for environmental policy implementation and enforcement. For instance, beyond alerting officials about the need to address particular issues through their advocacy efforts, these organisations can take the lead in a wide range of monitoring and implementation activities. NGOs offer the EU and CEE states an untapped source of energy, insight and action. It is likely that contestation and disagreement will always be present. However, working with these groups as partners rather than assuming that they are antagonists may help orient their actions and efforts towards promoting the realisation of national and international environmental policy.

Information and Ideas in a Multi-Dimensional Europe

Many contributions in this volume maintain that the accession process has been largely taken up with a one-way transfer of institutions and expertise from 'West to East', in general, and from Brussels to CEE capitals, in particular. Fifteen years of 'transition' combined with impending eastern enlargement have made some limitations of this unidirectional transfer of policy institutions and information more apparent. For example, a critical issue raised throughout the volume is the fact that a top-down approach does not allow knowledge and environmentally friendly practices present in the CEE countries to trickle 'up' to Brussels. Furthermore, EU dominance of CEE environmental policymaking agendas has marginalised both non-EU-sanctioned ideas and institutions and reflections about what policies and practices from the state socialist past could be kept or adapted. Although state socialism had an abysmal environmental record, several authors note that the formerly communist countries can bring some useful forms of environmental sensibility to the Union, such as bottle reuse, minimal packaging and a reliance on public transportation. In other words, from environmental protection and sustainability perspectives, not every aspect of the pre-1989 past need be pronounced environmentally damaging or politically outmoded.

Although applicant states are required to adopt the *acquis*, great differences exist across the 15 (pre-2004) EU member states in their organisation of environmental policymaking, monitoring and implementation [*Holzinger and Knoepfel, 2000; Jordan, Liefferink and Fairbrass, 2004; Jordan and Liefferink, 2004; Knill and Lenschow, 2000*]. Factors such as national policy style, risk perception, and the structure of various domestic institutions also vary greatly and these differences have important implications for national choices in environmental policy [see *O'Neill, 2000; Skjaerseth, 2002*]. In many respects, while more flexible, newer environmental policy instruments are highly circumscribed, serious discussions of the pros and cons associated with centralisation in Brussels and in national capitals has yet to take place [*Golub, 1998a, 1998b; Holzinger and Knoepfel, 2000*]. As the EU grows larger and more diverse, it is increasingly evident that CEE states, like the present EU states, need organisational structures best suited to their particular circumstances, sizes, capabilities and collective preferences. Just as 'one size does not fit all' in the Western European EU member states, so too must a balance be found between wholesale implementation of EU institutions and policies and those that will be effective and appropriate in the new CEE member states [*Holzinger and Knoepfel, 2000*].

As a number of contributors point out, the near single-minded pursuit of EU membership for CEE states on the part of CEE officials (and NGOs) means that many domestic issues have been marginalised or subsumed in the EU accession process. As membership is attained, domestic priorities may come to the fore, receiving more attention in national as well as in international political arenas. Further, the CEE region is not a single 'bloc' or a group of 'blocs'. The region encompasses as much political, economic and cultural difference as is seen among Western European states and peoples. CEE officials engage regional, pan-European and global politics with their own understandings of their national interests. As a result, CEE officials and publics, like their Western counterparts, are likely to choose different ways to accomplish the same environmental goals. Or they may have quite different environmental priorities, as seen in debates about nuclear power across Europe, for example. As a result, not only may CEE–EU communication need to become less unidirectional, it also may need to become more varied in response to individual state needs.

Several of the contributors to this volume argue that the EU is changing, not simply getting bigger, as its membership expands. As Schreurs notes [*this volume*], much of the debate about how EU institutions must change to accommodate additional member states has been taken up with national 'voting' rules for the larger EU. These concerns have focused on the potential for EU-level paralysis resulting from minorities of member states being able to block policy changes and on ways to protect some areas of sovereign authority so that smaller states are not dominated by larger nations [see *Homeyer, this volume*]. However, variability in institutions as well as in individual state priorities suggests that after membership is realised, CEE countries will likely take action to address their own concerns and interests rather than simply 'voting' as a bloc in Brussels.

Even with their focus on adopting and implementing the *acquis*, and subsequently being influenced by the EU, CEE states also may be influenced by a wide range of international organisations and institutions. One limitation of the nearly exclusive focus by Europeanisation scholars and European policymakers on the relationship between the EU and member countries is that they ignore the impact that other multilateral forums may have on states and domestic spheres. International environmental politics is populated by numerous organisations and multilateral arrangements. EU members are party to over 60 international environmental treaties [*VanDeveer, 2003*], most of which influence domestic environmental policies [*Keohane and Levy, 1996; Victor, Raustiala and Skolnikoff, 1998; Weiss and Jacobsen, 1998*]. Multilateral environmental efforts including the Organisation for Security and Cooperation in Europe (OSCE), the United Nations Economic Commission for Europe (UNECE), the 'Environment for Europe' processes, and a number of regional seas management arrangements also influence CEE environmental policy development in ways that are generally (but not always) consistent with EU policy [*Gutner and VanDeveer, 2001; Selin and VanDeveer, 2003, forthcoming; VanDeveer, 2000; Carius, 2002*]. The presence of multiple international environmental negotiating forums often encourages state officials and activists to 'venue shop' for international cooperation regimes most amenable to their policy goals [*VanDeveer, forthcoming*]. Such multi-dimensional international environmental politics are likely, over time, to further alter the bilateral dynamics between Brussels and CEE states witnessed during the accession process.

Conventional Wisdom Reconsidered

The contributions in this volume reveal a kind of paradox: greater diversity and more European unity both will result from EU enlargement. Consequently, it is clear that simply asking 'how do domestic institutions and actors change in response to EU action?' is too narrow a question to capture the diversity of factors shaping environmental policy change and environmental outcomes across the CEE region. European politics is changing, not just 'expanding' its territory. The contributors demonstrate that CEE states and NGO actors are responding to many revolutionary changes, not merely those engendered by the *acquis*. Changes in environmental institutions and outcomes as driven by EU accession cannot be explained without attention to larger changes in political and economic structure, as well as to shifts in the roles and expectations of state officials, citizens and civil society groups. In other words, while EU accession has driven the environmental policy agendas in CEE states for much of the last decade, it is certainly not the only factor that significantly shapes each policy outcome.

Europeanisation in a Larger and More Diverse Europe

The three pathways of EU-driven domestic change, discussed in detail in the first chapter [*Carmin and VanDeveer, this volume*], call attention to the strategic behaviour of various actors, including that of government officials, NGOs and firms, and to potential changes in values, norms and expectations. The contributions to this volume focus attention, in particular, on the role of domestic environmental leaders (officials and NGOs) and the ongoing diffusion of ideas and material resources among domestic and international actors in the CEE region. While EU officials have attempted to exercise power directly over CEE states during accession negotiations and via international aid programmes, incentives for domestic actors and these actors' values and beliefs also change in response to EU actions and to transformations engendered by political and economic transitions. As property rights and ownership patterns change across the region, so have the incentives for private sector and state actors. Furthermore, civil society actors have continued to push for greater participation in policymaking as well as greater access to information.

Consistent with the first Europeanisation pathway, the top-down institutional model, CEE officials have moved to enhance national legal, administrative and human resource capacities associated with environmental policy in response to EU directives and regulations that require a host of specific laws, procedures, and regulatory functions and standards. As many of the authors note, a number of CEE actions in environmental policy have been carried out because the EU has required them as part of accession. The second pathway, changing domestic incentives, is reflected in a variety of ways including the impacts of EU assistance programmes and the co-financing requirements of most international donors, the retention of environmental actors (state officials and civil society groups) in domestic policymaking, and the changing practices of consumers and private sector actors in response to liberalisation and privatisation. As the contributions demonstrate, such changes can lead to more or less environmentally friendly outcomes, such as reductions in industrial pollution emissions and increases in household wastes, traffic and urban sprawl. In a sense, these variable environmental outcomes all make CEE states and societies 'more European'.

The third pathway, alterations in beliefs and expectations of domestic actors as a result of EU decisions and discourses, is also represented in the contributions to this volume. For example, the institutionalisation of norms associated with NGO involvement in policymaking and attempts to apply concepts such as pollution prevention, the precautionary principle, and polluter pays approaches (manifest in pollution taxes, for example), all suggest changing norms and ideas about environmental governance across the CEE region. The contributions reinforce the notion that examination of both the material and ideational aspects of Europeanisation (or EU-driven change) are essential for explaining outcomes.

At one level, this volume demonstrates that Europeanisation processes affect applicant states through the same three mechanisms or pathways as they impact on member states. At the same time, the authors demonstrate that merely asking how domestic institutions adjust to accommodate EU policy, as is common in contemporary 'top-down' Europeanisation research, misses a number of important dynamics. First, it may underestimate the influence of domestic agents and institutions on policy and behavioural outcomes. The contributions highlight numerous instances in which domestic NGOs and public sector actors – sometimes in cooperation with a non-EU international actor such as a foundation – are important determinants of domestic CEE policy and environmental quality outcomes. Second, 'top-down' Europeanisation approaches appear to underestimate the interactive learning processes that may take place under the guise of EU harmonisation and implementation. For example, as discussed further below, the Europeanisation of CEE environmental policy has revealed problems and limitation with EU policy in numerous areas of environmental, transport, agricultural, investment and energy policies. In other words, in the process of trying to apply EU policy in CEE states and societies, change in EU policy may be engendered.

Environmental Policy and Quality

In addition to enriching perspectives on Europeanisation, this volume provides alternative views to the prevailing wisdom about the impacts that eastern enlargement will have on the European environment. Overall, the authors suggest that CEE accession will be neither the 'problem' for EU environmental policymaking that many have suggested, nor the saviour for CEE or pan-European environmental quality. In short, this is precisely because CEE policy, environmental and otherwise, looks increasingly similar to that of the EU and many of its current member states. Thus, the challenges facing CEE environmental policy look increasingly similar to the challenges facing EU environmental governance.

Among environmental analysts and advocates, debate about enlargement has often focused on the concern that CEE states would join the so-called environmental 'laggard' EU states to slow, or attempt to reverse the progressive nature of EU environmental policy [see *Holzinger and Knoepfel, 2000; Jehlièka and Tickle, this volume*]. This volume suggests that the picture is more complex and that blanket assumptions that CEE countries will become laggards are misguided. For instance, CEE states have established large protected areas that contain some of the greatest biodiversity in all of Europe and they have often been quick to enact environmental legislation. As the authors also note, with privatisation and international investment still in progress, industrial investments could be accompanied by clean technologies. Further, recent studies of the implementation of environmental policy commitments demonstrate that long-standing and wealthy EU member states may lack implementation capacity, as well [*McCormick, 2001; Knill and Lenshow, 2000; Skjaerseth, 2002*]. Contrary to commonly held notions of CEE states being environmental laggards *vis-à-vis* Western Europe, the implementation challenge increasingly looks like one that CEE accession states have in common with their fellow EU members on issues from wastewater treatment to the newer procedurally oriented EU policies, such as those focusing on environmental auditing, impact assessment and integrated pollution prevention.

Another conventional claim about enlargement is that adoption of the *acquis* will clean up the environment in CEE states. This volume suggests a more complex and nuanced picture. While EU environmental law is raising many environmental standards and opening more potential avenues for civil society actors, EU bodies send decidedly mixed or contradictory messages about environmental policy and quality. While environmentally sound behaviour is fostered in some arenas, so are many unsustainable practices. In fact, a 1999 report by the European Environment Agency (EEA) confirms that environmental conditions and indicators across a host of issue areas continue to deteriorate across the European continent [*EEA, 1999*].

Scholars, policy analysts and environmental advocates alike have tended to frame many aspects of CEE accession as a series of 'problems' that must be addressed. As Holzinger and Knoepfel [*2000: 31*] put it, 'There is no doubt that the Eastern enlargement of the EU represents a challenge for its environmental implementation.' These and other analysts have not seen many opportunities for environmental policy in enlargement. In contrast, the authors in this volume suggest that environmental gains have been achieved, that CEE countries are working to adopt and implement standards articulated in the *acquis*, and that some innovative policy ideas and practices to enhance EU environmental governance can be gleaned from both the communist and transition periods in the CEE region. In particular, the rapid harmonisation and transition processes across CEE states have revealed many of the limitations and contradictions across EU policy areas that work to undermine environmental quality, such as investments in transport and agriculture and the mixed messages sent around waste management and NGO involvement in policymaking. Furthermore, as Gille argues [*this volume*], while many environmental problems linger from the socialist legacy, this history can also be mined for policy lessons.

Taking advocates for greater policy flexibility and decentralisation seriously, one sees environmental policymakers and environmental programme managers as potential sources of more efficient and more innovative environmental policy implementation and goal achievement [*Rabe, 2002*]. Since implementation 'challenges' characterise EU environmental policy across all member states, older member states and EU policymakers alike might look to CEE accession states for ideas about how best to achieve significant policy changes in relatively short periods of time – and for less cost. In policy areas such as pollution and resource extraction taxes, and the operation of national environmental funds, some CEE states are leading and teaching some EU member states. While Western European and North American officials and environmental policy experts have talked about various green taxes, subsidies and market-style policy instruments for many years, such taxes and funds appear to be more widely practised in post-communist states [*Holzinger and Knoepfel, 2000; VanDeveer and Carmin, forthcoming*]. Similarly, despite years of rhetoric, the EU is hardly the place to look for leadership on sustainable agricultural policies. Taken together, these patterns suggest that the emergence of greater diversity in economic structure and ability, and in political and social institutions, will require greater intra-EU debate about the need for more flexibility and efficiency in EU environmental policy.

Beyond the EU's New Borders

The 2004 accession of CEE countries is setting precedents for later rounds of EU enlargement. With many more applicants in the pipeline, nearly all of them substantially poorer than the CEE accession states, the first wave of EU enlargement is framing the possibilities and choices available in future rounds of membership negotiations. For example, smaller non-member states of south-eastern Europe such as Albania and states from the former Yugoslavia are likely to look for lessons from the experiences of the smaller CEE accession states. Larger EU applicants may look to Poland. For their part, EU officials appear likely to hold to the general criteria for membership applied to CEE states, including the complete incorporation of the *acquis* prior to membership and the granting of few concessions to applicants. Because EU environmental policy continues to expand, the 'bar' may be raised even higher for future entrants.

In recent years, EU officials have become increasingly active in regional international forums such as those around the Baltic, Mediterranean and Black seas, and the conventions and protocols negotiated under the auspices of the United Nations Economic Commission for Europe (UNECE) [*Carius, 2002; VanDeveer, 2000; Selin and VanDeveer, 2003, forthcoming*]. The latter forum includes treaties on long-range air pollution, environmental impact assessment, management of transboundary rivers and lakes, the cross-border effects of industrial accidents, and public participation and access to information on environmental matters. European Commission officials have worked to harmonise many of these treaties with EU law and policy even as they pressure (and sometimes financially support) non-EU states across the post-Soviet space to ratify and implement them. As such, these regional forums may be used to reinforce EU influence in the new member states and to extend EU influence around and beyond the EU perimeter. A post-2004 EU of 25 member states likely will result in an increase in the EU role (and its number of national 'votes') in these regional treaties and organisations, many of them forums in which CEE officials have been relatively active in recent years. These regional cooperation forums may also serve as vehicles for lesson drawing and norm diffusion from new and old EU members to candidates and non-candidates around the EU perimeter.

An enlarged EU has ramifications for global environmental governance, as well. Garvey [*2002*] noted that a larger EU will act in global environmental politics in at least three potentially significant ways: unilaterally, through multilateral environmental agreements (MEAs), and through trade accords. EU unilateral environmental actions may also afford opportunities to lead by example or to construct and promulgate certain environmental standards, policy ideas, norms and principles. For example, environmental policy and product standard decisions made by an EU of 25 states, representing over 450 million people, will have ramifications well beyond its borders. First, many such decisions will have wide-ranging implications for global markets [*Mitchener, 2002*]. Markets can reinforce environmental norms and drive improved environmental protection. Global attention to the need to 'get the incentives right' in environmental policy has increased in the last decade. In conjunction with transnational private sector and non-governmental organisations, public demands – as articulated within markets – can improve the environmental performance of firms and governments [see *Andonova, 2003*]. Furthermore, well-designed public policy can shape markets in more environmentally sensitive ways. For example, 'green taxes' may shape demand and offer incentives to conserve or to increase the efficiency of resource use. Access to information about pollution emissions, or environmentally damaging practices, can shape public demand and preferences. Lastly, stringent product standards may drive how goods are produced globally, since many producers seek to export to the EU market. With a consumer market exceeding the size of the US market, EU product standards have already begun to achieve the *de facto* status of global standards, as producers manufacture products to meet the often higher standards of the world's largest market.

The design of EU foreign assistance programmes, if they are brought into line with EU environmental priorities, also offers opportunities to extend a kind of unilateral EU environmental policy influence beyond the border of the larger Union. For example, EU officials may seek to use EU foreign assistance to support implementation and expansion of facets of the Kyoto Protocol or other global accords. Aspects of the CEE transition and accession experiences may serve as models for non-European states and societies undergoing transitions away from state socialism, such as China and Vietnam, as well as those undergoing rapid industrialisation and the accompanying rapid rise in environmental problems. EU foreign assistance might seek to encourage this.

The EU has the potential to exercise more global influence and leadership within the context of international organisations, such as United Nations bodies, and multilateral negotiations around environmental and trade agreements. If the 25 EU states can negotiate as a block in global environmental and trade forums or use unilateral EU policy development to lead by example, they may increase their influence over global policy outcomes. Such influence might be exercised around issues such as climate change, biodiversity protection or trade law. Visibility in international forums and the growing size of EU membership and the EU marketplace offer opportunities for the Union to become a global environmental leader.

Conclusion

Throughout the accession processes, environmental issues remained important for EU officials and, therefore, for CEE countries alike. While some scepticism remains about the impact that the heavily polluted countries of Central and Eastern Europe will have on EU environmental policy and quality, it is clear that applicant states have enacted sweeping legal and regulatory changes as they seek EU membership. At various times, environmental protection and remediation moved from the centre to the fringes of the policy agenda. However, environmental issues have been important political priorities both in the early phases of the transition away from Soviet-style rule and in the final stages of EU accession preparations. In the last 15 years, CEE policymakers have developed new domestic institutions and formulated, enacted, and begun to implement a host of new policies for environmental protection.

CEE countries have focused on harmonisation and transposition of the environmental chapter of the *acquis*. Consequently, it appears that much of the environmental policy development in these states, particularly that which occurred in between the late 1990s and 2003, would not have occurred without EU pressure associated with the criteria for EU membership. Furthermore, foreign assistance programmes, especially those designed to build various types of capacity needed to enact and implement environmental policy have had discernible impacts on environmental policy outcomes in CEE states. Yet it is also clear that EU officials and policies send mixed signals and contradictory messages (and incentives) to CEE states and publics regarding environmental protection, policy priorities, consumer and firm behaviour, and NGO involvement and influence in policymaking and implementation. In a sense, these mixed messages confirm the extent of Europeanisation taking place in CEE states and societies because many of the opportunities and challenges faced by CEE environmental policymakers and civil society advocates look similar to those faced by their Western counterparts. Together, the chapters suggest that officials and NGOs in the accession states and the EU15 will have more in common in the enlarged EU than conventional wisdom suggests.

REFERENCES

Andonova, Liliana (2003), *Transnational Politics and the Environment; EU Integration and Environmental Policy in Central and Eastern Europe*, Cambridge: MIT Press.

Auer, Matthew R. (forthcoming), *Restoring Cursed Earth: Appraising Environmental Policy Reforms in Central and Eastern Europe and Russia*, Boulder, CO: Rowman & Littlefield Press.

Beckmann, Andreas, JoAnn Carmin and Barbara Hicks (2002), 'Catalysts for Sustainability: NGOs and Regional Development Initiatives in the Czech Republic', in Walter Leal Filho (ed.), *International Experiences on Sustainability*, Bern: Peter Lang Scientific Publishing, pp.159–77.

Börzel, Tanja A. (2002), 'Improving Compliance Through Domestic Mobilization? New Instruments and the Effectiveness of Implementation', in Christopher Knill and Andrea Lenschow (eds.), *Implementing EU Environmental Policy: New Directions and Old Problems*, Manchester: Manchester University Press, pp.222–50.

Carmin, JoAnn (2003), 'Resources, Opportunities, and Local Environmental Action in the Democratic Transition and Early Consolidation Periods in the Czech Republic', *Environmental Politics*, Vol.12, No.3, pp.42–64.

Carmin, JoAnn, Barbara Hicks and Andreas Beckmann (2003), 'Leveraging Local Action: Grassroots Initiatives and Transnational Collaboration in the Formation of the White Carpathian Euroregion', *International Sociology*, Vol.18, No.4, pp.703–25.

Carius, Alexander (2002), 'Challenges for Governance in a Pan-European Environment: Transborder Cooperation and Institutional Coordination', in Sabina Crisen and JoAnn Carmin (eds.), *EU Enlargement and Environmental Quality: Central and Eastern Europe & Beyond*, Washington, DC: Woodrow Wilson International Center for Scholars, pp.21–31.

Carius, Alexander, Ingmar von Homeyer and Stefani Bär (2000), 'Eastern Enlargement of the European Union and Environmental Policy: Challenges, Expectations, Multiple Speeds and Flexibility', in K. Holzinger and P. Knoepfel (eds.), *Environmental Policy in a European Union of Variable Geometry? The Challenge of the Next Enlargement*, Basel: Helbing & Lichtenhahn, pp.141–80.

EEA (European Environment Agency] (1999), *Environment and the European Union at the Turn of the New Century*, Copenhagen: EEA.

Garvey, Tom (2002), 'EU Enlargement: Is It Sustainable?', in Sabina Crisen and JoAnn Carmin (eds.), *EU Enlargement and Environmental Quality: Central and Eastern Europe & Beyond*, Washington, DC: Woodrow Wilson International Center for Scholars, pp.53–62.

Golub, Jonathon (1998a), *New Instruments for Environmental Policy in the EU*, London: Routledge.

Golub, Jonathon (1998b), *Global Competition and EU Environmental Policy*, London: Routledge.

Grindle, Merilee S. (1997), *Getting Good Government: Capacity Building in the Public Sector of Developing Countries*, Cambridge, MA: Harvard University Press.

Gutner, Tamar L. (2002), *Banking of the Environment: Multilateral Development Banks and Their Environmental Performance in Central and Eastern Europe*, Cambridge, MA: MIT Press.

Gutner, Tamar L. and Stacy D. VanDeveer (2001), 'Networks, Coalitions and Communities: Capacity Building in the Baltic Region', paper presented at the International Studies Association Annual Convention, Chicago, IL. 20–24 Feb. 2001.

Holzinger, Katharina and Peter Knoepfel (2000), *Environmental Policy in a European Union of Variable Geometry? The Challenge of the Next Enlargement*, Basel: Helbing & Lichtenhahn.

Jordan, Andrew, Duncan Liefferink and J. Fairbrass (2004), 'The Europeanization of National Environmental Policy: A Comparative Analysis', in J. Barry, B. Baxter and R. Dunphy (eds.), *Europe, Globalisation and Sustainable Development*, London: Routledge.

Jordan, Andrew and Duncan Liefferink (2004). *Environmental Policy in Europe: The Europeanization of Environmental Policy in Europe*, London: Routledge.

Keohane, Robert O. and Marc A. Levy (eds.) (1996), *Institutions for Environmental Aid: Pitfalls and Promise*, Cambridge: MIT Press.

KPMG (1998), *Environmental Policy and the Role of Foreign Assistance in Central and Eastern Europe*, Copenhagen: KPMG.

Knill, Christopher and Andrea Lenschow (2000), *Implementing EU Environmental Policy: New Directions and Old Problems*, Manchester: Manchester University Press.

Legro, Susan and Matthew R. Auer (forthcoming), 'Environmental Reform in the Czech Republic: Uneven Progress after 1989', in Mathew Auer (ed.), *Restoring Cursed Earth: Appraising Environmental Policy Reforms in Central and Eastern Europe and Russia*, Boulder, CO: Rowman & Littlefield Press.

McCormick, John (2001), *Environmental Policy in the European Union*, New York: Palgrave.

Mendelson, Sarah E. and John K. Glenn (2002), *The Power and Limits of NGOs: A Critical Look at Building Democracy in Eastern Europe and Eurasia*, New York: Columbia University Press.

Mitchener, Brandon (2002), 'Standard Bearers: Increasingly, Rules of Global Economy Are Set in Brussels', *Wall Street Journal*, 23 Apr., p.A1, A10.

O'Neill, Kate (2000), *Waste Trading Among Rich Nations: Building a New Theory of Environmental Regulation*, Cambridge, MA: MIT Press.

Princen, Thomas, Michael Maniates and Ken Conca (2002), *Confronting Consumption*, Cambridge, MIT Press.

Rabe, Barry G. (2002), 'Power to the States: The Promise and Pitfalls of Decentralization', in N. Vig and M. Kraft (eds.), *Environmental Policy: New Directions for the Twenty-first Century*, 5th edition, Washington, DC: CQ Press, pp.33–56.

Sagar, Ambuj (2000), 'Capacity Development for the Environment: A View from the South, A View from the North', *Annual Review of Energy and the Environment*, Vol.25, pp.377–439.

Selin, Henrik and Stacy D. VanDeveer (2003), 'Mapping Institutional Linkages in European Air Pollution Politics', *Global Environmental Politics*, Vol.3, No.3, pp.14–47.

Selin, Henrik and Stacy D. VanDeveer (forthcoming), 'Baltic Hazardous Substances Management: Results and Challenges', *AMBIO: Journal of the Human Environment.*

Skjaerseth, Jon Birger (2002), *North Sea Cooperation: Linking International and Domestic Pollution Control*, Manchester: Manchester University Press.

VanDeveer, Stacy, D. (1997), 'Normative Force: The State, Transnational Norms and International Environmental Regimes', PhD dissertation, University of Maryland, College Park, MD.

VanDeveer, Stacy D. (2000), 'Protecting Europe's Seas: Lessons after 25 Years', *Environment*, Vol.42, No.6, pp.10–26.

VanDeveer, Stacy D. (2003), 'Green Fatigue', *Wilson Quarterly*, Autumn, pp.55–9.

VanDeveer, Stacy D. (forthcoming), 'Assessment Information in European Politics: East and West', in Ronald B. Mitchell, William C. Clark, David W. Cash and Frank Alcock (eds.), *Global EnvironmentalAssessments: Information, Institutions and Influence,* Cambridge: MIT Press.

VanDeveer, Stacy D. and JoAnn Carmin (forthcoming), 'Sustainability and EU Accession: Capacity Development and Environmental Reform in Central and Eastern Europe', in Gary B. Cohen and Zbignew Bochniarz (eds.), *Sustainability in the New Central Europe.*

VanDeveer, Stacy D. and Geoffrey D. Dabelko (1999), *Protecting Regional Seas; Developing Capaicty and Fostering Environmental Cooperation in Europe*, Conference Proceedings, Washington, DC: Woodrow Wilson International Center for Scholars.

VanDeveer, Stacy D. and Geoffrey D. Dabelko (2001), 'It's Capacity Stupid: National Implementation and International Assistance', *Global Environmental Politics*, Vol.1, No.2, pp.18–29.

VanDeveer, Stacy D. and Ambuj Sagar (forthcoming), 'Capacity Building for the Environment: North and South', in E. Corell, A. Churie Kallhauge and G. Sjöstedt (eds), *Furthering Consensus: Meeting the Challenges of Sustainable Development Beyond 2000*, London: Greenleaf.

Victor, David G., Kal Raustiala and Eugene B. Skolnikoff (1998), *The Implementation and Effectiveness of International Environmental Commitments: Theory and Practice*, Cambridge: MIT Press.

Wedel, Janine (1998), *Collision and Collusion: The Strange Case of Western Aid to Eastern Europe 1989–1998*, New York: St Martin's Press.

Weiss, Edith Brown and Harold K. Jacobson. (1998), *Engaging Countries: Strengthening Compliance with International Environmental Accords*, Cambridge: MIT Press.

Wolchik, Sharon L. (1991), *Czechoslovakia in Transition: Politics, Economics, & Society*, London: Pinter Publishers.

INDEX